William Smoult Playfair

A Treatise on the Science and Practice of Midwifery

Fifth Edition. Vol. 1

William Smoult Playfair

A Treatise on the Science and Practice of Midwifery
Fifth Edition. Vol. 1

ISBN/EAN: 9783337034979

Printed in Europe, USA, Canada, Australia, Japan

Cover: Foto ©berggeist007 / pixelio.de

More available books at **www.hansebooks.com**

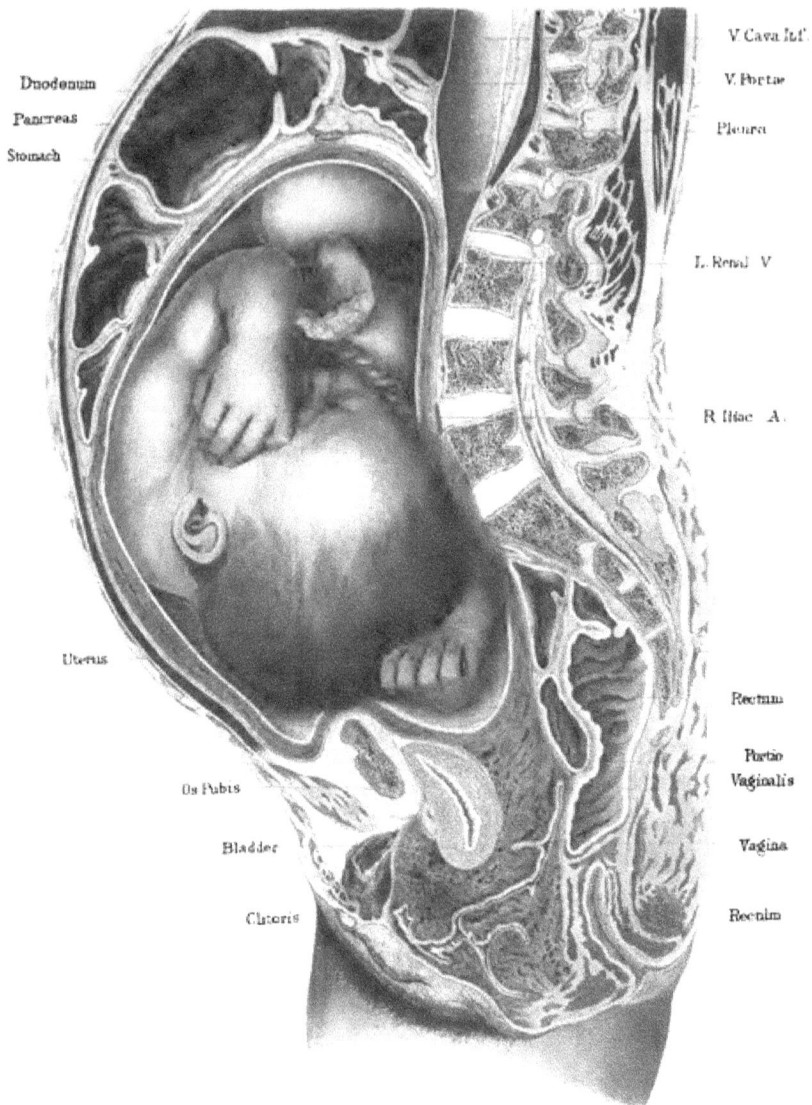

Duodenum
Pancreas
Stomach

V Cava Inf.
V. Portæ
Pleura

L. Renal V

R Iliac A.

Uterus

Os Pubis

Bladder

Clitoris

Rectum

Portio
Vaginalis

Vagina

Rectum

Section of a frozen body in the last month of pregnancy (after Braune.)
illustrating the relations of the Uterus to the surrounding parts, and the attitude
of the foetus, which is lying in the second cranial position

A TREATISE

ON

THE SCIENCE AND PRACTICE

OF

MIDWIFERY

BY

W. S. PLAYFAIR, M.D., F.R.C.P.

PHYSICIAN-ACCOUCHEUR TO H. I. AND R. H. THE DUCHESS OF EDINBURGH ; PROFESSOR OF OBSTETRIC
MEDICINE IN KING'S COLLEGE ; PHYSICIAN FOR THE DISEASES OF WOMEN AND CHILDREN TO
KING'S COLLEGE HOSPITAL ; CONSULTING PHYSICIAN TO THE GENERAL LYING-IN
HOSPITAL, AND TO THE EVELINA HOSPITAL FOR CHILDREN ; LATE
PRESIDENT OF THE OBSTETRICAL SOCIETY OF LONDON ; EXAMINER
IN MIDWIFERY TO THE UNIVERSITY OF LONDON, AND
TO THE ROYAL COLLEGE OF PHYSICIANS

IN TWO VOLUMES

VOL. I.

FIFTH EDITION

LONDON

SMITH, ELDER, & CO., 15 WATERLOO PLACE

1884

PREFACE

TO

THE FIFTH EDITION.

———◦◦◦———

A FIFTH edition being called for, the Author has endeavoured to make his work more worthy of the kind reception it continues to receive from the Profession, by subjecting it to a careful revision, and by adding several new illustrations. He has, more particularly, rewritten in great part the chapter on 'Conception and Generation,' and in so doing he has received valuable assistance from Dr. W. TYRRELL BROOKS, lately of the Physiological Laboratory in King's College, now of Oxford, for which he has to express his obligation. His special thanks are also due to Dr. JOHN PHILLIPS, for his aid in passing the work through the press.

31 GEORGE STREET, HANOVER SQUARE:
July 1884

PREFACE

THE FIRST EDITION.

————◆————

THOSE who have studied the progress of Midwifery know that there is no department of medicine in which more has been done of late years, and none in which modern views of practice differ more widely from those prevalent only a short time ago. The Author's object has been to place in the hands of his readers an epitome of the science and practice of midwifery which embodies all recent advances. He is aware that on certain important points he has recommended practice which not long ago would have been considered heterodox in the extreme, and which, even now, will not meet with general approval. He has, however, the satisfaction of knowing that he has only done so after very deliberate reflection, and with the profound conviction that such changes are right, and that they will stand the test of experience. He has endeavoured to dwell especially on the practical part of the subject, so as to make the work a useful guide in this most anxious and responsible branch of the profession. It is admitted by all, that emergencies and difficulties arise more often in this than in any other branch of practice; and there is no part of the practitioner's work which requires more thorough knowledge or greater experience. It is, moreover, a lamentable fact that students generally leave their schools more ignorant of obstetrics than of any other subject. So long as the absurd regulations exist which oblige the lecturer on midwifery to attempt the impossible task of teaching obstetrics in a short three months' course—an absurdity which has over and over again been pointed out—such must of necessity be the case. This must be the Author's excuse for dwelling on many topics at greater length than some will doubtless think their importance merits, since he desires to place in the hands of his students a work which may in some measure supply the inevitable defects of his lectures.

Many of the illustrations are copied from previous authors, while some are original. The following quotation from the preface to Tyler Smith's 'Manual of Obstetrics' will explain why the source of the copied woodcuts has not been in each instance acknowledged : 'When I began to publish, I determined to give the authority for every woodcut copied from other works. I soon found, however, that obstetric authors of all countries, from the time of Mauriceau downwards, had copied each other so freely without acknowledgment as to render it difficult or impossible to trace the originals.'

The Author has to express his acknowledgments to many friends for their kind assistance by the loan of illustrations and otherwise, and more especially to his colleague, Dr. HAYES, for his valuable aid in passing the work through the press.

31 GEORGE STREET, HANOVER SQUARE:
March 1876.

CONTENTS

OF

THE FIRST VOLUME.

——◆——

PART I.

*ANATOMY AND PHYSIOLOGY OF THE ORGANS
CONCERNED IN PARTURITION.*

———

CHAPTER I.

ANATOMY OF THE PELVIS.

CHAPTER II.

THE FEMALE GENERATIVE ORGANS.

CHAPTER III.

OVULATION AND MENSTRUATION.

PART II.

PREGNANCY.

CHAPTER I.

CONCEPTION AND GENERATION.

CHAPTER II.

THE ANATOMY AND PHYSIOLOGY OF THE FŒTUS.

CHAPTER III.

PREGNANCY.

CHAPTER IV.

SIGNS AND SYMPTOMS OF PREGNANCY.

CHAPTER IX.

PATHOLOGY OF THE DECIDUA AND OVUM.

CHAPTER X.

ABORTION AND PREMATURE LABOUR.

PART III.

LABOUR.

CHAPTER I.

THE PHENOMENA OF LABOUR.

CHAPTER II.

MECHANISM OF DELIVERY IN HEAD PRESENTATIONS.

CHAPTER III.

MANAGEMENT OF NATURAL LABOUR.

CHAPTER IV.

ANÆSTHESIA IN LABOUR.

CHAPTER V.

PELVIC PRESENTATIONS.

CHAPTER VI.

PRESENTATIONS OF THE FACE.

CHAPTER VII.

DIFFICULT OCCIPITO-POSTERIOR POSITIONS.

CHAPTER VIII.

PRESENTATIONS OF SHOULDER, ARM, OR TRUNK—COMPLEX PRESENTATIONS—PROLAPSE OF THE FUNIS.

ILLUSTRATIONS

IN THE

FIRST VOLUME.

————•◦•————

VOL. I.   a

THE SCIENCE AND PRACTICE

OF

MIDWIFERY.

PART I.

ANATOMY AND PHYSIOLOGY OF THE ORGANS CONCERNED IN PARTURITION.

CHAPTER I.

ANATOMY OF THE PELVIS.

THE *pelvis* is the bony basin situated between the trunk and the lower extremities. To the obstetrician its study is of paramount importance, for it not only contains, in the unimpregnated state, all the organs connected with the function of reproduction, but through its cavity the fœtus has to pass in the process of parturition. An accurate knowledge, therefore, of its anatomical formation may be said to be the very alphabet of obstetrics, without which no one can practise midwifery, either with satisfaction to himself or safety to his patient.

In a treatise on obstetrics, however, any detailed account of the purely descriptive anatomy of the pelvis would be out of place. A knowledge of that must be taken for granted, and it is only necessary to refer to those points which have a more or less direct bearing on the study of its obstetrical relations.

The *pelvis* is formed of four bones. On either side are

The pelvis.

Its importance in obstetrics.

Formation of pelvis.

the *ossa innominata*, joined together by the *sacrum*; to the inferior extremity of the sacrum is attached the *coccyx*, which is, in fact, its continuation.

Os innominatum: its three divisions.

The *os innominatum* (fig. 1) is an irregularly shaped bone originally formed of three distinct portions, the *ilium*, the *ischium*, and the *pubes*, which remain separated from each other up to and beyond the period of puberty. They are united at the acetabulum by a Y-shaped cartilaginous junction, which does not, as a rule, become ossified until about the twentieth year. The consequence is that the pelvis, during the period of growth, is subject to the action

Fig. 1.

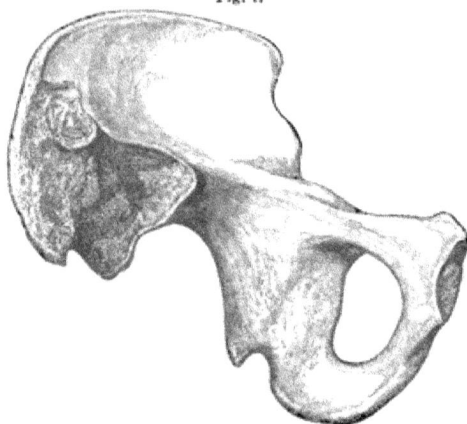

OS INNOMINATUM.

of various mechanical influences to a far greater extent than in adult life ; and these, as we shall presently see, have an important effect in determining the form of the bones. The external surface and borders of the os innominatum are chiefly of obstetric interest from giving attachment to muscles, many of which have an important accessory influence on parturition, such as the muscles forming the abdominal wall, which are attached to its crest, and those closing its outlet and forming the perinæum, which are attached to the tuberosity of the ischium. On the anterior and posterior extremities of the crest of the ilium are two prominences (the anterior and posterior spinous processes) which are points from which certain measurements are sometimes taken. The internal surface of the upper fan-shaped portion of the os innominatum gives attachment to the iliacus muscle, and contributes to the support of the abdominal contents; along with its fellow of the opposite side it forms the *false* pelvis. The false is separated from the *true* pelvis by the ilio-pectineal line, which, with the upper margin of the sacrum, forms the brim of the pelvis. This is of especial obstetric importance, as it is the first part of the pelvic cavity through which the child passes, and that in which osseous deformities are most often met with. At one portion of the ilio-pectineal line, corresponding with the junction of the ilium and pubes, is situated a prominence, which is known as the ilio-pectineal eminence. *Separation between the true and false pelvis. Importance of pelvic brim.*

The internal smooth surface of the innominate bone below the linea ilio-pectinea forms the greater portion of the pelvis proper. In front, with the corresponding portions of the opposite bone, it forms the arch of the pubes, under which the head of the child passes in labour. *Internal surface.*

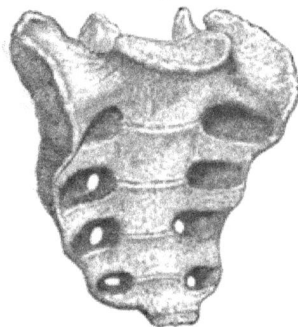

Fig. 2.

SACRUM AND COCCYX.

Behind this we observe the oval obturator foramen, and below that the tuberosity and spine of the ischium, the latter separating the great and lesser sciatic notches, and giving attachment to ligaments of importance. The rough articulating surface posteriorly, by which the junction with the sacrum is effected, may be noted, and above this the prominence to which the powerful ligaments joining the sacrum and os innominatum are attached. *Articulating surfaces.*

The *sacrum* (fig. 2) is a triangular and somewhat spongy bone forming the continuation of the spinal column, and binding together the ossa innominata. It is originally composed of five separate portions, analogous to the vertebræ, which ossify and unite about the period of puberty, leaving on its internal surface four prominent ridges at the points of junction. The upper of these is sometimes so well marked as to be mistaken, on vaginal examination, for the promontory of the sacrum itself. *Sacrum.*

The base of the sacrum is about 4½ inches in width, and its sides rapidly approximate until they nearly meet at its apex, giving the whole bone a triangular or wedge shape. The anterior and posterior surfaces also approximate in the same way, so that the bone is much thicker at the base than at the apex. The sacrum, in the erect position of the body, is directed from above downwards and from before backwards. At its upper edge it is joined, the lumbo-sacral cartilage intervening, with the fifth lumbar vertebra. The

Promontory of sacrum.

point of junction, called the promontory of the sacrum, is of great importance, as on its undue projection many deformities of the brim of the pelvis depend. The anterior surface of the bone is concave, and forms the curve of the sacrum ; more marked in some cases than in others. There is also more or less concavity from side to side. On it we observe four apertures on each side, the intervertebral foramina, giving exit to nerves. The posterior surface is convex, rough and irregular for the attachment of ligaments and muscles, and showing a ridge of vertical prominences, corresponding to the spinous processes of the vertebræ.

Mechanical relations of sacrum.

The sacrum is generally described as forming a keystone to the arch constituted by the pelvic bones, and transmitting the weight of the body, in consequence of its wedge-like shape, in a direction which tends to thrust it downwards and backwards, as if separating the ossa innominata. Dr. Duncan,[1] however, has shown, from a careful consideration of its mechanical relations, that it should rather be regarded as a strong transverse beam, curved on its anterior surface, the extremities of which are in contact with the corresponding articular surfaces of the ossa innominata. The weight of the body is thus transmitted to the innominate bones, and through them to the acetabula and the femora (fig. 3). There counter-pressure is applied, and the result is, as we shall subsequently see, an important modifying influence on the development and shape of the pelvis.

The coccyx (fig. 2) is composed of four small separate bones, which eventually unite into one, but not until late in life. The uppermost of these articulates with the apex of the sacrum. On its posterior surface are two small cornua, which unite with corresponding points at the tip of the

[1] Researches in Obstetrics, p. 67.

sacrum. The bones of the coccyx taper to a point. To it are attached various muscles which have the effect of imparting considerable mobility. During labour, also, it yields to the mechanical pressure of the presenting part, so as to increase the antero-posterior diameter of the pelvic outlet to the extent of an inch or more. *Its mobility.*

If, through disease or accident, as sometimes happens, the articular cartilages of the coccyx become prematurely ossified, the enlargement of the pelvic outlet during labour may be prevented, and considerable difficulty may thus arise. This is most apt to happen in aged primiparæ, or in women who have followed sedentary occupations ; and not infrequently, under such circumstances, the bone fractures under the pressure to which it is subjected by the presenting part. *Ossification of coccyx.*

The pelvic bones are firmly joined together by various articulations and ligaments. The latter are arranged so as to complete the canal through which the fœtus has to pass, and which is in great part formed by the bones. On its internal surface, where the absence of obstruction is of importance, they are everywhere smooth ; while externally, where strength is the desideratum, they are arranged in larger masses, so as to unite the bones firmly together. The pelvic articulations have been generally described as symphyses or amphiarthrodia, a term which is properly applied to two articulating surfaces, united by fibrous tissue in such a way as to prevent any sliding motion. It is certain, however, that this is not the case with the joints of the female pelvis during pregnancy and parturition. Lenoir found that in 22 females, between the ages of 18 and 35, there was a distinct sliding motion. Therefore, the pelvic articulations are, strictly speaking, to be considered examples of the class of joints termed arthrodia. *Pelvic articulations.*

The last lumbar vertebra is united to the sacrum by ligamentous union similar to that which joins the vertebræ to each other. The intervening fibro-cartilage forms a disk, which is thicker in front than behind, and this, in connection with a similar peculiarity of the fifth lumbar vertebra, tends to increase the sloped position of the sacrum, and the angle which it forms with the vertebral column. It constitutes the most prominent portion of the promontory of the *Lumbo-sacral joint.*

sacrum, and is the part on which the finger generally impinges in vaginal examinations. The anterior common vertebral ligament passes over the surface of the joints, and we also find the ligamenta sub-flava and the inter-spinous ligaments, as in the other vertebræ. The articular processes are joined together by a fibrous capsule, and there is also a peculiar ligament, the lumbo-sacral, extending from the transverse process of the vertebra on each side, and attaching itself to the sides of the sacrum and the sacro-iliac synchondrosis.

Ligaments of coccyx.

The sacrum is joined to the coccyx, and, in some cases at least, the separate bones of the coccyx to each other, by small cartilaginous disks like that connecting the sacrum with the last lumbar vertebra. They are further united by anterior and posterior common ligaments, the latter being much the thicker and more marked. In the adult female a synovial membrane is found between the sacrum and coccyx, and it is supposed that this is formed under the influence of the movements of the bones on each other.

Sacro-iliac synchondrosis.

The opposing articular surfaces of the sacrum and ilium are each covered by cartilages, that of the sacrum being the thicker. These are firmly united, but, in the female, according to Mr. Wood,[1] they are always more or less separated by an intervening synovial membrane. Posterior to these cartilaginous convex surfaces there are strong interosseous ligaments, passing directly from bone to bone, filling up the interspace between them, and uniting them firmly. There are also accessory ligaments, such as the superior and anterior sacro-iliac, which are of secondary consequence.

Posterior sacro-iliac ligaments.

The posterior sacro-iliac ligaments, however, are of great obstetric importance. They are the very strong attachments which unite the rough surfaces on the posterior iliac tuberosities to the posterior and lateral surfaces of the sacrum. They pass obliquely downwards from the former points, and suspend, as it were, the sacrum from them. According to Duncan, the sacrum has nothing to prevent its being depressed by the weight of the body but these ligaments, and it is mainly through them that the weight of the body is transmitted to the sacro-cotyloid beams and the heads of the femora.

[1] Todd's *Cyclopædia of Anatomy and Physiology*, article 'Pelvis,' p. 123.

The sacro-sciatic ligaments are instrumental in complet- Sacro-
ing the canal of the pelvis. The greater sacro-sciatic ligament sciatic
is attached by a broad base to the posterior-inferior spine ligaments.
of the ilium, and to the posterior surfaces of the sacrum and
coccyx. Its fibres unite into a thick cord, cross each other
in an X-like manner, and again expand at their insertion
into the tuberosity of the ischium. The lesser sacro-
sciatic ligament is also attached with the former to the

Fig. 3.

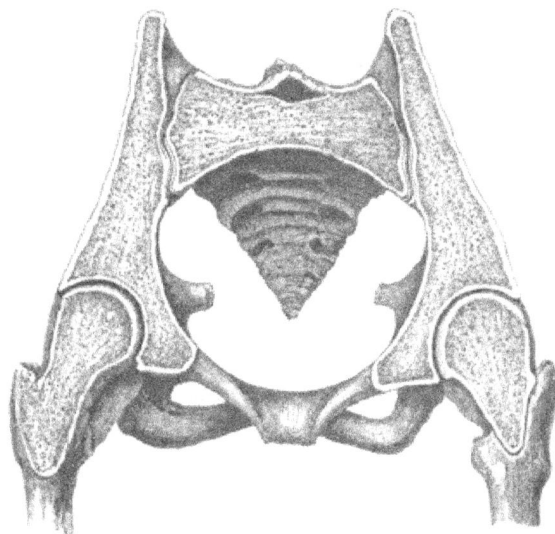

SECTION OF PELVIS AND HEADS OF THIGH-BONES, SHOWING THE SUSPENSORY ACTION OF
THE SACRO-ILIAC LIGAMENTS. (After Wood.)

back parts of the sacrum and coccyx, its fibres passing to
their much narrower insertion at the spine of the ischium,
and converting the sacro-sciatic notch into a complete
foramen.

The obturator membrane is the fibrous aponeurosis that Obturator
closes the large obturator foramen. Joulin[1] supposes that, mem-
along with the sacro-sciatic ligaments, it may, by yielding brane.
somewhat to the pressure of the fœtal head, tend to prevent
the contusion to which the soft parts would be subjected if
they were compressed between two entirely osseous surfaces.

[1] *Traité d'Accouchements*, p. 11.

Symphysis pubis.

The junction of the pubic bones **in front** is effected by means of two oval plates of fibro-cartilage, attached to each articular surface by nipple-shaped projections, which fit into corresponding depressions in the bones. There is a greater separation between the bones in front than behind, where the numerous fibres of the cartilaginous plates intersect, and unite the bones firmly together. At the upper and back part of the articulation there is an interspace between the cartilages, which is lined by a delicate membrane. In pregnancy this space often increases in size, so as to extend even to the front of the joint. The juncture is further strengthened by four ligaments, the anterior, the posterior, the superior, and the sub-pubic. Of these the last is the largest, connecting together the pubic bones and forming the upper boundary of the pubic arch.

Movements of pelvic joints.

The close apposition of the bones of the pelvis might not unreasonably lead to the supposition that no movement took place between its component parts; and this is the opinion which is even yet held by many anatomists. It is tolerably certain, however, that even in the unimpregnated condition there is a certain amount of mobility. Thus Zaglas has pointed out [1] that in man there is a movement in an antero-posterior direction of the sacro-iliac joints which has the effect, in certain positions of the body, of causing the sacrum to project downwards to the extent of about a line, thus narrowing the pelvic brim, tilting up the point of the bone, and thereby enlarging the outlet of the pelvis. This movement seems habitually brought into play in the act of straining during defæcation.

Observations in the lower animals.

During pregnancy in some of the lower animals there is a very marked movement of the pelvic articulations, which materially facilitates the process of parturition. This, in the case of the guinea-pig and cow, has been especially pointed out by Dr. Matthews Duncan.[2] In the former, during labour the pelvic bones separate from each other to the extent of an inch or more. In the latter the movements are different, for the symphysis pubis is fixed by bony anchylosis, and is immovable; but the sacro-iliac joints become swollen during pregnancy, and extensive movements in an antero-posterior

---

[1] *Monthly Journal of Med. Science*, Sept. 1851.
[2] *Researches in Obstetrics*, p. 19.

direction take place in them, which materially enlarge the pelvic canal during labour.

It is extremely probable that similar movements take place in women, both in the symphysis pubis and in the sacro-iliac joints, although to a less marked extent. These are particularly well described by Dr. Duncan. They seem to consist chiefly in an elevation and depression of the symphysis pubis, either by the ilia moving on the sacrum, or by the sacrum itself undergoing a forward movement on an imaginary transverse axis passing through it, thus lessening the pelvic brim to the extent of one or even two lines, and increasing, at the same time, the diameter of the outlet, by tilting up the apex of the sacrum. These movements are only an exaggeration of those which Zaglas describes as occurring normally during defæcation. The instinctive positions which the parturient woman assumes find an explanation in these observations. During the first stage of labour, when the head is passing through the brim, she sits, or stands, or walks about, and in these erect positions the symphysis pubis is depressed, and the brim of the pelvis enlarged to its utmost. As the head advances through the cavity of the pelvis, she can no longer maintain her erect position, and she lies down and bends her body forward, which has the effect of causing a nutatory motion of the sacrum, with corresponding tilting up of its apex, and an enlargement of the outlet.

Mode in which the movements are effected.

These movements during parturition are facilitated by the changes which are known to take place in the pelvic articulations during pregnancy. The ligaments and cartilages become swollen and softened, and the synovial membranes existing between the articulating surfaces become greatly augmented in size and distended with fluid. These changes act by forcing the bones apart, as the swelling of a sponge placed between them might do after it had imbibed moisture. The reality of these alterations receives a clinical illustration from those cases, which are far from uncommon, in which these changes are carried to so extreme an extent that the power of progression is materially interfered with for a considerable time after delivery.

Alterations in the pelvic joints during pregnancy.

They sometimes continue after delivery.

On looking at the pelvis as a whole, we are at once struck with its division into the true and false pelvis. The latter

Pelvis as a whole.

portion (all that is above the brim of the pelvis) is of comparatively little obstetric importance, except in giving attachments to the accessory muscles of parturition, and need
not be further considered.   The brim of the pelvis is a
heart-shaped opening, bounded by the sacrum behind, the
linea ilio-pectinea on either side, and the symphysis of the
pubes in front.   All below it forms the cavity, which is
bounded by the hollow of the sacrum behind, by the inner
surfaces of the innominate bones at the sides and in front,
and by the posterior surface of the symphysis pubis.   It is
in this part of the pelvis that the changes in direction which

Divisions of the true pelvis.

Fig. 4.

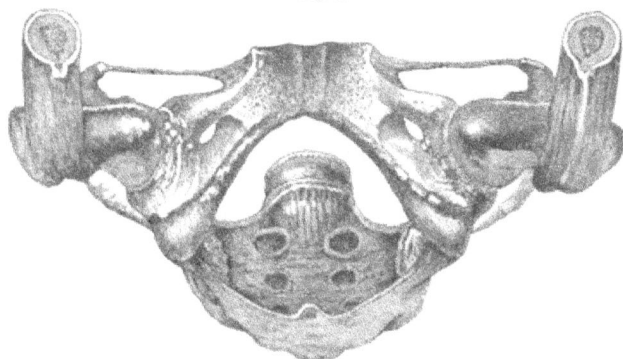

OUTLET OF PELVIS.

the fœtal head undergoes in labour are imparted to it.   The
lower border of this canal, or pelvic outlet (fig. 4), is lozenge-
shaped, is bounded by the ischiatic tuberosities on either
side, the tip of the coccyx behind, and the under-surface of
the pubic symphysis in front.   Posteriorly to the tuberosities
of the ischia the boundaries of the outlet are completed by
the sacro-sciatic ligaments.

Differences in the two sexes.

There is a very marked difference between the pelvis in
the male and the female, and the peculiarities of the latter
all tend to facilitate the process of parturition.   In the
female pelvis (fig. 5) all the bones are lighter in structure,
and have the points for muscular attachments much less
developed.   The iliac bones are more spread out, hence the
greater breadth which is observed in the female figure, and
the peculiar side-to-side movement which all females have

in walking.   The tuberosities of the ischia are lighter in
structure and further apart, and the rami of the pubes also
converge at a much less acute angle.   This greater breadth

Fig. 5.

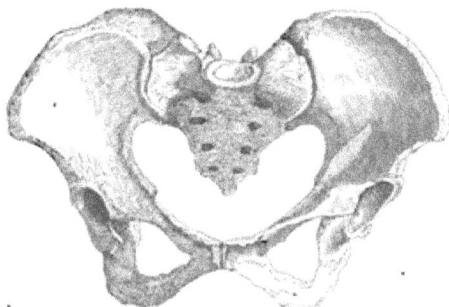

THE FEMALE PELVIS.

of the pubic arch gives one of the most easily appreciable
points of contrast between the male and the female pelvis; the
pubic arch in the female forms an angle of from 90° to 100°,
while in the male (fig. 6) it averages from 70° to 75°.   The
obturator foramina are more triangular in shape.

Fig. 6.

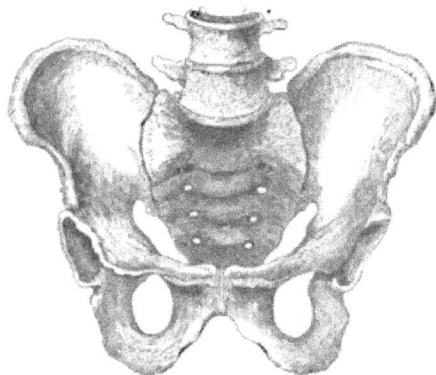

THE MALE PELVIS.

The whole cavity of the female pelvis is wider and less
funnel-shaped than in the male, the symphysis pubis is not
so deep, and, as the promontory of the sacrum does not
project so much, the shape of the pelvic brim is more oval
than heart-shaped.   These differences between the male and
female pelvis are probably due to the presence of the female

genital organs in the true pelvis, the growth of which increases its development in width.   In proof of this, Schroeder
states that in women with congenitally defective internal
organs, and in women who have had both ovaries removed
early in life, the pelvis has always more or less of the
masculine type.

The measurements of the pelvis that are of most importance from an obstetric point of view are taken between
various points directly opposite to each other, and are known
as the *diameters* of the pelvis.   Those of the true pelvis
are the diameters which it is especially important to fix in
our memories, and it is customary to describe three in works
on obstetrics—the antero-posterior or conjugate, the oblique,

Fig. 7.

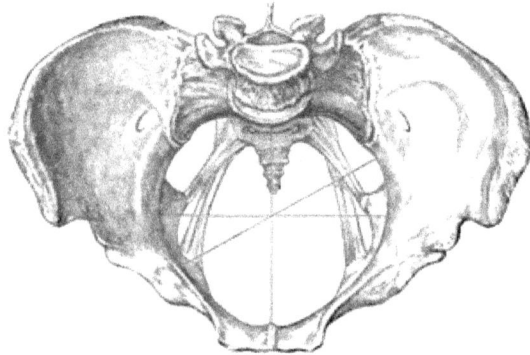

BRIM OF PELVIS, SHOWING ANTERO-POSTERIOR, OBLIQUE, AND CONJUGATE DIAMETERS.

and the transverse—although of course the measurements
may be taken at any opposing points in the circumference of
the bones.   The *antero-posterior* (sacro-pubic), at the brim
(fig. 7), is taken from the upper part of the posterior surface
of the symphysis pubis to the centre of the promontory of
the sacrum ; in the cavity, from the centre of the symphysis
pubis to a corresponding point in the body of the third piece
of the sacrum ; and at the outlet (coccy-pubic), from the
lower border of the symphysis pubis to the tip of the coccyx.
The *oblique*, at the brim, is taken from the sacro-iliac joint
on either side to a point of the brim corresponding with the
ilio-pectineal eminence (that starting from the right sacro-
iliac joint being called the right oblique, that from the left
the left oblique); in the cavity a similar measurement is

Marginal notes:
Causes of difference.

Measurements of the pelvis.

Points from which the diameters are measured.

Antero-posterior.

Oblique.

made at the same level as the conjugate; while at the outlet an oblique diameter is not usually measured. The *transverse* is taken at the brim, from a point midway between the sacro-iliac joint and the ilio-pectineal eminence to a corresponding point at the opposite side of the brim; in the cavity from points in the same plane as the conjugate and oblique diameters; and at the outlet from the centre of the inner border of one ischial tuberosity to that of the other. The measurements given by various writers differ considerably and vary somewhat in different pelves. Taking the average of a large number, the following may be given as the standard measurements of the female pelvis:—

Fig. 8.

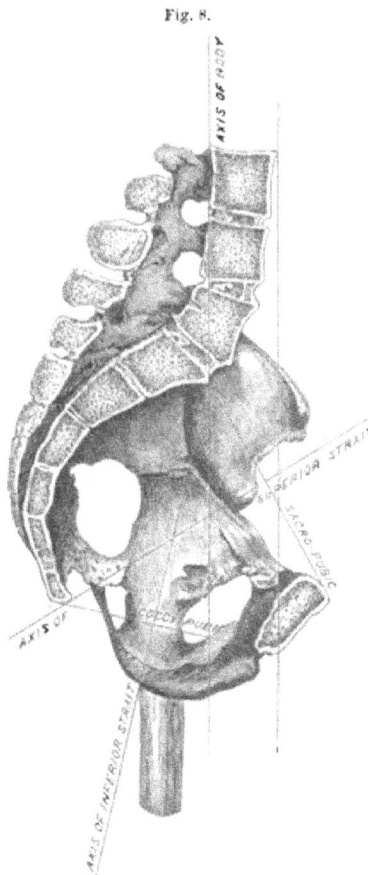

TRANSVERSE SECTION OF PELVIS, SHOWING THE DIAMETERS.

	Antero-posterior.	Oblique.	Transverse.
	In.	In.	In.
Brim . .	4·25	4·8	5·2
Cavity .	4·7	5·2	4·75
Outlet . .	5·0	—	4·2

It will be observed that the lengths of the corresponding diameters at different places vary greatly; thus while the transverse is longest at the brim, the oblique is longest in the cavity, and the antero-posterior at the outlet. It will be subsequently seen that this fact is of great practical importance in studying the mechanism of delivery, for the head in its descent through the pelvis alters its position in such a way as to adapt itself to the longest diameter of the pelvis; thus, as it passes through the cavity it lies in the oblique diameter, and then rotates so as to be expelled in the antero-posterior diameter of the outlet.

Diameters as altered by soft parts. In thinking of these measurements of the pelvis, it must not be forgotten that they are taken in the dried bones, and that they are considerably modified during life by the soft parts. This is especially the case at the brim, where the projection of the psoas and iliacus muscles lessens the transverse diameter about half an inch, while the antero-posterior diameter of the brim, and all the diameters of the cavity, are lessened by a quarter of an inch. The right oblique diameter of the brim is, even in the dried pelvis, found to be on an average slightly longer than the left, probably on account of the increased development of the right side of the pelvis from the greater use made of the right leg; but, in addition to this, the left oblique diameter is somewhat lessened during life by the presence of the rectum on the left side. The advantage gained by the comparatively frequent passage of the head through the pelvis in the right oblique diameter is thus explained.

Other measurements. There are one or two other measurements of the true pelvis which are sometimes given, but which are of secondary importance. One of these, the sacro-cotyloid diameter, is that between the promontory of the sacrum and a point immediately above the cotyloid cavity, and averages from 3·4 to 3·5 inches. Another, called by Wood the lower or inclined conjugate diameter, is that between the centre of the lower margin of the symphysis pubis and the promontory of the sacrum, and averages half an inch more than the antero-posterior diameter of the brim. These measurements are chiefly of importance in relation to certain pelvic deformities.

External measurements. The external measurements of the pelvis are of no real consequence in normal parturition, but they may help us, in certain cases, to estimate the existence and amount of deformities. Those which are generally given are: Between the anterior-superior iliac spines, 10 inches; between the central points of the crests of the ilia, 10½ inches; between the spinous process of the last lumbar vertebra and the upper part of the symphysis pubis (external conjugate), 7 inches.

Planes of the pelvis. By the planes of the pelvis are meant imaginary levels at any portion of its circumference. If we were to cut out a

piece of cardboard so as to fit the pelvic cavity, and place it
either at the brim or elsewhere, it would represent the
pelvic plane at that particular part, and it is obvious that
we may conceive as many planes as we desire.   Observation
of the angle which the pelvic planes form with the horizon
shows the great obliquity at which the pelvis is placed in
regard to the spinal column.   Thus the angle A B I (fig. 9)
represents the inclination to the horizon of the plane of the

Fig. 9.

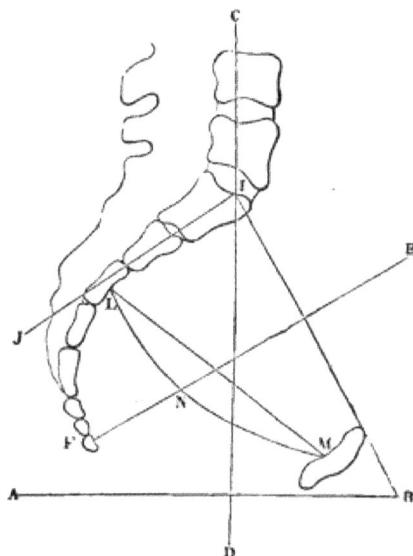

PLANES OF THE PELVIS WITH HORIZON.

A B. Horizon.   C D. Vertical line.
A B I. Angle of inclination of pelvis to horizon, equal to 60°.
B I C. Angle of inclination of pelvis to spinal column, equal to 150°.
C I J. Angle of inclination of sacrum to spinal column, equal to 130°.
E F. Axis of pelvic inlet.   L M. Mid plane in the middle line.
N. Lowest point of mid plane of ischium.

pelvic brim, I B, and is estimated to be about 60°, while the
angle which the same plane forms with the vertebral column
is about 150°.   The plane of the outlet forms, with the
coccyx in its usual position, an angle with the horizon of
about 11°, but which varies greatly with the movements of
the tip of coccyx, and the degree to which it is pushed back
during parturition.   These figures must only be taken as
giving an approximate idea of the inclination of the pelvis
to the spinal column, and it must be remembered that the

Inclina-
tion of the
pelvis
varies at
different
times.

degree of inclination varies considerably in the same female at different times, in accordance with the position of the body. During pregnancy especially, the obliquity of the brim is lessened by the patient throwing herself backwards in order to support more easily the weight of the gravid uterus. The height of the promontory of the sacrum above the upper margin of the symphysis pubis is on an average about 3¾ inches, and a line passing horizontally backwards from the latter point would impinge on the junction of the second and third coccygeal bones.

Fig. 10.

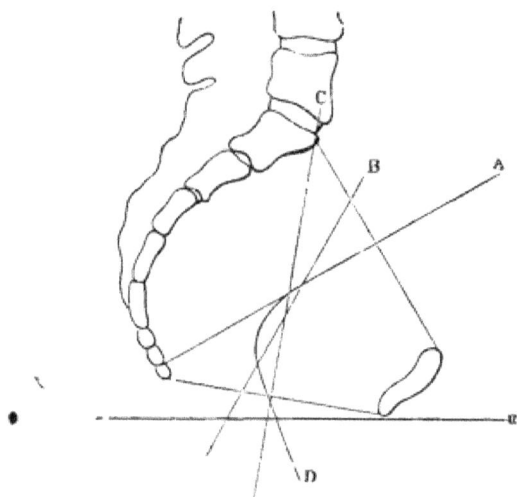

AXES OF THE PELVIS.

A. Axis of superior plane.   B. Axis of mid plane.   C. Axis of inferior plane.
D. Axis of canal.   E. Horizon.

Axes of the
parturient
canal.

By the axis of the pelvis is meant an imaginary line which indicates the direction which the fœtus takes during its expulsion. The axis of the brim (fig. 10) is a line drawn perpendicular to its plane, which would extend from the umbilicus to about the apex of the coccyx; the axis of the outlet of the bony pelvis intersects this, and extends from the centre of the promontory of the sacrum to midway between the tuberosities of the ischia. The axis of the entire pelvic canal is represented by the sum of the axes of an indefinite number of planes at different levels of the pelvic cavity, which forms an irregular parabolic line, as represented in the accompanying diagram (fig. 10, A D).

It must be borne in mind, however, that it is not the axis of the bony pelvis alone that is of importance in obstetrics. We must always, in considering this subject, remember that the general axis of the parturient canal (fig. 11) also includes that of the uterine cavity above, and of the soft parts below. These are variable in direction according to circumstances; and it is only the axis of

Axis of the whole parturient canal.

Fig. 11.

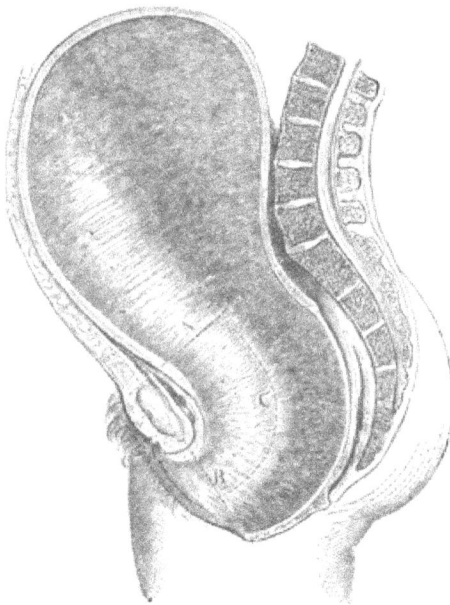

REPRESENTING GENERAL AXIS OF PARTURIENT CANAL, INCLUDING THE UTERINE
CAVITY AND SOFT PARTS.

that portion of the parturient canal extending between the plane of the pelvic brim and a plane between the lower edge of the pubic symphysis and the base of the coccyx that is fixed. The axis of the lower part of the canal will vary according to the amount of distension of the perinæum during labour; but when this is stretched to its utmost, just before the expulsion of the head, the axis of the plane between the edge of the distended perinæum and the lower border of the symphysis looks nearly directly forwards. The axis of the uterine cavity generally corresponds with that of the pelvic brim, but it may be much altered by abnormal positions of the uterus, such as anteversion from laxity of the abdominal walls. The fœtus, under such circumstances, will not enter the brim in its proper axis, and difficulties in the labour arise. A knowledge of the general direction of the parturient canal is of great importance in practical midwifery in guiding us to the introduction of the hand or instruments in obstetric operations, and in showing us how to obviate difficulties arising from such accidental deviations of the uterus as have been just alluded to.

Cavity of the pelvis. The arrangements of the bones in the interior of the pelvic canal (fig. 12) are important in relation to the mechanism of delivery. A line passing between the spine of the ischium and the ilio-pectineal eminence divides the inner surface of ischial bone into two smooth plane surfaces, which have received the name of the planes of the ischium. Two other planes are formed by the inner surfaces of the pubic bones in front and by the upper portion of the sacrum behind, both having a direction downwards and backwards. In studying the

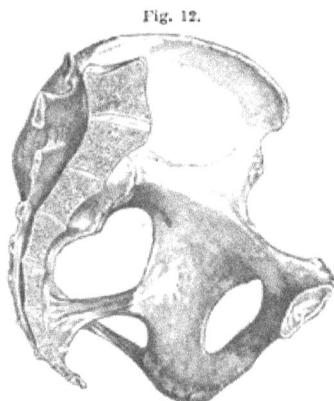

Fig. 12.

SIDE VIEW OF PELVIS.

mechanism of delivery, it will be seen that many obstetricians attribute to these planes, in conjunction with the spines of the ischia, a very important influence in effecting rotation of the fœtal head from the oblique to the antero-posterior diameter of the pelvis.

Development of the pelvis. The peculiarities of the pelvis during infancy and childhood are of interest as leading to a knowledge of the manner in which the form observed during adult life is impressed

Peculiarities of the infantile pelvis. upon it. The sacrum in the pelvis of the child (fig. 13) is less developed transversely, and is much less deeply curved than in the adult. The pubes is also much shorter from side to side, and the pubic arch is an acute angle. The result of this narrowness of both the pubes and sacrum is that the transverse diameter of the pelvic brim is shorter instead of longer than the antero-posterior. The sides of the pelvis have a tendency to parallelism, as well as the antero-posterior walls; and this is stated by Wood to be a peculiar characteristic of the infantile pelvis. The iliac bones are not spread out as in adult life, so that the centres of the crests of the ilia are not more distant from each other than the anterior superior spines. The cavity of the true pelvis is small, and the tuberosities of the ischia are proportionately nearer to each other than they afterwards become; the pelvic viscera are consequently crowded up into the abdominal cavity, which is, for this reason, much more prominent in children than in adults. The bones are soft

Fig. 13.

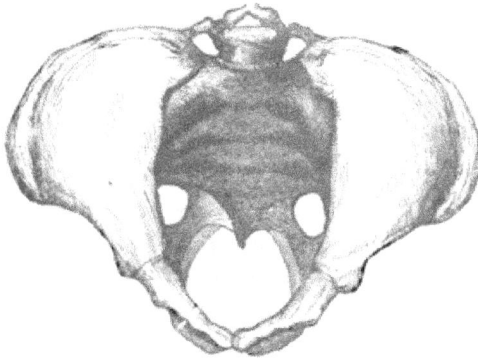

PELVIS OF A CHILD.

and semi-cartilaginous until after the period of puberty, and yield readily to the mechanical influences to which they are subjected; and the three divisions of the innominate bone remain separate until about the twentieth year.

As the child grows older the transverse development of the sacrum increases, and the pelvis begins to assume more and more of the adult shape. The mere growth of the bones, however, is not sufficient to account for the change in the shape of the pelvis, and it has been well shown by Duncan that this is chiefly produced by the pressure to which the bones are subjected during early life. The iliac bones are acted upon by two principal and opposing forces. One is the weight of the body above, which acts vertically

Mode in which the development of the pelvis is produced.

upon the sacral extremity of the iliac beam through the strong posterior sacro-iliac ligaments, and tends to throw the lower or acetabular ends of the sacro-cotyloid beams outwards. This outward displacement, however, is resisted, partly by the junction between the two acetabular ends at the front of the pelvis, but chiefly by the opposing force, which is the upward pressure of the lower extremities through the femurs.

Effects of pressure. The result of these counteracting forces is that the still soft bones bend near their junction with the sacrum; and thus the greater transverse development of the pelvic brim characteristic of adult life is established. In treating of pelvic deformities it will be seen that the same forces applied to diseased and softened bones explain the peculiarities of form that they assume.

Pelvis in different races. The researches that have been made on the differences of the pelvis in different races prove that these are not so great as might have been expected. Joulin pointed out that in all human pelves the transverse diameter was larger than the antero-posterior, while the reverse was the case in all the lower animals, even in the highest simiæ. This observation has been more recently confirmed by Von Franque,[1] who has made careful measurements of the pelvis in various races. In the pelvis of the gorilla, the oval form of the brim, resulting from the increased length of the conjugate diameter, is very marked. In certain races there is so far a tendency to animality of type that the difference between the transverse and conjugate diameters is much less than in European women, but is not sufficiently marked to enable us to refer any given pelvis to a particular race. Von Franque makes the general observation that the size of the pelvis increases from south to north, but that the conjugate diameter increases in proportion to the transverse in southern races.

Soft parts in connection with pelvis. In closing the description of the pelvis, the attention of the student must be directed to the muscular and other structures which cover it. It has already been pointed out that the measurements of the pelvic diameters are considerably lessened by the soft parts, which also influence parturition in other ways. Thus attached to the crests of the ilia are strong muscles which not only support the enlarged

---

[1] Scanzoni's *Beiträge,* 1867.

uterus during pregnancy, but are powerful accessory muscles in labour: in the pelvic cavity are the obturator and pyriformis muscles lining it on either side ; the pelvic cellular tissue and fasciæ ; the rectum and bladder ; the vessels and nerves, pressure on which often gives rise to cramps and pains during pregnancy and labour ; while below the outlet of the pelvis is closed, and its axis directed forwards by the numerous muscles forming the floor of the pelvis and perinæum. The structures closing the pelvis have been accurately described by Dr. Berry Hart,[1] who points out that they form a complete diaphragm stretching from the pelvis to the sacrum, in which are three ' faults ' or ' slits ' formed by the orifices of the urethra, vagina, and rectum. The first of these is a mere capillary slit, the last is closed by a strong muscular sphincter, while the vagina, in a healthy condition, is also a mere slit, with its walls in accurate apposition. Hence it follows that none of these apertures impairs the structural efficiency of the pelvic floor, or the support it gives to the structures above it.

The pelvic floor.

---

[1] *The Structural Anatomy of the Female Pelvic Floor.*

CHAPTER II.

THE FEMALE GENERATIVE ORGANS.

Division according to function.

1. External or copulative.

2. Internal or formative.

THE reproductive organs in the female are conveniently divided, according to their function, into: 1. The external or copulative organs, which are chiefly concerned in the act of insemination, and are only of secondary importance in parturition : they include all the organs situated externally which form the vulva; and the vagina, which is placed internally and forms the canal of communication between the uterus and the vulva. 2. The internal or formative organs : they include the ovaries, which are the most important of all, as being those in which the ovule is formed ; the Fallopian tubes, through which the ovule is carried to the uterus; and the uterus, in which the impregnated ovule is lodged and developed.

1. The external organs consist of :—

Mons veneris.

The *mons veneris* (fig. 14, *f*), a cushion of adipose and fibrous tissue which forms a rounded projection at the upper part of the vulva. It is in relation above with the lower part of the hypogastric region, from which it is often separated by a furrow, and below it is continuous with the labia majora on either side. It lies over the symphysis and horizontal rami of the pubes. After puberty it is covered with hair. On its integument are found the openings of numerous sweat and sebaceous glands.

Labia majora.

The *labia majora* (fig. 14, *a*) form two symmetrical sides to the longitudinal aperture of the vulva. They have two surfaces, one external, of ordinary integument, covered with hair, and another internal, of smooth mucous membrane, in apposition with the corresponding portion of the opposite labium, and separated from the external surface by a free convex border. They are thicker in front, where they run into the

mons veneris, and thinner behind, where they are united, in
front of the perinæum, by a thin fold of integument called
the fourchette, which is almost invariably ruptured in the first
labour.   In the virgin the labia are closely in apposition,
and conceal the rest of the generative organs.   After child-

Fig. 14.

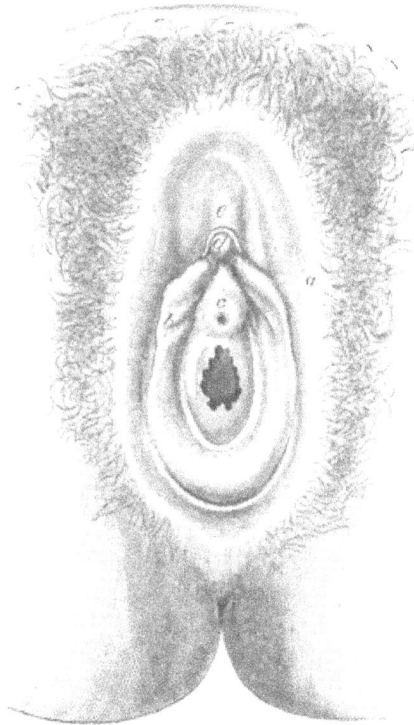

EXTERNAL GENITALS OF VIRGIN WITH DIAPHRAGMATIC HYMEN   (After Sappey.)

a. Labium majus. b. Labium minus. c. Præputium clitoridis. d. Glans
clitoridis. e. Vestibule just above urethral orifice. f. Mons veneris.

bearing they become more or less separated from each other,
and in the aged they waste, and the internal nymphæ
protrude through them.   Both their cutaneous and mucous
surfaces contain a large number of sebaceous glands, opening
either directly on the surface or into the hair follicles.   In
structure the labia are composed of connective tissue, containing a varying amount of fat, and parallel with their

external surface are placed tolerably close plexuses of elastic tissue, interspersed with regularly arranged smooth muscular fibres. These fibres are described by Broca as forming a membranous sac, resembling the dartos of the scrotum, to which the labia majora are analogous. Towards its upper and narrower end this sac is continuous with the external inguinal ring, and in it terminate some of the fibres of the round ligament. The analogy with the scrotum is further borne out by the occasional hernial protrusion of the ovary into the labium, corresponding to the normal descent of the testis in the male.

Labia minora.

The *labia minora*, or *nymphæ* (fig. 14, *b*), are two folds of mucous membrane, commencing below, on either side, about the centre of the internal surface of the labium externum ; they converge as they proceed upwards, bifurcating as they approach each other. The lower branch of this bifurcation is attached to the clitoris (fig. 14, *c*), while the upper and larger unites with its fellow of the opposite side, and forms a fold round the clitoris, known as its prepuce. The nymphæ are usually entirely concealed by the labia majora, but after childbearing and in old age they project somewhat beyond them; then they lose their delicate pink colour and soft texture, and become brown, dry, and like skin in appearance. This is especially the case in some of the negro races, in whom they form long projecting folds called the apron.

The surfaces of the nymphæ are covered with tesselated epithelium, and over them are distributed a large number of vascular papillæ, somewhat enlarged at their extremities, and sebaceous glands, which are more numerous on their internal surfaces. The latter secrete an odorous, cheesy matter, which lubricates the surface of the vulva, and prevents its folds adhering to each other. The nymphæ are composed of trabeculæ of connective tissue, containing muscular fibres.

The clitoris.

The *clitoris* (fig. 14, *d*) is a small erectile tubercle situated about half an inch below the anterior commissure of the labia majora. It is the analogue of the penis in the male, and is similar to it in structure, consisting of two corpora cavernosa, separated from each other by a fibrous septum. The crura are covered by the ischio-cavernous muscles, which serve the same

purpose as in the male. It has also a suspensory ligament.
The corpora cavernosa are composed of a vascular plexus
with numerous traversing muscular fibres. The arteries are
derived from the internal pudic artery, which gives a branch,
the cavernous, to each half of the organ; there is also a
dorsal artery distributed to the prepuce. According to
Gussenbauer these cavernous arteries pour their blood directly
into large veins, and a finer venous plexus near the surface receives arterial blood from small arterial branches. By these
arrangements the erection of the organ which takes place
during sexual excitement is favoured. The nervous supply
of the clitoris is large, being derived from the internal pudic
nerve, which supplies branches to the corpora cavernosa, and
terminates in the glands and prepuce, where Paccinian corpuscles and terminal bulbs are to be found. On this account
the clitoris has been supposed by some to be the chief seat
of voluptuous sensation in the female.

The *vestibule* (fig. 14, *e*) is a triangular space, bounded at
its apex by the clitoris, and on either side by the folds of the
nymphæ. It is smooth, and, unlike the rest of the vulva, is
destitute of sebaceous glands, although there are several groups
of muciparous glands opening on its surface. At the centre of
the base of the triangle, which is formed by the upper edge
of the opening of the vagina, is a prominence, distant about
an inch from the clitoris, on which is the orifice of the urethra.
This prominence can be readily made out by the finger, and
the depression upon it—leading to the urethra—is of importance as our guide in passing the female catheter. This
little operation ought to be performed without exposing
the patient, and it is done in several ways. The easiest is
to place the tip of the index finger of the left hand (the
patient lying on her back) on the apex of the vestibule,
and slip it gently down until we feel the bulb of the
urethra, and the dimple of its orifice, which is generally
readily found. If there is any difficulty in finding the
orifice, it is well to remember that it is placed immediately
below the sharp edge of the lower border of the symphysis
pubis, which will guide us to it. The catheter (and a male
elastic catheter is always the best, especially during labour,
when the urethra is apt to be stretched) is then passed
under the thigh of the patient, and directed to the orifice

The
vestibule.

Orifice
of the
urethra.

Passing of
the female
catheter.

of the urethra by the finger of the left hand, which is placed upon it. We must be careful that the instrument is really passed into the urethra, and not into the vagina. It is advisable to have a few feet of elastic tubing attached to the end of the catheter, so that the urine can be passed into a vessel under the bed without uncovering the patient. If the patient be on her side, in the usual obstetric position, the operation can be more readily performed by placing the tip of the finger in the vagina and feeling its upper edge. The orifice of the urethra lies immediately above this, and if the catheter be slipped along the palmar surface of the finger, it can generally be inserted without much trouble. If, however, as is often the case during labour, the parts are much swollen, it may be difficult to find the aperture, and it is then always better to look for the opening than to hurt the patient by long-continued efforts to feel it.

The urethra.

The *urethra* is a canal 1½ inches in length, and it is intimately connected with the anterior wall of the vagina, through which it may be felt. It is composed of muscular and erectile tissue, and is remarkable for its extreme dilatability, a property which is turned to practical account in some of the operations for stone in the female bladder.

Orifice of the vagina.

The orifice of the vagina is situated immediately below the bulb of the urethra. In virgins it is a circular opening, but in women who have borne children or practised sexual intercourse it is, in the undistended state, a fissure, running transversely, and at right angles to that between the labia.[1] In virgins it is generally more less blocked up by a fold of mucous membrane, containing some cellular tissue and muscular fibres, with vessels and nerves, which is known as the *hymen*. This is most often crescentic in shape, with the concavity of the crescent looking upwards; sometimes, however, it is circular with a central opening, or cribriform; or it may even be entirely imperforate, and this gives rise to the retention of the menstrual secretion. These varieties of form depend on the peculiar mode of development of the fold of vaginal mucous membrane which blocks up the orifice of the vagina in the fœtus, and from which the hymen is formed. The density of the membrane also varies in different

The hymen.

[1] Hart, *op. cit.*

individuals. Most usually it is very slight, so as to be ruptured in the first sexual approaches, or even by some accidental circumstance, such as stretching the limbs, so that its absence cannot be taken as evidence of want of chastity. A knowledge of this fact is of considerable importance from a medico-legal point of view. Sometimes it is so tough as to prevent intercourse altogether, and may require division by the knife or scissors before this can be effected ; and at others it rather unfolds than ruptures, so that it may exist even after impregnation has been effected, and it has been met with intact in women who have habitually led unchaste lives. In a few rare cases it has even formed an obstacle to delivery, and has required incision during labour.

The *carunculæ myrtiformes* are small fleshy tubercles, varying from two to five in number, situated round the orifice of the vagina, and which are generally supposed to be the remains of the ruptured hymen. Schroeder, however, maintains that they are only formed after childbearing, in consequence of parts of the hymen having been destroyed by the injuries received during the passage of the child. *Carunculæ myrti-formes.*

Near the posterior part of the vaginal orifice, and below the superficial perineal fascia, are situated two conglomerate glands which are the analogues of Cowper's glands in the male. Each of these is about the size and shape of an almond, and is contained in a cellular fibrous envelope. Internally they are of a yellowish-white colour, and are composed of a number of lobules separated from each other by prolongations of the external envelope. These give origin to separate ducts which unite into a common canal, about half an inch in length, which opens in front of the attached edge of the hymen in virgins, and in married women at the base of one of the carunculæ myrtiformes. According to Huguier, the size of the glands varies much in different women, and they appear to have some connection with the ovary, as he has always found the largest gland to be on the same side as the largest ovary. They secrete a glairy, tenacious fluid, which is ejected in jets during the sexual orgasm, probably through the spasmodic action of the perineal muscles. At other times their secretion serves the purpose of lubricating the vulva, and thus preserves the sensibility of its mucous membrane. *Vulvo-vaginal glands.*

Fossa na-
vicularis. Immediately behind the hymen in the unmarried, and between it and the perinæum, is a small depression called the *fossa navicularis,* which disappears after childbearing.

Perinæum. The *perinæum* separates the orifice of the vagina from that of the rectum. It is about 1½ inches in breadth, and is of great obstetric interest, not only as supporting the internal organs from below, but because of its action in labour. It is largely stretched and distended by the presenting part of the child, and if unusually tough and unyielding, may retard delivery, or it may be torn to a greater or less extent, thus giving rise to various subsequent troubles.

Vascular
supply of
the vulva. The structures described above together form the *vulva,* and they are remarkable for their abundant vascular and nervous supply. The former constitutes an erectile tissue, similar to that which has already been described in the clitoris, and which is especially marked about the bulb of the vestibule (fig. 15). From this point and extending on either side of the vagina, there is a well-marked plexus of convoluted veins, which, in their distended state, are likened by Dr. Arthur Farre, to a filled leech. The erection of the erectile tissue, as well as that of the clitoris, is brought about under excitement, as in the male, by the compression of the efferent veins by the contraction of the ischio-cavernous muscles, and by that of a thin layer of muscular tissues surrounding the orifice of the vagina, and described as the constrictor vaginæ.

The
vagina. The *vagina* is the canal which forms the communication between the external and internal generative organs, through which the semen passes to reach the uterus, the menses flow, and the fœtus is expelled. Roughly speaking it lies in the axis of the pelvis, but its opening is placed anterior to the axis of the pelvic outlet, so that its lower portion is curved forwards, so as to lie parallel to the pelvic brim. It is narrow below, but dilated above, where the cervix uteri is inserted into it, so that it is more or less conoidal in shape. Under ordinary circumstances, especially in the virgin, the anterior and posterior walls lie in close contact with each other (see Plate I.), and there is, strictly speaking, no vaginal canal, although they are capable of wide distension, as in copulation, and during the passage of the fœtus. The anterior wall of the vagina is shorter than the posterior,

the former measuring on an average 2½ inches, the latter 3
inches; but the length of the canal varies greatly in different
subjects and under certain circumstances. In front the
vagina is closely connected with the base of the bladder, so
that when the vagina is prolapsed, as often occurs, it drags
the bladder with it (fig. 17); behind, it is in relation with
the rectum, but less intimately; laterally with the broad
ligaments and pelvic fascia; and superiorly with the lower

Fig. 15.

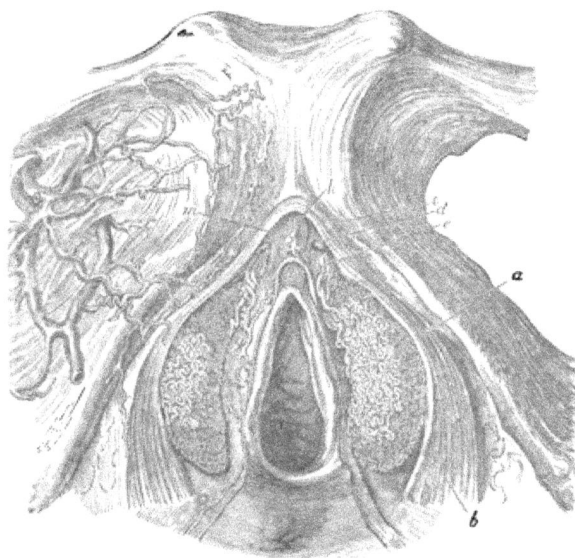

VASCULAR SUPPLY OF VULVA. (After Kobelt.)

a. Bulb of vestibule. *b.* Muscular tissue of vagina. *c, d, e, f.* The clitoris and muscles. *g, h, i, k, l, m, n,* Veins of the nymphæ and clitoris communicating with the epigastric and obturator veins.

portion of the uterus and folds of peritoneum both before
and behind. The vagina is composed of mucous, muscular,
and cellular coats. The mucous lining is thrown into
numerous folds. These start from longitudinal ridges which
exist on both the anterior and posterior walls, but most distinctly on the anterior. They are very numerous in the
young and unmarried, and greatly increase the sensitive
surface of the vagina (fig. 16). After childbearing, and in
the aged, they become atrophied, but they never completely
disappear, and towards the orifice of the vagina, where they

(margin) Composed of mucous, muscular, and cellu- lar coats.

exist in greatest abundance, they are always to be met with. The whole of the mucous membrane is lined with tesselated epithelium, and it is covered with a large number of papillæ, either conical or divided, which are highly vascular and project into the epithelial layer. Unlike the **vulvar** mucous membrane, that of the vagina seems to be destitute of glands. Beneath the epithelial layer is a submucous tissue containing a large number of elastic and some muscular fibres, derived from the muscular walls of the vagina. These are strong and well developed, especially towards the ostium vaginæ. They consist of two layers—an internal longitudinal,

Fig. 16.

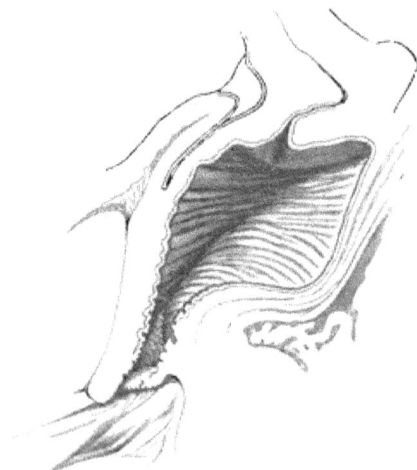

RIGHT HALF OF VIRGIN VAGINA, WITH WALLS HELD APART, SHOWING THE ABUNDANT TRANSVERSE RUGÆ, THE GREATER DEPTH OF THE VAGINA ABOVE THAN BELOW, AND THE HYMENEAL SEGMENT. (After Hart.)

and an external circular—with oblique decussating fibres connecting the two. Below they are attached to the ischio-pubic rami, and above they are continuous with the muscular coat of the uterus. The muscular tissue of the vagina increases in thickness during pregnancy, but to a much less degree than that of the uterus. Its vascular arrangements, like those of the vulva, are such as to constitute an erectile tissue. The arteries form an intricate network around the tube, and eventually end in a submucous capillary plexus from which twigs pass to supply the papillæ; these again give origin to venous radicles which unite into meshes freely

Its vascular arrange-ments.

interlacing with each other, and forming a well-marked venous plexus.

2. The internal organs of generation consist of the uterus, the Fallopian tubes, and the ovaries; and in connection with them we have to study the various ligaments and folds of peritoneum which serve to maintain the organs in position, along with certain accessory structures. Physiologically, the most important of all the generative organs are the ovaries, in which the ovules are formed, and which dominate the

Fig. 17.

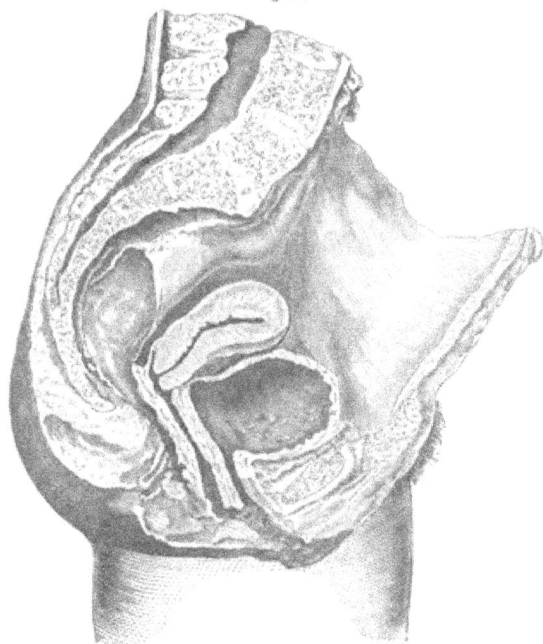

LONGITUDINAL SECTION OF BODY, SHOWING RELATIONS OF GENERATIVE ORGANS.

entire reproductive life of the female. The Fallopian tubes, which convey the ovule to the uterus, and the uterus itself—whose main function is to receive, nourish, and eventually expel the impregnated product of the ovary—may be said to be, in fact, accessory to these viscera. Practically, however, as obstetricians, we are chiefly concerned with the uterus, and may conveniently commence with its description.

The *uterus* is correctly described as a pyriform organ, flattened from before backwards, consisting of the body, with its rounded fundus, and the cervix, which projects into the upper part of the vaginal canal. In the adult female it is deeply situated in the pelvis, being placed between the bladder in front and the rectum behind, its fundus being below the plane of the pelvic brim (fig. 18). It only assumes this position, however, towards the period of puberty; and in the fœtus it is placed much higher, and lies, indeed, entirely within the cavity of the abdomen. It is maintained in this position partly by being slung by its ligaments, which

Fig. 18.

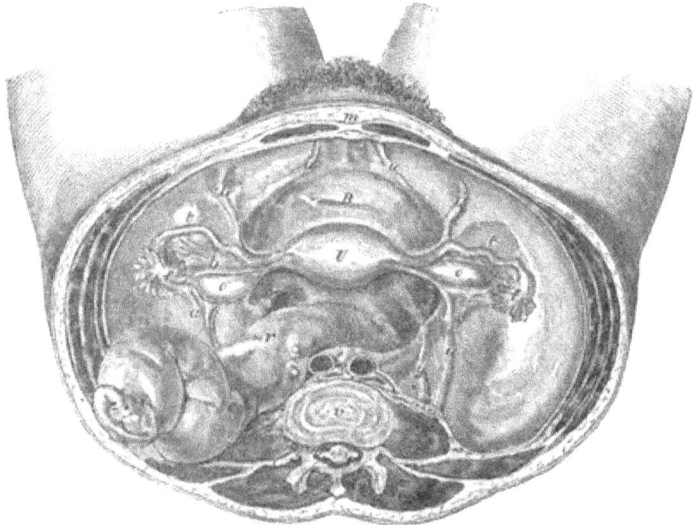

TRANSVERSE SECTION OF THE BODY, SHOWING RELATIONS OF THE FUNDUS UTERI.

m. Pubes. *a a* (in front). Remainder of hypogastric arteries. *a a* (behind). Spermatic vessels and nerves. *B.* Bladder. *L L.* Round ligaments. *U.* Fundus uteri. *t, t.* Fallopian tubes. *o, o.* Ovaries. *r.* Rectum. *g.* Right ureter, resting on the psoas muscle. *c.* Utero-sacral ligaments. *v.* Last lumbar vertebra.

we shall subsequently study, and partly by being supported from below by the pelvic cellular tissue and the fleshy column of the vagina. The result is that the uterus, in the healthy female, is a perfectly movable body, altering its position to suit the condition of the surrounding viscera, especially the bladder and rectum, which are subjected to variations of size according to their fulness or emptiness. When from any cause—as, for example, some peri-uterine inflammation producing adhesions to the surrounding textures—the mobility of the organ is interfered with, much

Is a perfectly movable organ.

distress ensues, and if pregnancy supervenes more or less serious consequences may result. Generally speaking, the uterus may be said to lie in a line roughly corresponding with the axis of the pelvic brim, its fundus being pointed forwards and its cervix lying in such a direction that a line drawn from it would impinge on the junction between the sacrum and coccyx. According to some authorities, the uterus in early life is more curved in the anterior direction, and is, in fact, normally in a state of ante-flexion. Sappey holds that this is not necessarily the case, but that the amount of anterior curvature depends on the emptiness or fulness of the bladder, on which the uterus, as it were, moulds itself in the unimpregnated state. It is believed also that the body of the uterus is very generally twisted

It lies in the axis of the pelvis.

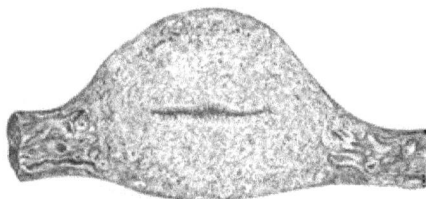

Fig. 19.

TRANSVERSE SECTION OF UTERUS.

somewhat obliquely, so that its interior surface looks a little towards the right side, this probably depending on the presence and frequent distension of the rectum in the left side of the pelvis. The anterior surface of the uterus is convex, and is covered in three-fourths of its extent by the peritoneum, which is intimately adherent to it. Below the reflexion of the membrane it is loosely connected by cellular tissue to the bladder, so that any downward displacement of the uterus drags the bladder along with it. The posterior surface is also convex, but more distinctly so than the anterior, as may be observed in looking at a transverse section of the organ (fig. 19). It is also covered by peritoneum, the reflection of which on the rectum forms the cavity known as Douglas's pouch. The fundus is the upper extremity of the uterus, lying above the points of entry by the Fallopian tubes. It is only slightly rounded in the virgin, but becomes more decidedly and permanently rounded in the woman who has borne children.

Its sur-
faces.

Until the period of puberty the uterus remains small and undeveloped (fig. 20); after that time it reaches the adult size, at which it remains until menstruation ceases, when it again atrophies. If the woman has borne children, it always remains larger than in the nullipara. In the virgin adult the uterus measures $2\frac{1}{4}$ inches from the orifice to the fundus, rather more than half being taken up by the cervix. Its greatest breadth is opposite the insertion of the Fallopian tubes; its greatest thickness, about 11 or 12 lines, opposite the centre of its body. Its average weight is about 9 or 10

Fig. 20.

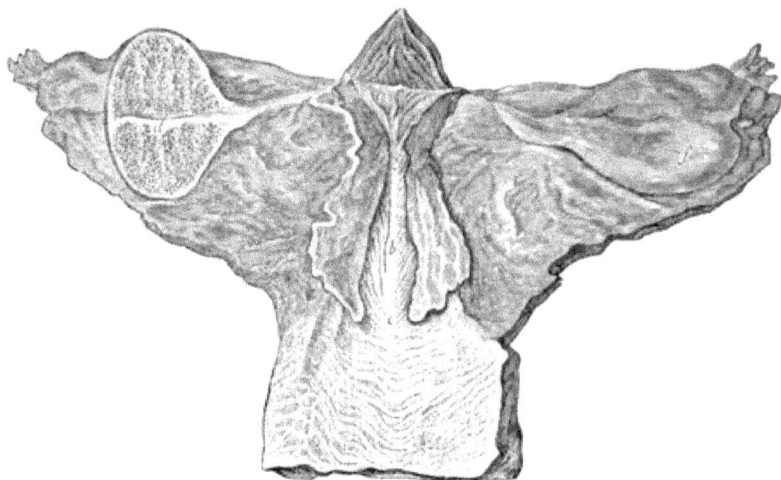

UTERUS AND APPENDAGES IN AN INFANT. (After Farre.)

drachms. Independently of pregnancy, the uterus is subject to great alterations of size towards the menstrual period, when on account of the congestion then present, it enlarges, sometimes, it is said, considerably. This fact should be borne in mind, as this periodical swelling might be taken for an early pregnancy.

Regional
division

For the purpose of description the uterus is conveniently divided into the *fundus*, with its rounded upper extremity, situated between the insertions of the Fallopian tubes; the *body*, which is bounded above by the insertions of the Fallopian tubes, and below by the upper extremity of the cervix, and which is the part chiefly concerned in the reception and growth of the ovum; and the *cervix*, which projects

into the vagina, and dilates during labour to give passage to
the child. The cervix is conical in shape, measuring 11 to
12 lines transversely at the base, and 6 or 7 in the antero-
posterior direction; while at the apex it measures 7 to 8
transversely, and 5 antero-posteriorly. It projects about 4
lines into the canal of the vagina, the remainder of the
cervix being placed above the reflection of the vaginal
mucous membrane. It varies much in form in the virgin
and nulliparous married woman, and in the woman who has
borne children; and the differences are of importance in the
diagnosis of pregnancy and uterine disease. In the virgin
it is regularly pyramidal in shape. At its lower extremity is
the opening of the external os uteri, forming a small transverse fissure, sometimes difficult to feel, and generally described as giving a sensation to the examining finger like
the extremity of the cartilage at the tip of the nose. It is
bounded by two lips, the anterior of which is apparently
larger on account of the position of the uterus. The surface
of the cervix and the borders of the os are very smooth and
regular.

In women who have borne children these parts become
considerably altered. The cervix is no longer conical, but is
irregular in form and shortened. The lips of the os uteri
become fissured and lobulated, on account of partial lacerations which have occurred during labour. The os is larger
and more irregular in outline, and is sometimes sufficiently
patulous to admit the tip of the finger. In old age the
cervix atrophies, and after the change of life it not uncommonly entirely disappears, so that the orifice of the os uteri
is on a level with the roof of the vagina. *Changes after childbirth.*

The internal surface of the uterus comprises the cavities
of the body and cervix—the former being rather less than
the latter in length in virgins, but about equal in women
who have borne children—separated from each other by a
constriction forming the upper boundary of the cervical
canal. The cavity of the body is triangular in shape, the
base of the triangle being formed by a line joining the openings of the Fallopian tubes, its apex by the upper orifice of
the cervix, or internal os, as it is sometimes called. In the
virgin its boundaries are somewhat convex, projecting inwards.
After child-bearing they become straight or slightly concave. *Internal surface of the uterus.* *The cavity of the body.*

D 2

The opposing surfaces of the cavity are always in contact in the healthy state, or are only separated from each other by a small quantity of mucus.

Cavity of the cervix. The cavity of the cervix is spindle-shaped or fusiform, narrower above and below, at the internal and external os uteri, and somewhat dilated between these two points. It is flattened from before backwards, and its opposing surfaces also lie in contact, but not so closely as those of the body. On the mucous lining of the anterior and posterior surfaces is a prominent perpendicular ridge, with a lesser one at each

Fig. 21.

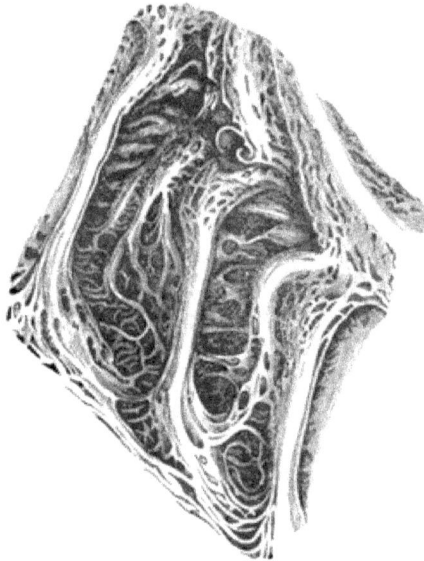

PORTION OF INTERIOR OF CERVIX. (Enlarged nine diameters.)
(After Tyler Smith and Hassall.)

side, from which transverse ridges proceed at more or less acute angles. They have received the name of the *arbor vitæ*. According to Guyon the perpendicular ridges are not exactly opposite, so that they fit into each other, and serve more completely to fill up the cavity of the cervix, especially towards the internal os (fig. 21). The arbor vitæ is most distinct in the virgin, and atrophies considerably after child-bearing.

The superior extremity of the cervical canal forms a

narrow isthmus separating it from the cavity of the body, and
measuring about ⅜ths of an inch in diameter. Like the external os, it contracts after the cessation of menstruation,
and in old age sometimes becomes entirely obliterated.

The uterus is composed of three principal structures—the
peritoneal, muscular, and mucous coats. The peritoneum
forms an investment to the greater part of the organ, extending downwards in front to the level of the os internum, and
behind to the top of the vagina, from which points it is
reflected upwards on the bladder and rectum respectively.
At the sides the peritoneal investment is not so extensive,
for a little below the level of the Fallopian tubes the peritoneal folds separate from each other, forming the broad
ligaments (to be afterwards described); here it is that the
vessels and nerves sup-
plying the uterus gain
access to it. At the upper
part of the organ the pe-
ritoneum is so closely adherent to the muscular
tissue that it cannot be
separated from it; below
the connection is more
loose. The mass of the
uterine tissue, both in the body and cervix, consists of
unstriped muscular fibres, firmly united together by
nucleated connective tissue and elastic fibres. The muscular
fibre cells are large and fusiform, with very attenuated extremities, generally containing in their centre a distinct
nucleus. These cells, as well as their nuclei, become greatly
enlarged during pregnancy (fig. 23); according to Stricker,
this is only the case with the muscular fibres which play an
important part in the expulsion of the fœtus, those of the
outermost and innermost layers not sharing in the increase
of size.[1] In addition to these developed fibres there are,
especially near the mucous coat, a number of round elementary corpuscles, which are believed by Dr. Farre[2] to be the
elementary form of the muscular fibres, and which he has
traced in various intermediate states of development. Dr.

Margin notes: Structure of the uterus. Its peritoneal investment. Its proper tissue is composed of unstriped muscular fibres.

Fig. 22.

MUSCULAR FIBRES OF UNIMPREGNATED
UTERUS. (After Farre.)
a. Fibres united by connective tissue.
b. Separate fibres and elementary corpuscles.

[1] *Comparative Histology*, vol. iii. *Syd. Soc. Trans.* p. 477.
[2] *The Uterus and its Appendages*, p. 632.

A great part of the tissue represents muscularis mucosæ.

John Williams [1] believes that a great part of the muscular tissue of the uterus, rather more indeed than three-fourths of its thickness, is an integral part of the mucous membrane, analogous to the muscularis mucosæ of the mucous membrane of the alimentary canal. This he describes as being separated from the rest of the muscular tissue by a layer of rather loose connective tissue, containing numerous vessels. In early fœtal life, and in the uteri of some of the lower

Fig. 23.

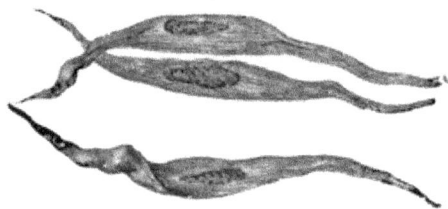

DEVELOPED MUSCULAR FIBRES FROM THE GRAVID UTERUS. (After Wagner.)

animals, this appearance is very distinct; in the adult female uterus, however, it cannot be readily made out.

Arrangement of the muscular fibres.

On examining the uterine tissue in an unimpregnated condition no definite arrangement of its muscular fibres can be made out, and the whole seem blended in inextricable confusion. By observation of their relations when hypertrophied during pregnancy, Helié [2] has shown that they may, speaking roughly, be divided into three layers: an external; a middle, chiefly longitudinal; and an internal, chiefly circular. Into the details of their distribution, as described by him, it is needless to enter at length. Briefly, however, he describes the external layer as arising posteriorly at the junction of the body and cervix, and spreading upwards and over the fundus. From this are derived the muscular fibres found in the broad and round ligaments, and more particularly described by Rouget. The middle layer is made up of strong fasciculi, which run upwards, but decussate and unite with each other in a remarkable manner, so that those which are at first superficial become most deeply seated, and vice versâ. The muscular fasciculi which form this coat curve in

[1] 'On the Structure of the Mucous Membrane of the Uterus,' *Obstet. Journ.* 1875.

[2] *Recherches sur la disposition des Fibres musculaires de l'Utérus.* Paris, 1869.

a circular manner round the large veins, so as to form a species of muscular canal through which they run. This arrangement is of peculiar importance, as it affords a satis-factory explanation of the mechanism by which hæmorrhage after delivery is prevented. The internal layer is mainly composed of circular rings of muscular fibres, beginning round the openings of the Fallopian tubes, and forming wider and wider circles which eventually touch and interlace with each other. They surround the internal os, to which they form a kind of sphincter. In addition to these circular fibres on the internal uterine surface, both anteriorly and posteriorly, there is a well-marked triangular layer of longitudinal fibres, the base being above and the apex below, which sends mus-cular fasciculi into the mucous membrane.

The anatomy of the lining membrane of the uterus has been the subject of considerable discussion. Its existence

Its mucous membrane.

Fig. 24.

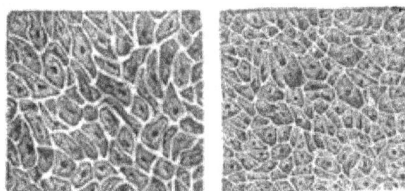

LINING MEMBRANE OF UTERUS, SHOWING NETWORK OF CAPILLARIES AND ORIFICES OF UTERINE GLANDS. (After Farre.)

From the body.   From orifice of Fallopian tube.

has been denied by many authorities, most recently by Snow Beck,[1] who maintains that it is in no sense a mucous mem-brane, but only a softened portion of true uterine tissue. It is, however, pretty generally admitted by the best authorities that it is essentially a mucous membrane, differing from others only in being more closely adherent to the subjacent structures, in consequence of not possessing any definite con-nective tissue framework.

It is a pale pink membrane of considerable thickness, most marked at the centre of the body, where it forms from $\frac{1}{8}$th to $\frac{1}{4}$th of the thickness of the whole uterine walls. At the internal os uteri it terminates by a distinct border, which

---

[1] *Obst. Trans.* vol. xiii. p. 294.

separates it from the mucous membrane lining the cervical cavity.

The utricular glands.

On the surface of the mucous membrane may be observed a multitude of little openings, about $\frac{1}{30}$th of a line in width, (fig. 24). These are the orifices of the utricular glands which are found in immense numbers all over the cavity of the uterus, and very closely agglomerated together. They are little cul-de-sacs, narrower at their mouths than in their length, the blind extremities of which are found in the subjacent tissues (fig. 26). Williams describes them as running obliquely towards the surface at the lower third of the cavity, perpendicularly at its middle, while towards the fundus they are at first perpendicular, and then oblique in their course (fig. 25). By others they are described as being often twisted and corkscrew-like. One or more may unite to form a common orifice, several of which may open together in little pits or depressions on the surface of the mucous membrane. These glands are composed of structureless membrane lined with epithelium, the precise character of which is doubtful. By some it is described as columnar, by others tesselated, and by some again as ciliated. The most generally received opinion is that it is columnar, but not ciliated; therein differing from the epithelium covering the surface of the membrane, which is undoubtedly ciliated, the movements of the cilia being from within outwards. Williams, however, has observed cilia in active movement on the columnar epithelium lining the glands, and also states that at the deep-seated extremities of the glands, which penetrate between the muscular fibres for some distance, the columnar epithelium is replaced by rounded cells. The capillaries of

Fig. 25.

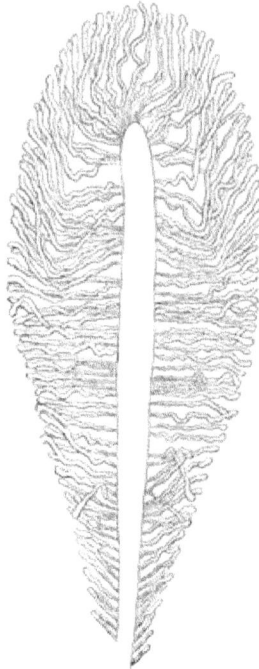

THE COURSE OF THE GLANDS IN THE FULLY DEVELOPED MUCOUS MEMBRANE OF THE UTERUS, VIZ., JUST BEFORE THE ONSET OF A MENSTRUAL PERIOD. (After Williams.)

the mucous membrane run down between the tubes, forming a lacework on their surfaces, and round their orifices.
No true papillæ exist in the membrane lining the uterine
cavity. The mucous membrane of the uterus is peculiar in
being always in a state of change and alteration, being thrown
off at each menstrual period in the form of débris, in conse-

Fig. 26.

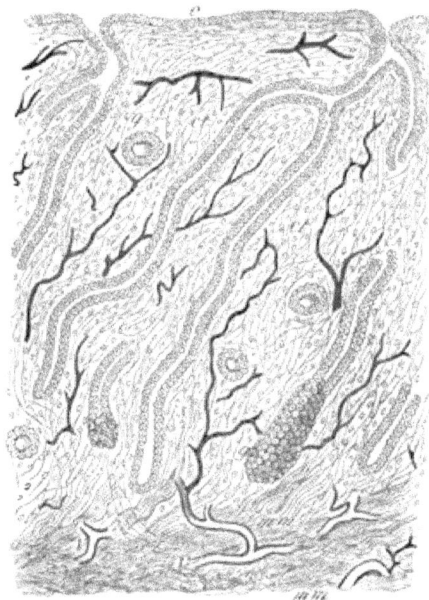

VERTICAL SECTION THROUGH THE MUCOUS MEMBRANE OF THE HUMAN UTERUS.
(After Turner.)

e. Columnar epithelium, the Cilia are not represented; g.g. Utricular
glands; ct. ct. Interglandular connective tissue; r.r. Blood-vessels;
m.m. Muscularis mucosæ (¹⁄₁₀.)

quence of fatty degeneration of its structures, and reformed
afresh by proliferation of the cells of the muscular and connective tissues, probably from below upwards, the new
membrane commencing at the internal os. Hence its appearance and structure vary considerably according to the time at
which it is examined. The subject, however, will be more
particularly studied in connection with menstruation.

The mucous membrane of the cervix is much thicker
and more transparent than that of the body of the uterus,
from which it also differs in certain structural peculiarities.

The general arrangements of its folds and surface have
already been described.　The lower half of the membrane
lining the cavity of the cervix, and the whole of that covering its external or vaginal portion, are closely set with a
large number of minute filiform, or clavate papillæ (fig. 27),
Their structure is similar to that of the mucous membrane

Fig. 27.

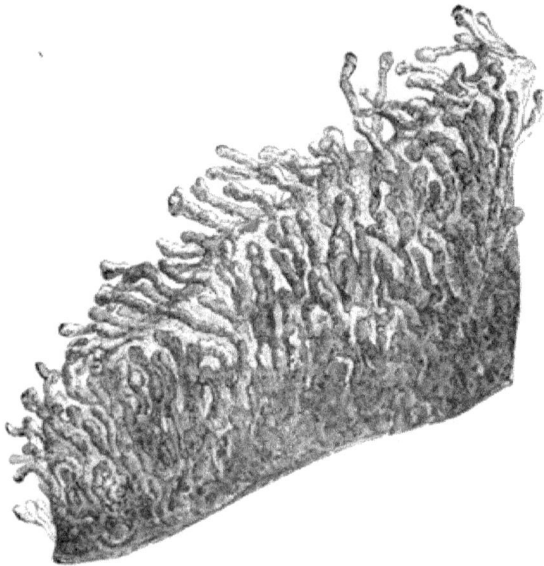

VILLI OF OS UTERI STRIPPED OF EPITHELIUM. (After Tyler Smith and Hassall.)

itself, of which they seem to be merely elevations.　They
each contain a vascular loop (fig. 28), and they are believed
by Kilian and Farre to be mainly concerned in giving sensibility to this part of the generative tract.　All over the
interior of the cervix, both on the ridges of the mucous
membrane and between their folds, are a very large
number of mucous follicles, consisting of a structureless
membrane lined with cylindrical epithelium, and intimately
united with connective tissue.　They cease at the external
orifice of the cervix, and they secrete the thick, tenacious,
and alkaline mucus which is generally found filling the
cervical cavity.　The transparent follicles, known as the
'ovula Nabothii,' which are sometimes found in considerable numbers in the cavity of the cervix, consist of mucous

follicles the mouths of which have become obstructed, and their canals distended by mucous secretion. The lower third of the cervical canal, as well as the exterior of the cervix, is covered with pavement epithelium; while on its upper portion is found a columnar and ciliated epithelium similar to that lining the uterine cavity.

Bandl[1] describes the cervical mucous membrane as ex-

Fig. 28.

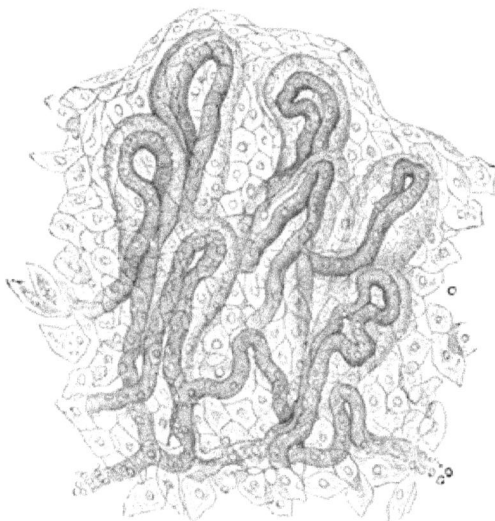

VILLI OF UTERUS, COVERED WITH PAVEMENT EPITHELIUM, AND CONTAINING LOOPED VESSELS. (After Tyler Smith and Hassall.)

tending much higher in the virgin than in women who have borne children, being traceable in the former nearly to the middle of the body of the uterus. During the first pregnancy he believes that the upper portion of the cervix is taken up into the body of the uterus, its mucous membrane never regaining the arrangement peculiar to that of the cervical canal.

Peculiarities of the cervical mucous membrane in virgins.

The arteries of the uterus are derived from the internal iliac, and from the ovarian. They enter the uterus between the folds of the broad ligaments, and, penetrating its muscular coat, anastomose freely with each other and with the corresponding vessels of the opposite side. Their walls are

Vessels of the uterus.

[1] *Arch. f. Gyn.* B. xiv. s. 237.

thick and well developed, and they are remarkable for their
very tortuous course, forming spiral curves, especially in the
upper part of the uterus. They end in minute capillaries
which form the fine meshes surrounding the glands, and in
the cervix give off the loops entering the papillæ. Beneath
the uterine mucous membrane these capillaries form a plexus,
terminating in veins without valves, which unite with each
other to form the large veins traversing the substance of the
uterus, known during pregnancy as the uterine sinuses, the
walls of which are closely adherent to the uterine tissues.
These veins, freely anastomosing with each other, pass outwards to the folds of the broad ligaments, where they unite
to form, with the ovarian and vaginal veins, a large and well-
developed venous network, known as the *pampiniform
plexus*.

The lym-
phatics of
the uterus.

The lymphatics of the uterus are large and well developed,
and they have recently, and with much probability, been
supposed to play an important part in the production of
certain puerperal diseases. A more minute knowledge than
we at present possess of their course and distribution will
probably throw much light on their influence in this respect.
According to the researches of Leopold,[1] who has studied
their minute anatomy carefully, they originate in lymph
spaces between the fine bundles of connective tissue forming
the basis of the mucous lining of the uterus. Here they
are in intimate contact with the utricular glands and the
ultimate ramifications of the uterine bloodvessels. As they
pass into the muscular tissue they become gradually nar-
rowed into lymph-vessels and spaces, which have a very complicated arrangement, and which eventually unite together
in the external muscular layer, especially on the sides of the
uterus, to form large canals which probably have valves.
Immediately under this peritoneal covering these lymph-
vessels form a large and characteristic network covering the
anterior and posterior surfaces of the uterus, and present, in
various parts of their course, large ampullæ. They then
spread over the Fallopian tubes. The lymphatics of the
body of the uterus unite with the lumbar glands, those of
the cervix with the pelvic glands.

[1] *Arch. f. Gynak.* Bd. vi. Heft 1.

The distribution and arrangement of the nerves of the
uterus have been the subject of much controversy. They
are derived mainly from the ovarian and hypogastric plexuses,
inosculating freely with each other between the folds of the
broad ligament, from which they enter the muscular tissue
of the uterus, generally, but not invariably, following the
course of the arteries. They are chiefly derived from the
sympathetic ; but, as the hypogastric plexus is connected
with the sacral nerves, it is probable that some fibres from
the cerebro-spinal system are distributed to the cervix. It is
now generally admitted that nervous filaments are distributed

Fig. 29.

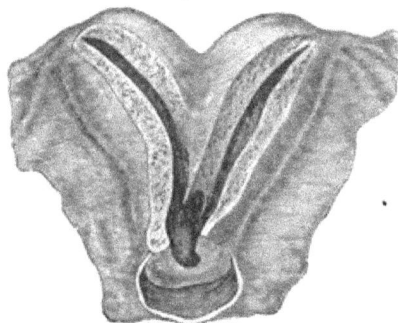

BIFID UTERUS.   (After Farre.)

to the cervix, even as far as the external os, although their
existence in this situation has been denied by Jobert and
other writers. The ultimate distribution of the nerves is not
yet made out. Polle describes a nerve filament as entering
the papillæ of the cervical mucous membrane along with the
capillary loop, and Frankenhauser says the nerve fibres
surround the muscles of the uterus in the form of plexuses,
and terminate in the nuclei of the muscle cells.

Various abnormal conditions of the uterus and vagina are
occasionally met with, which it is necessary to mention, as
they may have an important practical bearing on parturition.
The most frequent of these is the existence of a double, or
partially double, uterus (fig. 29), similar to that found
normally in many of the lower animals. This abnormality
is explained by the development of the organ during fœtal
life. The uterus is formed out of structures existing only in

early fœtal life, known as the Wolffian bodies. These consist of a number of tubes, situated on either side of the vertebral column, and opening externally into an excretory duct. Along their external border a hollow canal is formed, termed the canal of Müller, which, like the excretory ducts, proceeds to the common cloaca of the digestive and urinary organs which then exists. The canal of Müller unites with its fellow of the opposite side to form the uterus and Fallopian tubes in the female, and subsequently the central partition at their point of junction disappears. If, however, the progress of development be in any way checked, the central partition may remain. Then we have produced either a complete double uterus or the uterus bicornis, which is bifid at its upper extremity only ; or a double vagina, each leading to a separate uterus.

Pregnancy in cases of bifid uterus.

If pregnancy occur in any of these anomalous uteri, and many such cases are recorded, serious troubles may follow. It may happen that one horn of the double uterus is not sufficiently large to admit of pregnancy going on to term, and rupture may occur. It is supposed that some cases, presumed to be tubal gestation, were really thus explicable. Impregnation may also occur in the two cornua at different times, leading to superfœtation. It is, however, quite possible that impregnation may occur in one horn of a bifid uterus, and labour be completed without anything unusual being observed. A remarkable case of this sort has been recorded by Dr. Ross, of Brighton,[1] in which a patient miscarried of twins on July 16, 1870, and on October 31, fifteen weeks later, she was delivered of a healthy child. Careful examination showed the existence of a complete double uterus, each side of which had been impregnated. Curiously enough, this patient had formerly given birth to six living children at term, nothing remarkable having been observed in her labours. It can only rarely happen that, under such circumstances, so favourable a result will follow, and more or less difficulty and danger may generally be expected. Occasionally the vagina only is double, the uterus being single. Dr. Matthews Duncan has recorded some cases of this kind,[2] in which the vaginal septum

---

[1] *Lancet*, August, 1871.

[2] *Researches in Obstetrics*, p. 443.

formed an obstacle to the birth of the child, and required division.

The various folds of peritoneum which invest the uterus serve to maintain it in position, and they are described as its ligaments. They are the broad, the vesico-uterine, and sacro-uterine ligaments; the round ligaments are not peritoneal folds like the others. Ligaments of the uterus.

The *broad ligaments* extend from either side of the uterus, where their laminæ are separated from each other, transversely across to the pelvic wall, and thus divide the The broad ligaments.

Fig. 30.

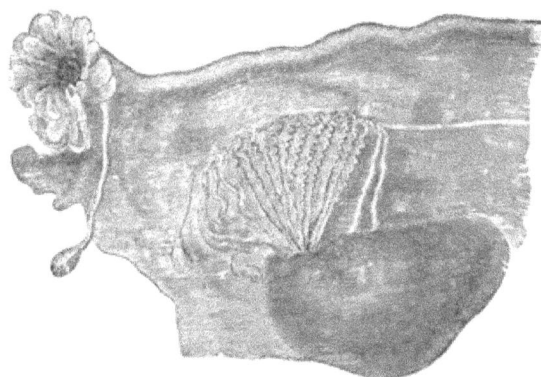

ADULT PAROVARIUM, OVARY, AND FALLOPIAN TUBE. (After Kobelt.)

cavity of the pelvis into two parts; the anterior containing the bladder, the posterior the rectum. Their upper borders are divided into three subsidiary folds, the anterior of which contains the round ligament, the middle the Fallopian tube, and the posterior the ovary. The arrangement has received the name of the *ala vespertilionis*, from its fancied resemblance to a bat's wing. Between the folds of the broad ligaments are found the uterine vessels and nerves, and a certain amount of loose cellular tissue continuous with the pelvic fasciæ. Here is situated that peculiar structure called the organ of Rosenmüller, or the *parovarium* (fig. 30), which is the remains of the Wolffian body, and corresponds to the epididymus in the male. This may best be seen in young subjects, by holding up the broad ligaments and looking through them by transmitted light; but it exists at Structures between the folds of the broad ligaments. The parovarium.

all ages.   It consists of several tubes (eight or ten according
to Farre, eighteen or twenty according to Bankes [1]), which
are tortuous in their course.   They are arranged in a pyramidal form, the base of the pyramid being towards the
Fallopian tube, its apex being lost on the surface of the
ovary.   They are formed of fibrous tissue, and lined with

Fig. 31.

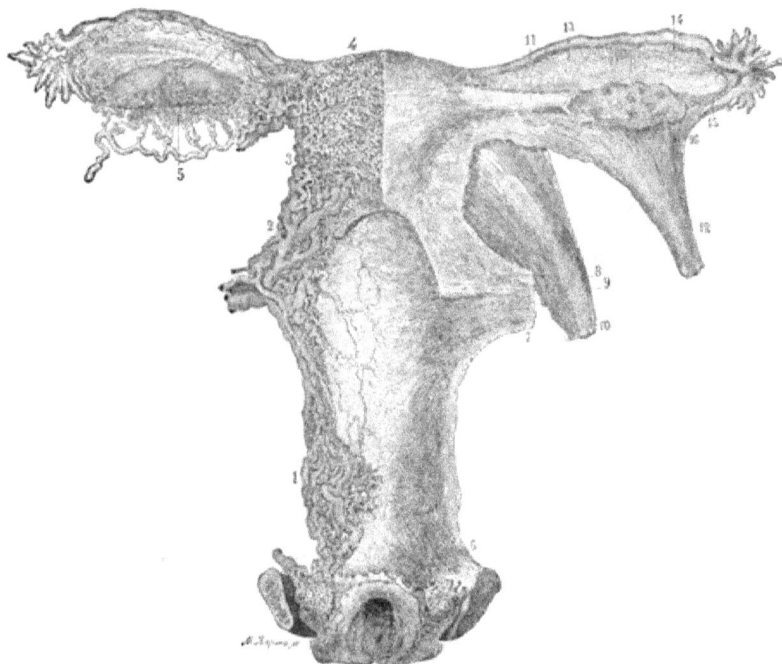

POSTERIOR VIEW OF MUSCULAR AND VASCULAR ARRANGEMENTS.   (After Rouget.)

Vessels—1, 2, 3. Vaginal, cervical, and uterine plexuses.   4. Arteries of body
of uterus.   5. Arteries supplying ovary.   *Muscular fasciculi.*—6, 7. Fibres
attached to vagina, symphysis pubis, and sacro-iliac joint.   8. Muscular
fasciculi from uterus and broad ligaments.   9, 10, 11, 12. Fasciculi attached
to ovary and Fallopian tubes.

pavement epithelium.   They have no excretory duct or
communication with either the uterus or ovary, and their
function, if they have any, is unknown.

Muscular fibres between its folds.   A number of muscular fibres are also found in this
situation, lying between the meshes of the connective tissue.
They have been particularly studied by Rouget, who describes them as interlacing with each other, and forming an
open network, continuous with the muscular tissues of the

[1] Bankes *On the Wolffian Bodies.*

uterus (fig. 31). They are divisible into two layers, the anterior of which is continuous with the muscular fibres of the anterior surface of the uterus, and goes to form part of the round ligament; the posterior arises from the posterior wall of the uterus, and proceeds transversely outwards, to become attached to the sacro-iliac synchondrosis. A continuous muscular envelope is thus formed, which surrounds the whole of the uterus, Fallopian tubes, and ovaries. Its function is not yet thoroughly established. It is supposed to have the effect of retracting the stretched folds of peritoneum after delivery, and more especially of bringing the entire generative organs into harmonious action during menstruation and the sexual orgasm; in this way explaining, as we shall subsequently see, the mechanism by which the fimbriated extremity of the Fallopian tube grasps the ovary prior to the rupture of a Graafian follicle.

The *round ligaments* are essentially muscular in structure. They extend from the upper border of the uterus, with the fibres of which their muscular fibres are continuous, transversely and then obliquely downwards, until they reach the inguinal rings, where they blend with the cellular tissue. In the first part of their course the muscular fibres are solely of the unstriped variety, but soon they receive striped fibres from the transversalis muscles, and the columns of the inguinal ring, which surround and cover the unstriped muscular tissue. In addition to these structures they contain elastic and connective tissue, and arterial, venous, and nervous branches; the former from the iliac or cremasteric arteries, the latter from the genito-crural nerve. According to Mr. Rainey, the principal function of these ligaments is to draw the uterus towards the symphysis pubis during sexual intercourse, and thus to favour the ascent of the semen.

The *vesico-uterine ligaments* are two folds of peritoneum passing in front from the lower part of the body of the uterus to the fundus of the bladder.

The *utero-sacral ligaments* consist of folds of peritoneum of a crescentic form, with their concavities looking inwards: they start from the lower part of the posterior surface of the uterus, and curve backwards to be attached to the third and fourth sacral vertebræ. Within their folds exist bundles of muscular fibres, continuous with those of the uterus, as

The round ligaments.

The vesico-uterine and utero-sacral ligaments.

well as connective tissue, vessels, and nerves. The experiments of Savage, as well as of other anatomists, show that these ligaments have an important influence in preventing downward displacement of the womb.

Alterations during pregnancy. During pregnancy all these ligaments become greatly stretched and unfolded, rising out of the pelvic cavity and accommodating themselves to the increased size of the gravid uterus; and they again contract to their natural size, possibly through the agency of the muscular fibres contained within them, after delivery has taken place.

The *Fallopian tubes*, the homologues of the vasa de-

Fig. 32.

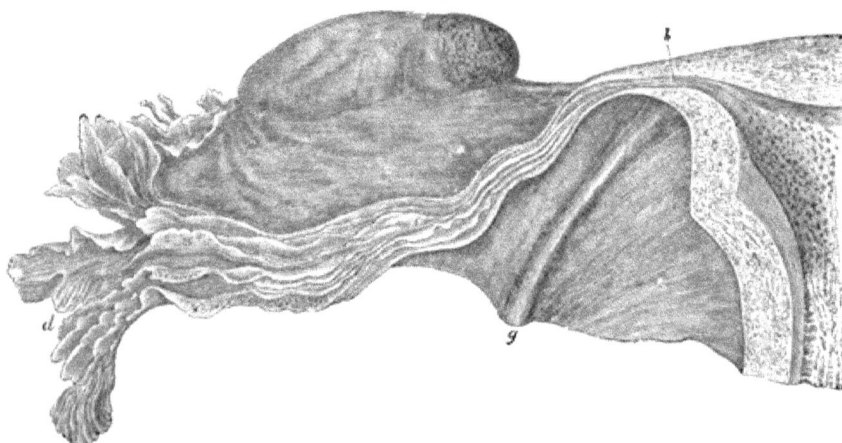

FALLOPIAN TUBE LAID OPEN. (After Richard.)

a, b. Uterine portion of tube. *c, d.* Plicæ of mucous membrane. *e.* Tubo-ovarian ligaments and fringes. *f.* Ovary. *g.* Round ligaments.

The Fallopian tubes. ferentia in the male, are structures of great physiological interest. They serve the double purpose of conveying the semen to the ovary, and of carrying the ovule to the uterus. From the latter function they may be looked on as the excretory ducts of the ovaries; but, unlike other excretory ducts, they are movable, so that they may apply themselves to the part of the ovaries from which the ovule is to come; and so great is their mobility that there is reason to believe that a Fallopian tube may even grasp the ovary of the opposite side. Each tube proceeds from the upper angle of the uterus at first transversely outwards, and then downwards, backwards, and inwards, so as to reach the neighbourhood of the ovary. In the first part of its course it is straight, afterwards it becomes flexuous and twisted on itself. It is contained in the upper part of the broad ligament, where it may be felt as a hard cord. It commences at the uterus by a narrow opening, admitting only the passage of a bristle, known as *ostium uterinum*. As it passes through the muscular walls of the uterus the tube takes a somewhat curved course, and opens into the uterine cavity by a dilated aperture. From its uterine attachment the tube expands gradually until it terminates in its trumpet-shaped extremity; just before its distal end, however, it again contracts slightly. The ovarian end of the tube is surrounded by a number of remarkable fringe-like processes. These consist of longitudinal membranous fimbriæ, surrounding the aperture of the tube, like the tentacles of a polyp, varying considerably in number and size, and having their edges cut and subdivided. On their inner surface are found both transverse and longitudinal folds of mucous membrane, continuous with those lining the tube itself (fig. 32). One of these fimbriæ is always larger and more developed than the rest, and is indirectly united to the surface of the ovary by a fold of peritoneum proceeding from its external surface. Its under surface is grooved so as to form a channel, open below. The function of this fringe-like structure is to grasp the ovary during the menstrual nisus; and the fimbria which is attached to the ovary would seem to guide the tentacles to the ovary which they are intended to seize. One or more supplementary series of fimbriæ sometimes exist, which have an aperture of communication with the canal of the Fallopian tube, beyond its ovarian extremity. His has recently shown that the fimbriated extremity of the tube, after running over the upper part of the ovary, turns down along its free border; so that its aperture lies below it, ready to receive the ovule when expelled from the Graafian follicle.[1]

The fimbriated extremities.

The tubes themselves consist of peritoneal, muscular, and mucous coats. The peritoneum surrounds the tube for three-fourths of its calibre, and comes into contact with the mucous lining at its fimbriated extremity, the only instance in the body where such a junction occurs. The muscular coat is principally composed of circular fibres, with a few longi-

Their structure.

[1] His, *Archiv. fur Anat. und Phys.* 1881.

tudinal fibres interspersed. Its muscular character has been
doubted by Robin and Richard, but Farre had no difficulty in
demonstrating the existence of muscular fibres, both in the
human female and many of the lower animals. According to
Robin, the muscular tissue of the Fallopian tubes is entirely
distinct from that of the uterus, from which he describes it as
being separated by a distinct cellular septum. The mucous
lining is thrown into a number of remarkable longitudinal
folds, each of which contains a dense and vascular fibrous
septum, with small muscular fibres, and is covered with
columnar and ciliated epithelium. The apposition of these
produces a series of minute capillary tubes, along which the
ovules are propelled, the action of the cilia, which is towards
the uterus, apparently favouring their progress.

The
ovaries.

The *ovaries* are the bodies in which the ovules are formed,
and from which they are expelled, and the changes going on
in them in connection with the process of ovulation, during
the whole period between the establishment of puberty and
the cessation of menstruation, have an enormous influence
on the female economy. Normally, the ovaries are two in
number; in some exceptional cases a supplementary ovary
has been discovered; or they may be entirely absent. They
are placed in the posterior folds of the broad ligaments,
usually below the brim of the pelvis, behind the Fallopian
tubes, the left in front of the rectum, the right in front of
some coils of the small intestine. Their situation varies,
however, very much under different circumstances, so that
they can scarcely be said to have a fixed and normal position;

Their
position.

most probably, however, as has been recently shown by His,[1]
they are normally placed close below the brim of the pelvis,
with their long diameters almost vertical, and immediately
above the aperture of the distal extremity of the Fallopian
tubes. In pregnancy they rise into the abdominal cavity
with the enlarging uterus; and in certain conditions they
are dislocated downwards into Douglas's space, where they
may be felt through the vagina as rounded and very tender
bodies.

Their con-
nections.

The folds of the broad ligament, between which the
ovaries are placed, form for them a kind of loose mesentery.
Each of them is united to the upper angle of the uterus by a

[1] *Op. cit.*

special ligament called the utero-ovarian. This is a rounded *Their form* band of organic muscular fibres, about an inch in length, *and di-* continuous with the superficial muscular fibres, of the poste- *mensions.* rior wall of the uterus, and attached to the inner extremity of the ovary. It is surrounded by peritoneum, and through it the muscular fibres, which form an important integral part in the structure of the ovaries, are conveyed to them. The ovary is also attached to the fimbriated extremity of the Fallopian tube in the manner already described.

The ovary is of an irregular oval shape (fig. 33), the upper border being convex, the lower—through which the

Fig. 33.

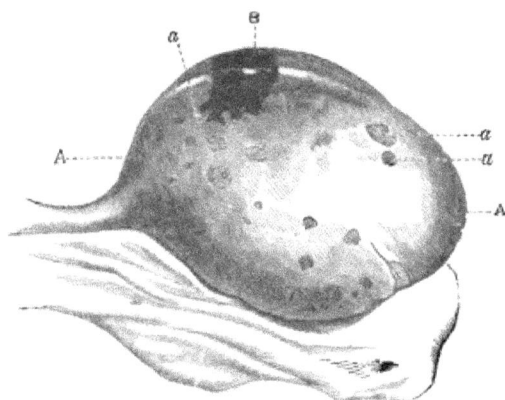

A A. Ovary enlarged under menstrual nisus.   B. Ripe follicle projecting on its surface.
a, a, a, Traces of previously ruptured follicles.

vessels and nerves enter—being straight. The anterior surface, like that of the uterus, is less convex than the posterior. The outer extremity is more rounded and bulbous than the inner, which is somewhat pointed and eventually lost in its proper ligament. By these peculiarities it is possible to distinguish the left from the right ovary, after they have been removed from the body. The ovary varies much in size under different circumstances. On an average, in adult life, it measures from one to two inches in length, three-quarters of an inch in width, and about half an inch in thickness. It increases greatly in size during each menstrual period—a fact which has been demonstrated in certain cases of ovarian hernia, in which the protruded ovary has been seen to swell as menstruation commenced; also during

pregnancy, when it is said to be double its usual size. After
the change of life it atrophies, and becomes rough and
wrinkled on its surface. Before puberty, the surface of the
ovary is smooth and polished, and of a whitish colour. After
menstruation commences, its surface becomes scarred by the
rupture of the Graafian follicles (fig. 33, *a a*), each of which
leaves a little linear or striated cicatrix, of a brownish colour ;
and the older the patient the greater are the number of these
cicatrices.

Their structure. Epithelial investment.   The structure of the ovary has been made the subject of
many important observations. It has an external covering of
epithelium, originally continuous with the peritoneum, called
by some the germ-epithelium, in consequence of the ovules
being formed from it in early fœtal life. In the adult it is
separated from the peritoneum at the base of the organ by a
circular white line, and it consists of columnar epithelium,
differing only from the epithelium lining the Fallopian tubes,
with which it is sometimes continuous through the attached
fimbria uniting the tube and the ovary, in being destitute of
cilia. Immediately beneath this covering is the dense coat
Tunica albuginea.   known as the *tunica albuginea*, on account of its whitish
colour. It consists of short connective-tissue fibres, arranged
in laminæ, among which are interspersed fusiform muscular
fibres. At the point where the vessels and nerves enter the

Fig. 34.

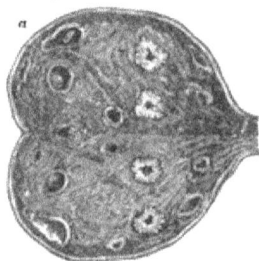

LONGITUDINAL SECTION OF ADULT
OVARY. (After Farre.)

ovary this membrane is raised into
a ridge, which is continuous with
the utero-ovarian ligament, and is
called the *hilum*. The tunica albuginea is so intimately blended
with the stroma of the ovary as to
be inseparable on dissection ; it
does not, however, exist as a distinct
lamina, but is merely the external
part of the proper structure of the
ovary, in which more dense connective tissue is developed than elsewhere.

The stroma.   On making a longitudinal section of the ovary (fig. 34),
it will be seen to be composed of two parts, the more internal
of which is of a reddish colour from the number of vessels
The medullary substance.   that ramify in it, and is called the *medullary* or vascular
zone ; while the external, of a whitish tint, receives the

name of the *cortical* or parenchymatous substance. The
former consists of loose connective tissue interspersed with
elastic, and a considerable number of muscular fibres. According to Rouget[1] and His[2] the muscular structure forms
the greater part of the ovarian stroma. The latter describes
it as consisting essentially of interwoven muscular fibres,
which he terms the ' fusiform tissue,' and which he believes
to be continuous with the muscular layers of the ovarian
vessels. The former believes that the muscular fasciculi

Fig. 35.

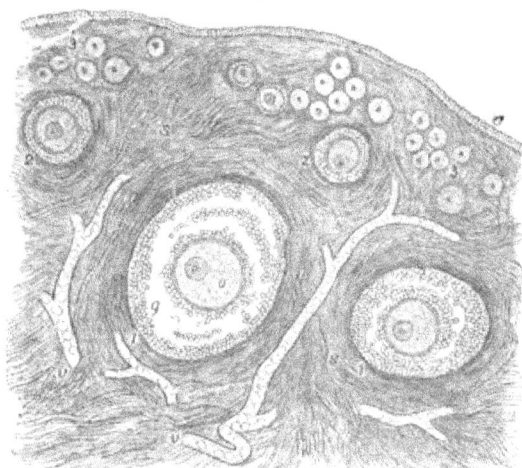

SECTION THROUGH THE CORTICAL PART OF THE OVARY.

e. Surface epithelium. *s. s.* Ovarian stroma. 1 1. Large-sized Graafian follicles.
2 2. Middle-sized; and 3 3. Small-sized Graafian follicles. *o.* Ovule within
Graafian follicle. *v v.* Blood-vessels in the stroma. *g.* Cells of the membrana
granulosa. (After Turner.)

accompany the vessels in the form of sheaths, as in erectile
tissues. Both attribute to the muscular tissues an important
influence in the expulsion of the ovules, and in the rupture
of the Graafian follicles. Waldeyer and other writers, however, do not consider it to be so extensively developed as
Rouget and His believe. The cortical substance is the more   The cor-
                                                              tical sub-
important as that in which the Graafian follicles and ovules   stance.
are formed. It consists of interlaced fibres of connective
tissue, containing a large number of nuclei. The muscular
fibres of the medullary substance do not seem to penetrate

[1] *Journal de Physiol.* i. p. 737.
[2] Schultze's *Arch. f. Mikroscop. Anat.* 1865.

into it in the human female. In it are found the Graafian follicles, which exist in enormous numbers from the earliest periods of life, and in all stages of development (fig. 35).

The Graafian follicles.

According to the researches of Pflüger, Waldeyer, and other German writers, the Graafian follicles are formed in early fœtal life by cylindrical inflections of the epithelial covering of the ovary, which dip into the substance of the gland. These tubular filaments anastomose with each other, and in them are formed the ovules, which are originally the

Fig. 36.

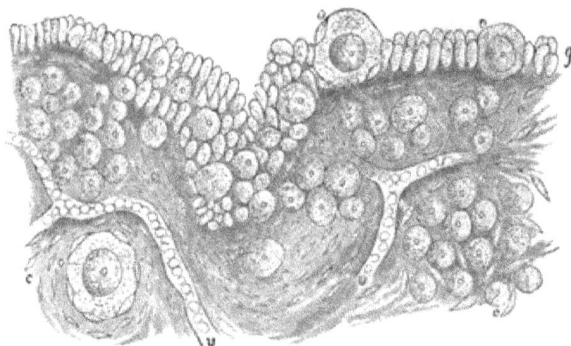

VERTICAL SECTION THROUGH THE OVARY OF THE HUMAN FŒTUS.

g g. Germ-epithelium, with *o o.* Developing ovules in it. *s s.* Ovarian stroma, containing *c c c.* Fusiform connective-tissue corpuscles. *r r.* Capillary blood-vessels. In the centre of the figure an involution of the germ epithelium is shown; and at the left lower side a primordial ovule, with the connective-tissue corpuscles ranging themselves round it. (After Foulis.)

Formation of the ovules and Graafian follicles.

epithelial cells lining the tubes. Portions become shut off from the rest of the filaments, and form the Graafian follicles. The ovules, on this view, are highly developed epithelial cells, originally derived from the surface of the ovary, and not developed in its stroma. These tubular filaments disappear shortly after birth, but they have recently been detected by Slavyansky [1] in the ovaries of a woman thirty years of age. These observations have been modified by Dr. Foulis.[2] He recognises the origin of the ovules from the germ-epithelium covering the surface of the ovary, which is itself derived from the Wolffian body. He believes all the ovules to be formed from the germ-epithelium corpuscles,

[1] *Annales de Gynec.* Feb. 1871.

[2] *Proceedings of the Royal Soc. of Edinb.* April 1875, and *Journ. of Anat. and Phys.* vol. xiii. 1879.

which become embedded in the stroma of the ovary, by the outgrowth of processes of vascular connective tissue, fresh germ-epithelial corpuscles being constantly produced on the surface of the organ up to the age of $2\frac{1}{2}$ years, to take the place of those already embedded in its stroma. He believes the Graafian follicles to be formed by the growth of delicate processes of connective tissue between and around the ovules, but not from tubular inflections of the epithelium covering the gland, as described by Waldeyer (fig. 36). This view is supported by the researches of Balfour,[1] who arrives at the conclusion that the whole egg-containing part of the ovary is really the thickened germinal epithelium, broken up into a

Fig. 37.

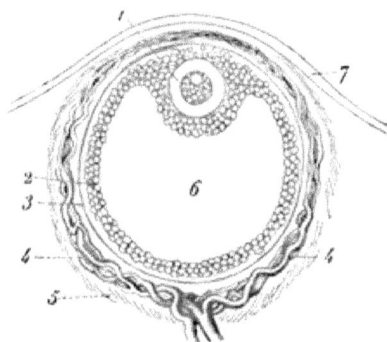

DIAGRAMMATIC SECTION OF GRAAFIAN FOLLICLE.

1. Ovum. 2. Membrana granulosa. 3. External membrane of Graafian follicle. 4. Its vessels. 5. Ovarian stroma. 6. Cavity of Graafian follicle. 7. External covering of ovary.

kind of meshwork by growths of vascular stroma. According to this theory, Pflüger's tubular filaments are merely trabeculæ of germinal epithelium, modified cells of which become developed into ovules.

The greater proportion of the Graafian follicles are only visible with the high powers of the microscope, but those which are approaching maturity are distinctly to be seen by the naked eye. The quantity of these follicles is immense. Foulis estimates that at birth each human ovary contains not less than 30,000. No fresh follicles appear to be formed after birth, and as development goes on some only grow, and,

[1] F. M. Balfour, 'Structure and Development of Vertebrate Ovary,' Quarterly Journal of Microscopical Science, vol. xviii. 1878

by pressure on the others, destroy them. Of those that grow, of course only a few ever reach maturity; they are scattered through the substance of the ovary, some developing in the stroma, others on the surface of the organ, where they eventually burst, and are discharged into the Fallopian tube.

Structure of the Graafian follicle

A ripe Graafian follicle has an external investing membrane (fig. 37), which is generally described as consisting of two distinct layers; the external, or *tunica fibrosa*, highly vascular, and formed of connective tissue; the internal, or *tunica propria*, composed of young connective tissue, containing a large number of fusiform or stellate cells, and numerous oil-globules. These layers, however, appear to be essentially formed of condensed ovarian stroma. Within this capsule is the epithelial lining called the *membrana granulosa*, consisting of stratified columnar epithelial cells, which, according to Foulis, are originally formed from the nuclei of the fibro-nuclear tissue of the stroma of the ovary. At one part of the circumference of the ovisac is situated the ovule, around which the epithelial cells are congregated in greater quantity, constituting the projection known as the *discus proligerus*. The remainder of the cavity of the follicle is filled with a small quantity of transparent fluid, the *liquor folliculi*, traversed by three or four minute bands, the retinacula of Barry, which are attached to the opposite walls of the follicular cavity, and apparently serve the purpose of suspending the ovule and maintaining it in a proper position. In many young follicles this cavity does not at first exist, the follicle being entirely filled by the ovule. According to Waldeyer, the liquor folliculi is formed by the disintegration of the epithelial cells, the fluid thus produced collecting, and distending the interior of the follicle.

The ovule.

The ovule is attached to some part of the internal surface of the Graafian follicle. It is a rounded vesicle about $\frac{1}{120}$th of an inch in diameter, and is surrounded by a layer of columnar cells, distinct from those of the discus proligerus, in which it lies. It is invested by a transparent elastic membrane, the *zona pellucida*, or vitelline membrane. In most of the lower animals the zona pellucida is perforated by numerous very minute pores, only visible under the highest powers of the microscope; in others there is a distinct aperture of a

larger size, the micropyle, allowing the passage of the spermatozoa into the interior of the ovule. It is possible that
similar apertures may exist in the human ovule, but they
have not been demonstrated. Within the zona pellucida
some embryologists describe a second fine membrane, the
existence of which has been denied by Bischoff. The cavity
of the ovule is filled with a viscid yellow fluid, the *yelk*,
containing numerous granules. It entirely fills the cavity,
to the walls of which it is non-adherent. In the centre of
the yelk in young, and at some portion of its periphery in
mature ovules, is situated the *germinal vesicle*, which is

Fig. 34.

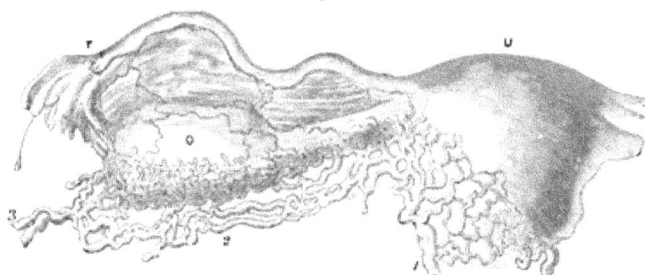

BULB OF OVARY.

U. Uterus. O. Ovary and utero-ovarian ligament. F. Fallopian tube. 1. Utero-
ovarian vein. 2. Pampiniform ovarian plexus. 3. Commencement of spermatic
vein.

a clear circular vesicle, refracting light strongly, and about
$\frac{1}{60}$th of a line in diameter. It contains a few granules, and
a nucleolus, or *germinal spot*, which is sometimes double.

From within outwards, therefore, we find—

1. The *germinal* spot; round this

2. The *germinal* vesicle, contained in

3. The *yelk*, which is surrounded by the

4. *Zona pellucida*, with its layers of columnar epithelial
cells.

These constitute the ovule.

The ovule is contained in—

The *Graafian follicle*, and lies in that part of its epithelial lining called the—

Discus proligerus, the rest of the follicle being occupied
by the *liquor folliculi*. Round these we have the epithelial
lining or *membrana granulosa*, and the external coat consisting of the *tunica propria* and the *tunica fibrosa*.

The vessels and nerves of the ovary.

The vascular supply of the ovary is complex. The arteries enter at the hilum, penetrating the stroma in a spiral curve, and are ultimately distributed in a rich capillary plexus to the follicles. The large veins unite freely with each other, and form a vascular and erectile plexus, continuous with that surrounding the uterus, called the bulb of the ovary (fig. 38). Lymphatics and nerves exist, but their mode of termination is unknown.

The mammary glands.

To complete the consideration of the generative organs of the female we must study the *mammary glands*, which secrete the fluid destined to nourish the child. In the human subject they are two in number, and instead of being placed upon the abdomen, as in most animals, they are situated on either side of the sternum, over the pectorales majora muscles, and extend from the third to the sixth ribs.

Their position and dimensions.

This position of the glands is obviously intended to suit the erect position of the female in suckling. They are convex anteriorly, and flattened posteriorly where they rest on the muscles. They vary greatly in size in different subjects, chiefly in proportion to the amount of adipose tissue they contain. In man, and in girls, previous to puberty, they are rudimentary in structure; while in pregnant women they increase greatly in size, the true glandular structures becoming much hypertrophied. Anomalies in shape and position are sometimes observed. Supplementary mammæ, one or more in number, situated on the upper portion of the mammæ, are sometimes met with, identical in structure with the normally situated glands; or, more commonly, an extra nipple is observed by the side of the normal one. In some races, especially the African, the mammæ are so enormously developed that the mother is able to suckle her child over her shoulder.

Their structure.

The skin covering the gland is soft and supple, and during pregnancy often becomes covered with fine white lines, while large blue veins may be observed coursing over. Underneath it is a quantity of connective tissue, containing a considerable amount of fat, which extends between the true glandular structure. This is composed of from fifteen to twenty lobes, each of which is formed of a number of lobules. The lobules are produced by the aggregation of the terminal acini in which the milk is formed. The acini are

minute cul-de-sacs opening into little ducts, which unite
with each other until they form a large duct for each lobule;
the ducts of each lobule unite with each other, until they
end in a still larger duct common to each of the fifteen
or twenty lobes into which the gland is divided, and eventually open on the surface of the nipple. These terminal
canals are known as the galactophorous ducts (fig. 39).
They become widely dilated as they approach the nipple, so

Fig. 39.

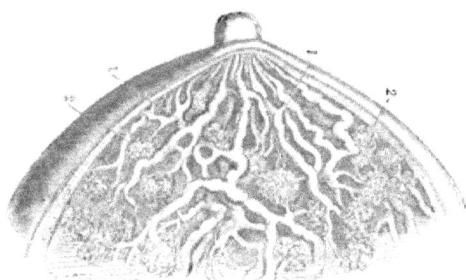

1. Galactophorous ducts.   2. Lobuli of the mammary gland.

as to form reservoirs in which milk is stored until it is
required, but when they actually enter the nipple they again
contract. Sometimes they give off lateral branches, but,
according to Sappey, they do not anastomose with each
other, as some anatomists have described. These excretory
ducts are composed of connective tissue, with numerous
elastic fibres, on their external surface. Sappey and Robin
describe a layer of muscular fibres, chiefly developed near their
terminal extremities. They are lined with columnar epithelium, continuous with that in the acini ; and it is by the distension of its cells with fatty matter, and their subsequent
bursting, that the milk is formed.

The *nipple* is the conical projection at the summit of the
mamma, and it varies in size in different women. Not very
unfrequently, from the continuous pressure to which it has
been subjected by the dress, it is so depressed below the
surface of the skin as to prevent lactation. It is generally
larger in married than in single women, and increases in
size during pregnancy. Its surface is covered with numerous
papillæ, giving it a rugous aspect, and at their bases the

The
nipple.

orifices of the lactiferous ducts open. Here are also the openings of numerous sebaceous follicles, which secrete an unctuous material supposed to protect and soften the integument during lactation. Beneath the skin are muscular fibres, mixed with connective and elastic tissues, vessels, nerves, and lymphatics. When the nipple is irritated it contracts and hardens, and by some this is attributed to its erectile properties. The vascularity, however, is not great, and it contains no true erectile tissue : the hardening is, therefore, due to muscular contraction. Surrounding the nipple is the *areola*, of a pink colour in virgins, becoming dark from the development of pigment cells during pregnancy, and always remaining somewhat dark after child-bearing. On its surface are a number of prominent tubercles, sixteen to twenty in number, which also become largely developed during gestation. They are supposed by some to secrete milk, and to open into the lactiferous tubes ; most probably they are composed of sebaceous glands only. Beneath the areola is a circular band of muscular fibres, the object of which is to compress the lactiferous tubes which run through it, and thus to favour the expulsion of their contents. The mammæ receive their blood from the internal mammary and intercostal arteries, and they are richly supplied with lymphatic vessels, which open into the axillary glands. The nerves are derived from the intercostal and thoracic branches of the brachial plexus.

The secretion of milk in women who are nursing is accompanied by a peculiar sensation, as if milk were rushing into the breast, called the ' draught,' which is excited by the efforts of the child to suck, and by various other causes. The sympathetic relations between the mammæ and the uterus are very well marked, as is shown in the unimpregnated state by the fact of the frequent occurrence of sympathetic pains in the breast in connection with various uterine diseases, and, after delivery, by the well-known fact that suction produces reflex contraction of the uterus, and even severe after-pains.

Marginal notes:

The areola.

Their vessels, nerves, and lymphatics.

Their sympathetic relations with the uterus.

CHAPTER III.

OVULATION AND MENSTRUATION.

THE main function of the ovary is to supply the female generative element, and to expel it, when ready for impregnation, into the Fallopian tube, along which it passes into the uterus. This process takes place spontaneously in all viviparous animals, and without the assistance of the male. In the lower animals this periodical discharge receives the name of the œstrum or rut, at which time only the female is capable of impregnation and admits the approach of the male. In the human female the periodical discharge of the ovule, in all probability, takes place in connection with menstruation, which may therefore be considered to be the analogue of the rut in animals. Between each menstrual period Graafian follicles undergo changes which prepare them for rupture and the discharge of their contained ovules. After rupture certain changes occur which have for their object the healing of the rent in the ovarian tissue through which the ovule has escaped, and the filling up of the cavity in which it was contained. This results in the formation of a peculiar body in the substance of the ovary, called the *corpus luteum*, which is essentially modified should pregnancy occur, and is of great interest and importance. During the whole of the child-bearing epoch the periodical maturation and rupture of the Graafian follicles are going on. If impregnation does not take place, the ovules are discharged and lost; if it does, ovulation is stopped, as a general rule, during gestation and lactation.

Functions of the ovary.

This, broadly speaking, is an outline of the modern theory of menstruation, which was first broached in the year 1821 by Dr. Power, and subsequently elaborated by Negrier, Bischoff, Raciborski, and many other writers. Although the sequence of events here indicated may be taken to be the

Theory of menstruation.

rule, it must be remembered that it is one subject to many exceptions, for undoubtedly ovulation may occur without its outward manifestation, menstruation, as in cases in which impregnation takes place during lactation, or before menstruation has been established, of which many examples are recorded. These exceptions have led some modern writers to deny the ovular theory of menstruation, and their views will require subsequent consideration.

In order to understand the subject properly it will be necessary to study the sequence of events in detail.

Changes in the Graafian follicle. The changes in the Graafian follicle which are associated with the discharge of the ovules comprise—① *Maturation.* As the period of puberty approaches, a certain number of 1. Maturation. the Graafian follicles, fifteen to twenty in number, increase in size, and come near the surface of the ovary. Amongst these one becomes especially developed, preparatory to rupture, and upon it for the time being all the vital energy of the ovary seems to be concentrated. A similar change in one, sometimes in more than one, follicle takes place periodically during the whole of the child-bearing epoch, in connection with each menstrual period, and an examination of the ovary will show several follicles in different stages of development. The maturing follicle becomes gradually larger, until it forms a projection on the surface of the ovary, from five to seven lines in breadth, but sometimes even as large as a nut (fig. 33). This growth is due to the distension of the follicle by the increase of its contained fluid, which causes it so to press upon the ovarian structures covering it that they become thinned, separated from each other, and partially absorbed, until they eventually readily lacerate. The follicle also becomes greatly congested, the capillaries coursing over it become increased in size and loaded with blood, and being seen through the attenuated ovarian tissue, give it, when mature, a bright red colour. At this time some of these distended capillaries in its inner coat lacerate, and a certain quantity of blood escapes into its cavity. This escape of blood takes place before rupture, and seems to have for its principal object the increase of the tension of the follicle, of which it has been termed the menstruation. Pouchet was of opinion that the blood collects behind the ovule, and carries it up to the surface of the follicle. By

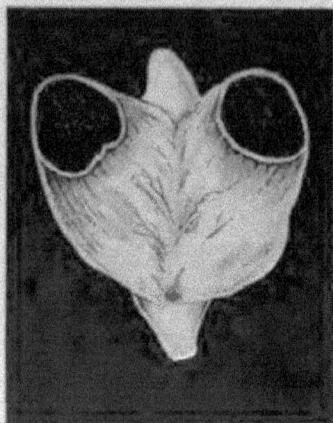

Plate II

Fig: 1.
A recently ruptured and bloody Graafian
follicle, just developing into a Corpus luteum.

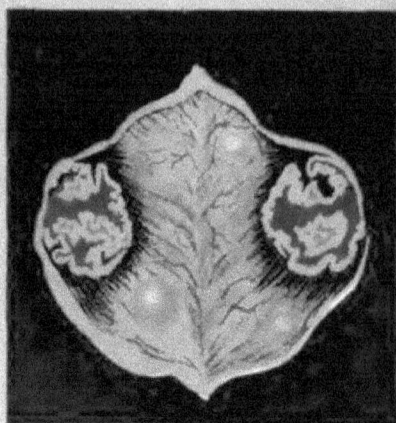

Fig: 2.
Corpus luteum two days after menstruation
with three older obsolete Corpora lutea,
diminishing successively in size.

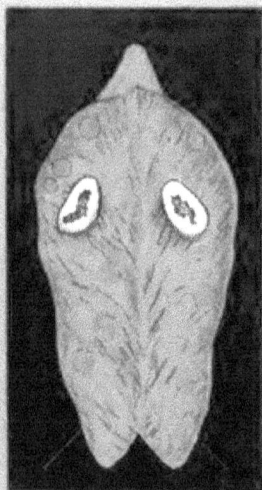

Fig: 3.
Degenerated Graafian follicle
which has never ruptured.
(The "false corpus luteum of Dalton.")

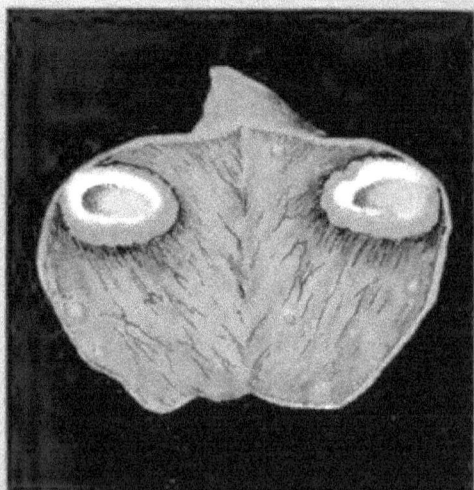

Fig: 4.
Corpus luteum of Pregnancy.

ILLUSTRATIONS OF THE CORPUS LUTEUM. (AFTER DALTON)
Mintern Bros. lith

these means the follicle is more and more distended, until at last it ruptures (Plate II., fig. 1), either spontaneously or, it may be, under the stimulus of sexual excitement. Whether the laceration takes place during, before, or after the menstrual discharge is not yet positively known; from the results of post-mortem examination in a number of women who died shortly before or after the period, Williams believes that the ovules are expelled before the monthly flow commences.[1] In order that the ovule may escape, the laceration must, of course, involve not only the coats of the Graafian follicles, but also the superincumbent structures.

Laceration seems to be aided by the growth of the internal layer of the follicle, which increases in thickness before rupture, and assumes a characteristic yellow colour from the number of oil-globules it then contains. It is also greatly facilitated, if it be not actually produced, by the turgescence of the ovary at each menstrual period, and by the contraction of the muscular fibres in the ovarian stroma. As soon as the rent in the follicular walls is produced, the ovule is discharged, surrounded by some of the cells of the membrana granulosa, and is received into the fimbriated extremity of the Fallopian tube, which grasps the ovary over the site of the rupture. By the vibratile cilia of its epithelial lining, it is then conducted into the canal of the tube, along which it is propelled, partly by ciliary action and partly by muscular contraction in the walls of the tube.

After the ovule has escaped, certain characteristic changes occur in the empty Graafian follicle, which have for their object its cicatrisation and obliteration. There are great differences in the changes which occur when impregnation has followed the escape of the ovule, and they are then so remarkable that they have been considered certain signs of pregnancy. They are, however, differences of degree rather than of kind. It will be well, however, to discuss them separately.

As soon as the ovule is discharged, the edges of the rent through which it has escaped become agglutinated by exudation, and the follicle shrinks, as is generally believed, by the inherent elasticity of its internal coat, but according to Robin, who denies the existence of this coat, from compression

Proceedings of the Royal Society, 1875.

2. Escape of the ovule.

Obliteration of **the** Graafian follicle.

Changes undergone by the follicle when impregnation does not occur.

by the muscular fibres of the ovarian stroma.   In proportion
to the contraction that takes place, the inner layer of
the follicle, the cells of which have become greatly hypertrophied and loaded with fat-granules previous to rupture, is
thrown into numerous folds (Plate II. fig. 2).   The greater
the amount of contraction the deeper these folds become,
giving to a section of the follicle an appearance similar to
that of the convolutions of
the brain (fig. 40).   These
folds in the human subject
are generally of a bright
yellow colour, but in some
of the mammalia they are
of a deep red.   The tint was
formerly ascribed by Raciborski to absorption of the
colouring matter of the
blood-clot contained in the
follicular cavity, a theory
he has more recently abandoned in favour of the view
maintained by Coste that
it is due to the inherent

Fig. 40.

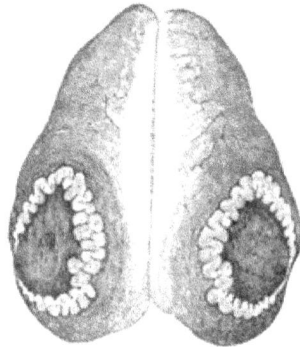

SECTION OF OVARY, SHOWING CORPUS LUTEUM
THREE WEEKS AFTER MENSTRUATION.
(After Dalton.)

colour of the cells of the lining membrane of the follicle,
which, though not well marked in a single cell, becomes
very apparent en masse.   The existence of a contained blood-
clot is also denied by the latter physiologist, except as an
unusual pathological condition ; and he describes the cavity
as containing a gelatinous and plastic fluid, which becomes absorbed as contraction advances.   The more recent researches
of Dalton,[1] however, show the existence of a central blood-
clot in the cavity of the follicle, and he considers its occasional absence to be connected with disturbance or cessation
of the menstrual function.   The folds into which the membrane has been thrown continue to increase in size, from the
proliferation of their cells, until they unite and become adherent, and eventually fill the follicular cavity.   By the time
that another Graafian follicle is matured and ready for rupture the diminution has advanced considerably, and the
empty ovisac is reduced to a very small size.   The cavity is

[1] 'Report on the Corpus Luteum,' American Gynæc. Trans. vol. ii. 1878.

now nearly obliterated, the yellow colour of the convolutions is altered into a whitish tint, and on section the corpus luteum has the appearance of a compact white stellate cicatrix, which generally disappears in less than forty days from the period of rupture. The tissue of the ovary at the site of laceration also shrinks, and this, aided by the contraction of the follicle, gives rise to one of those permanent pits or depressions which mark the surface of the adult ovary. Slavyansky[1] has shown that only a few of the immense number of Graafian follicles undergo these alterations. The greater proportion of them seem never to discharge their ovules, but, after increasing in size, undergo retrogressive changes exactly similar in their nature, but to a much less extent, to those which result in the formation of a corpus luteum. The sites of these may afterwards be seen as minute striae in the substance of the ovary.

Should pregnancy occur, all the changes above described take place; but, inasmuch as the ovary partakes of the stimulus to which all the generative organs are then subjected, they are much more marked and apparent (Plate II. fig. 4). Instead of contracting and disappearing in a few weeks, the corpus luteum continues to grow until the third or fourth month of pregnancy; the folds of the inner layer of the ovisac become large and fleshy, and permeated by numerous capillaries, and ultimately become so firmly united that the margins of the convolutions thin and disappear, leaving only a firm fleshy yellow mass, averaging from 1 to 1½ inches in thickness, which surrounds a central cavity, often containing a whitish fibrillated structure, believed to be the remains of a central blood-clot. This was erroneously supposed by Montgomery to be the inner layer of the follicle itself, and he conceived the yellow substance to be a new formation between it and the external layer, while Robert Lee thought it was placed external to both the external and internal layers.

Changes undergone by the follicle when impregnation has taken place.

Between the third and fourth months of pregnancy, when the corpus luteum has attained its maximum of development (fig. 41), it forms a firm projection on the surface of the ovary, averaging about 1 inch in length, and rather more than ½ an inch in breadth. After this it commences

[1] *Archiv. de Phys.* March 1874.

to atrophy (fig. 42), the fat-cells become absorbed, and the capillaries disappear. Cicatrisation is not complete until from one to two months after delivery.

Its value as a sign of pregnancy.   On account of the marked appearance of the corpus luteum it was formerly considered to be an infallible sign of pregnancy ; and it was distinguished from the corpus luteum of the non-pregnant state by being called a 'true' as opposed to a 'false' corpus luteum.   From what has been said it will be obvious that this designation is essentially wrong, as the difference is one of degree only.   Dalton[1] applies the term 'false corpus luteum' to a degenerated condition

Fig. 41.   Fig. 42.

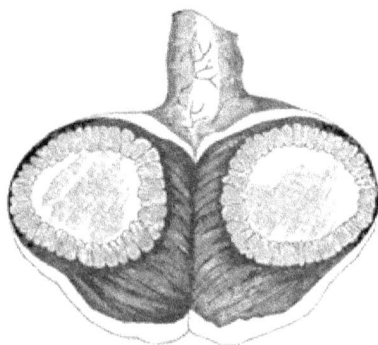

CORPUS LUTEUM OF THE FOURTH MONTH OF PREGNANCY. (After Dalton.)

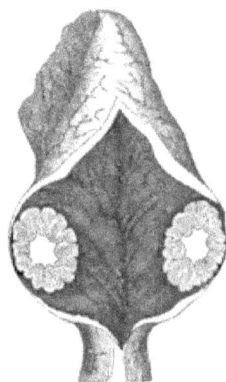

CORPUS LUTEUM OF PREGNANCY AT TERM. (After Dalton.)

sometimes met with in an unruptured Graafian follicle consisting in reabsorption of its contents and thickening of its walls (Plate II. fig. 3).   It differs from the 'true' corpus luteum in being deeply seated in the substance of the ovary, in having no central clot, and in being unconnected with a cicatrix on the surface of the ovary.   Nor do obstetricians attach by any means the same importance as they did formerly to the presence of the corpus luteum as indicating impregnation ; for even when well marked, other and more reliable signs of recent delivery, such as enlargement of the uterus, are sure to be present, especially at the time when the corpus luteum has reached its maximum of development ; while after delivery at term it has no longer a sufficiently characteristic appearance to be depended on.

[1] *Op. cit.* p. 64.

By the term *menstruation* (catamenia, periods, etc.) is meant the periodical discharge of blood from the uterus which occurs, in the healthy woman, every lunar month, except during pregnancy and lactation, when it is, as a rule, suspended.

The first appearance of menstruation coincides with the establishment of puberty, and the physical changes that accompany it indicate that the female is capable of conception and childbearing, although exceptional cases are recorded in which pregnancy occurred before menstruation had begun. In temperate climates it generally commences between the 14th and 16th years, the largest number of cases being met with in the 15th year. This rule is subject to many exceptions, it being by no means very rare for menstruation to become established as early as the 10th or 11th years, or to be delayed until the 18th or 20th. Beyond these physiological limits a few cases are from time to time met with in which it has begun in early infancy, or not until a comparatively late period of life.

Various accidental circumstances have much to do with its establishment. As a rule it occurs somewhat earlier in tropical, and later in very cold than in temperate climates. The influence of climate has been unduly exaggerated. It used to be generally stated that in the Arctic regions women did not menstruate until they were of mature age, and that in the tropics girls of 10 or 12 years of age did so habitually. The researches of Robertson, of Manchester,[1] first showed that the generally received opinions were erroneous; and the collection of a large number of statistics has corroborated his opinion. There can be no doubt, however, that a larger proportion of girls menstruate early in warm climates. Joulin found that in tropical climates, out of 1,635 cases, the largest proportion began to menstruate between the 12th and 13th years; so that there is an average difference of more than two years between the period of its establishment in the tropics and in temperate countries. Harris[2] states that among the Hindoos 1 to 2 per cent. menstruate as early as nine years of age; 3 to 4 per cent. at ten; 8 per cent. at eleven; and 25 per cent. at twelve; while in London or Paris

---

[1] *Edin. Med. and Surg. Journ.* 1832.
[2] *Amer. Journ. of Obstet.* 1871.   R. P. Harris on early puberty.

probably not more than one girl in 1,000 or 1,200 does so at nine years. The converse holds true with regard to cold climates, although we are not in possession of a sufficient number of accurate statistics to draw very reliable conclusions on this point; but out of 4,715 cases, including returns from Denmark, Norway and Sweden, Russia and Labrador, it was found that menstruation was established on an average a year later than in more temperate countries. It is probable that the mere influence of temperature has much to do in producing these differences, but there are other factors, the action of which must not be overlooked. Raciborski attributes considerable importance to the effect of race; and he has quoted Dr. Webb, of Calcutta, to the effect that English girls in India, although subjected to the same climatic influence as the Indian races, do not, as a rule, menstruate earlier than in England; while in Austria, girls of the Magyar race menstruate considerably later than those of German parentage.[1] The surroundings of girls, and their manner of education and living, have probably also a marked influence in promoting or retarding its establishment. Thus, it will commence earlier in the children of the rich, who are likely to have a highly developed nervous organisation, and are habituated to luxurious living, and a premature stimulation of the mental faculties by novel-reading, society, and the like; while amongst the hard-worked poor, or in girls brought up in the country, it is more likely to begin later. Premature sexual excitement is said also to favour its early appearance, and the influence of this among the factory girls of Manchester, who are exposed in the course of their work to the temptations arising from the promiscuous mixing of the sexes, has been pointed out by Dr. Clay.[2]

Changes occurring at puberty.

The first appearance of menstruation is accompanied by certain well-marked changes in the female system, on the occurrence of which we say that the girl has arrived at the period of puberty. The pubes become covered with hair, the breasts enlarge, the pelvis assumes its fully developed form, and the general contour of the body fills out. The mental qualities also alter; the girl becomes more shy and retiring, and her whole bearing indicates the change that has taken place. The menstrual discharge is not established regularly

[1] *Op. cit.* p. 227.   [2] *Brit. Record of Obstet. Med.* vol. i.

at once. For one or two months there may be only premonitory symptoms—a vague sense of discomfort, pains in the
breasts, and a feeling of weight and heat in the back and
loins. There then may be a discharge of mucus tinged with
blood, or of pure blood, and this may not again show itself
for several months. Such irregularities are of little consequence on the first establishment of the function, and need
give rise to no apprehension.

As a rule, the discharge recurs every twenty-eight days,   Period of
and with some women with such regularity that they can   duration
foretell its appearance almost to the hour. The rule is,   and re-
however, subject to very great variations. It is by no means   currence.
uncommon, and strictly within the limits of health, for it
to appear every twentieth day, or even with less interval ;
while in other cases, as much as six weeks may habitually
intervene between two periods. The period of recurrence
may also vary in the same subject. I am acquainted with
patients who sometimes only have twenty-eight days, at
others as many as forty-eight days, between their periods,
without their health in any way suffering. Joulin mentions
the case of a lady who only menstruated two or three times
in the year, and whose sister had the same peculiarity.

The duration of the period varies in different women, and
in the same woman at different times. In this country its
average is four or five days, while in France, Dubois and
Brierre de Boismont fix eight days as the most usual length.
Some women are only unwell for a few hours, while in others
the period may last many days beyond the average without
being considered abnormal.

The quantity of blood lost varies in different women.   Quantity
Hippocrates puts it at ℥xviij, which, however, is much too   of blood
high an estimate. Arthur Farre thinks that from ℥ij to ℥iij   lost.
is the full amount of a healthy period, and that the quantity
cannot habitually exceed this without producing serious constitutional effects. Rich diet, luxurious living, and anything
that unhealthily stimulates the body and mind, will have an
injurious effect in increasing the flow, which is, therefore,
less in hard-worked countrywomen than in the better classes
and residents in towns.

It is more abundant in warm climates, and our countrywomen in India habitually menstruate over-profusely, be-

coming less abundantly unwell when they return to England.
The same observation has been made with regard to American
women residing in the Gulf States, who improve materially
by removing to the Lake States. Some women appear to
menstruate more in summer than in winter. I am acquainted
with a lady who spends the winter in St. Petersburg, where
her periods last eight or ten days, and the summer in England, where they never exceed four or five. The difference
is probably due to the effect of the over-heated rooms in
which she lives in Russia.

The daily loss is not the same during the continuance
of the period. It generally is at first slight, and gradually
increases so as to be most profuse on the second or third
day, and as gradually diminishes. Towards the last days it
sometimes disappears for a few hours, and then comes on
again, and is apt to recur under any excitement or emotion.

Quality of
menstrual
blood.
As the menstrual fluid escapes from the uterus it consists of pure blood, and if collected through the speculum, it
coagulates. The ordinary menstrual fluid does not coagulate
unless it is excessive in amount. Various explanations of
this fact have been given. It was formerly supposed either to
contain no fibrine, or an unusually small amount. Retzius
attributes its non-coagulation to the presence of free lactic
and phosphoric acids. The true explanation was first given
by Mandl, who proved that even small quantities of pus or
mucus in blood were sufficient to keep the fibrine in solution ; and mucus is always present to greater or less amount
in the secretions of the cervix and vagina, which mix with
the menstrual blood in its passage through the genital tract.
If the amount of blood be excessive, however, the mucus
present is insufficient in quantity to produce this effect, and
coagula are then formed.

On microscopic examination the menstrual fluid exhibits
blood corpuscles, mucous corpuscles, and a considerable
amount of epithelial scales, the last being the débris of the
epithelium lining the uterine cavity. According to Virchow,
the form of the epithelium often proves that it comes from
the interior of the utricular glands. The colour of the blood
is at first dark, and as the period progresses it generally
becomes lighter in tint. In women who are in bad health
it is often very pale. These differences doubtless depend

upon the amount of mucus mingled with it. . The menstrual
blood has always a characteristic, faint, and heavy odour,
which is analogous to that which is so distinct in the lower
animals during the rut. Raciborski mentions a lady who
was so sensitive to this odour that she could always tell to a
certainty when any woman was menstruating. It is attributed either to decomposing mucus mixed with the blood,
which, when partially absorbed, may cause the peculiar odour
of the breath often perceptible in menstruating women ; or
to the mixture with the fluid of the sebaceous secretion from
the glands of the vulva. It probably gave rise to the old
and prevalent prejudices as to the deleterious properties of
menstrual blood, which, it is needless to say, are altogether
without foundation.

It is now universally admitted that the source of the men- Source of
the blood.
strual blood is the mucous membrane lining the interior of
the uterus, for the blood may be seen oozing through the os
uteri by means of the speculum, and in cases of prolapsus
uteri ; while in cases of inverted uterus it may be actually
observed escaping from the exposed mucous membrane,
and collecting in minute drops upon its surface. During the
menstrual nisus the whole mucous lining becomes congested
to such an extent that, in examining the bodies of women
who have died during menstruation, it is found to be thicker,
larger, and thrown into folds, so as to completely fill the
uterine cavity. The capillary circulation at this time becomes very marked, and the mucous membrane assumes a
deep red hue, the network of capillaries surrounding the
orifices of the utricular glands being especially distinct.
These facts have an unquestionable connection with the
production of the discharge, but there is much difference of
opinion as to the precise mode in which the blood escapes
from the vessels. Coste believed that the blood transudes
through the coats of the capillaries without any laceration
of their structure. Farre inclines to the hypothesis that the
uterine capillaries terminate by open mouths, the escape of
blood through these, between the menstrual periods, being
prevented by muscular contraction of the uterine walls.
Pouchet believed that during each menstrual epoch the
entire mucous membrane is broken down and cast off in
the form of minute shreds, a fresh mucous membrane being

developed in the interval between two periods. During this process the capillary network would be laid bare and ruptured, and the escape of blood readily accounted for. Tyler Smith, who adopted this theory, states that he has frequently seen the uterine mucous membrane, in women who have died during menstruation, in a state of dissolution, with the broken loops of the capillaries exposed. The phenomena attending the so-called membranous dysmenorrhœa, in which the mucous membrane is thrown off in shreds, or as a cast of the uterine cavity—the nature of which was first pointed out by Simpson and Oldham—have been supposed to corroborate this theory. This view is, in the main, corroborated by the recent researches of Engelman,[1] Williams,[2] and others. Williams describes the mucous lining of the uterus as undergoing a fatty degeneration before each period, which commences near the inner os, and extends over the whole mucous membrane, and down to the muscular wall. This seems to bring on a certain amount of muscular contraction, which drives the blood into the capillaries of the mucosa, and these, having become degenerated, readily rupture, and permit the escape of the blood. The mucous membrane now rapidly disintegrates, and is cast off in shreds with the menstrual discharge, in which masses of epithelial cells may always be detected. Engelman, however, holds that the fatty degeneration is limited to the superficial layers, and that a portion only of the epithelial investment is thrown off. As soon as the period is over, the formation of a new mucous membrane is begun, from proliferation of the elements of the muscular coat, and at the end of a week the whole uterine cavity is lined by a thin mucous membrane. This grows until the advent of another period, when the same degenerative changes occur unless impregnation has taken place, in which case it becomes further developed into the decidua.

Theory of menstruation.   That there is an intimate connection between ovulation and menstruation is admitted by most physiologists, and it is held by many that the determining cause of the discharge is the periodic maturation of the Graafian follicles. There is

[1] American Journal of Obstetrics, May 1875.

[2] 'On the Structure of the Mucous Membrane of the Uterus,' Obst. Journ. 1875.

abundant evidence of this connection, for we know that when, at the change of life, the Graafian follicles cease to develop, menstruation is arrested; and when the ovaries are removed by operation, of which there are now numerous cases on record, or when they are congenitally absent, menstruation does not generally take place.   A few cases, however, have been observed in which menstruation continued after double ovariotomy, or the removal of the ovaries by Battey's operation, and these have been used as an argument by those physiologists who doubt the ovular theory of menstruation.   Slavyanski has particularly insisted on such cases, which, however, are probably susceptible of explanation.   It may be that the habit of menstruation may continue for a time even after the removal of the ovaries; and it has not been shown that menstruation has continued permanently after double ovariotomy, although it certainly has occasionally, although quite exceptionally, done so for a time.   It is possible, also, that in such cases a small portion of ovarian tissue may have been left unremoved, sufficient to carry on ovulation.   Roberts, a traveller quoted by Depaul and Gueniot in their article on Menstruation in the 'Dictionnaire des Sciences Médicales,' relates that in certain parts of Central Asia it is the custom to remove both ovaries in young girls who act as guards to the harems. These women, known as 'hedjeras,' subsequently assume much of the virile type, and never menstruate.   The same close connection between ovulation and the rut of animals is observed, and supports the conclusion that the rut and menstruation are analogous.   The chief difference between ovulation in man and the lower animals is that in the latter the process is not generally accompanied by a sanguineous flow.   To this there are exceptions, for in monkeys there is certainly a discharge analogous to menstruation occurring at intervals.   Another point of distinction is that in animals connection never takes place except during the rut, and that it is then only that the female is capable of conception; while in the human race conception only occurs in the interval between the periods.   This is another argument brought against the ovular theory, because, it is said, if menstruation depend on the rupture of a Graafian follicle and the emission of an ovule, then impregnation should only take place during or immediately after menstruation.   Coste explains this by

Menstruation does not occur in the absence of the ovaries. Exceptions to this rule.

Similarity between menstruation and the rut of animals.

supposing that it is the *maturation* and not the *rupture* of the follicle which determines the occurrence of menstruation ; and that the follicle may remain unruptured for a considerable time after it is mature, the escape of the ovule being subsequently determined by some accidental cause, such as sexual excitement. However this may be, there is good reason to believe that the susceptibility to conception is greater during the menstrual epochs. Raciborski believes that in the large proportion of cases impregnation occurs in the first half of the menstrual interval, or in the few days immediately preceding the appearance of the discharge. There are, however, very numerous exceptions, for in Jewesses, who almost invariably live apart from their husbands for eight days after the cessation of menstruation, impregnation must constantly occur at some other period of the interval, and it is certain that they are not less prolific than other people. This rule with them is very strictly adhered to, as will be seen by the accompanying interesting letter from a medical friend who is a well-known member of that community, and which I have permission to publish.[1] This

Susceptibility to conception.

---

[1] 10, Bernard Street, Russell Square : July 21, 1873.

MY DEAR SIR,

1. To the best of my knowledge and belief, the law which prohibits sexual intercourse among Jews for seven clear days after the cessation of menstruation, is almost universally observed ; the exceptions not being sufficient to vitiate statistics. The law has perhaps fewer exceptions on the Continent—especially Russia and Poland, where the Jewish population is very great—than in England. Even here, however, women who observe no other ceremonial law observe this, and cling to it after everything else is thrown overboard. There are doubtless many exceptions, especially among the better classes in England, who keep only three days after the cessation of the menses.

2. The law is—as you state—that should the discharge last only an hour or so, or should there be only one gush or one spot on the linen, the five days during which the period *might* continue are observed ; to which must be superadded the seven clear days—twelve days per mensem in which connection is disallowed. Should any discharge be seen in the intermenstrual period, seven days would have to be kept, but not the five, for such *irregular* discharge.

3. The ' bath of purification,' which must contain at least eighty gallons, is used on the last night of the seven clear days. It is not used till after a bath for cleansing purposes ; and, from the night when such ' purifying ' bath is used, Jewish women are accustomed to calculate the commencement of pregnancy. That you should not have heard it is not strange : its mention would be considered highly indelicate.

4. Jewish women reckon their pregnancy to last nine calendar or ten

fact is of itself sufficient to disprove the theory advanced by Dr. Avrard,[1] that impregnation is impossible in the latter half of the menstrual interval. This, and the other reasons referred to, undoubtedly throw some doubt on the ovular theory, but they do not seem to be sufficient to justify the conclusion that menstruation is a physiological process altogether independent of the development and maturation of the Graafian follicles. All that they can be fairly held to prove is that the escape of the ovules may occur independently of menstruation, but the weight of evidence remains strongly in favour of the theory which is generally received. It should be stated that Lawson Tait attributes considerable influence in menstruation to the Fallopian tubes themselves; but his views on this point, based on observations made after the removal of the ovaries for certain morbid conditions, cannot yet be taken as proved. *The ovular theory has the weight of evidence in its favour.*

The cause of the monthly periodicity is quite unknown, and will probably always remain so. Goodman[2] has suggested what he calls the 'cyclical theory of menstruation,' which refers the phenomena to a general condition of the vascular system, specially localising itself in the generative organs, and connected with rhythmical changes in their nerve-centres. It does not seem to me, however, that he has satisfactorily proved the recurrence of the conditions which his ingenious theory assumes. The purpose of the loss of so much blood is also somewhat obscure. To a certain extent it must be considered an accident or complication of ovulation, produced by the vascular turgescence. Nor is it essential to fecundation, because women often conceive during lactation, when menstruation is suspended; or before the function has become established. It may, however, serve the negative purpose of relieving the congested uterine *Purpose of the menstrual loss.*

lunar months - 270 to 280 days. There are no special data on which to reckon an average, nor do I know of any books on the subject, except some Talmudic authorities, which I will look up for you if you desire it. Pray make no apologies for writing to me: any information I possess is at your service.

I am, dear Sir, yours very truly,

Dr. Playfair.   A. ASHER.

P.S.—The Biblical foundation for the law of the seven clear days is Leviticus xv., verse 19 till the end of the chapter—especially verse 28.

[1] *Rev. de Thérap. Méd. Chir.*, 1867.
[2] *American Journal of Obstetrics*, Oct. 1878.

capillaries which are periodically filled with a supply of
blood for the great growth which takes place when conception has occurred. Thus immediately before each period the
uterus may be considered to be placed by the afflux of blood
in a state of preparation for the function it may be suddenly
called upon to perform. That the discharge relieves a state
of vascular tension which accompanies ovulation is proved by
Vicarious the singular phenomenon of vicarious menstruation which
menstrua-
tion. is occasionally, though rarely, met with. It occurs in cases
in which, from some unexplained cause, the discharge does
not escape from the uterine mucous membrane. Under such
circumstances a more or less regular escape of blood may
take place from other sites. The most common situations
are the mucous membranes of the stomach, of the nasal
cavities, or of the lungs; the skin, not uncommonly that of
the mammæ, probably on account of their intimate sympathetic relation with the uterine organs ; from the surface of
an ulcer ; or from hæmorrhoids. It is a noteworthy fact that
in all these cases the discharge occurs in situations where its
external escape can readily take place. This strange deviation of the menstrual discharge may be taken as a sign of
general ill-health, and it is usually met with in delicate
young women of highly mobile nervous constitution. It
may, however, begin at puberty, and it has even been
observed during the whole sexual life. The recurrence is
regular, and always in connection with the menstrual nisus,
although the amount of blood lost is much less than in
ordinary menstruation.

Cessation   After a certain time changes occur, showing that the
of men-
struation. woman is no longer fitted for reproduction ; menstruation
ceases, Graafian follicles are no longer matured, and the
ovary becomes shrivelled and wrinkled on its surface.
Analogous alterations take place in the uterus and its appendages. The Fallopian tubes atrophy, and are not unfrequently obliterated. The uterus decreases in size. The
cervix undergoes a remarkable change, which is readily
detected on vaginal examination ; the projection of the
cervix into the vaginal canal disappears, and the orifice of
the os uteri in old women is found to be flush with the roof
of the vagina. In a large number of cases there is, after the
cessation of menstruation, an occlusion both of the external

and internal os; the canal of the cervix between them, however, remains patulous, and is not unfrequently distended
with a mucous secretion.

The age at which menstruation ceases varies much in Period of
different women. In certain cases it may cease at an un- cessation.
usually early age, as between 30 and 40 years, or it may
continue far beyond the average time, even up to 60 years;
and exceptional, though perhaps hardly reliable, instances
are recorded, in which it has continued even to 80 or 90
years. These are, however, strange anomalies, which, like
cases of unusually precocious menstruation, cannot be considered as having any bearing on the general rule. Most
cases of so-called protracted menstruation will be found to
be really morbid losses of blood depending on malignant or
other forms of organic disease, the existence of which, under
such circumstances, should always be suspected.

In this country menstruation usually ceases between 40
and 50 years of age. Raciborski says that the largest number of cases of cessation are met with in the 46th year. It
is generally said that women who commence to menstruate
when very young cease to do so at a comparatively early
age, so that the average duration of the function is about
the same in all women. Cazeaux and Raciborski, whose
opinion is strengthened by the observations of Guy in 1,500
cases,[1] think, on the contrary, that the earlier menstruation
commences the longer it lasts, early menstruation indicating
an excess of vital energy which continues during the whole
childbearing life. Climate and other accidental causes do
not seem to have as much effect on the cessation as on the
establishment of the function. It does not appear to cease
earlier in warm than in temperate climates. The change of
life is generally indicated by irregularities in the recurrence
of the discharge. It seldom ceases suddenly, but it may be
absent for one or more periods, and then occur irregularly;
or it may become profuse or scanty, until eventually it
entirely stops. The popular notions as to the extreme danger
of the menopause are probably much exaggerated; although
it is certain that at that time various nervous phenomena
are apt to be developed. So far from having a prejudical

---

[1] *Med. Times and Gaz.* 1845.

effect on the health, however, it is not an uncommon
observation to see an hysterical woman, who has been for years
a martyr to uterine and other complaints, apparently take
a new lease of life when her uterine functions have ceased
to be in active operation, and statistical tables abundantly
prove that the general mortality of the sex is not greater at
this than at any other time.

PART II.

PREGNANCY.

CHAPTER I.

CONCEPTION AND GENERATION.

GENERATION in the human female, as in all mammals, re- quires the congress of the two sexes, in order that the semen, the male element of generation, may be brought into contact with the ovule, the female element of generation.

The semen secreted by the testicle of an adult male is a viscid, opalescent fluid, forming an emulsion when mixed with water, and having a peculiar faint odour, which is attributed to the secretions which are mixed with it, such as those from the prostate and Cowper's glands. On analysis it is found to be an albuminous fluid, holding in solution various salts, principally phosphates and chlorides, and an animal substance, spermatine, analogous to fibrine. Examined under a magnifying power of from 400 to 500 diameters, it consists of a transparent and homogeneous fluid, in which are floating a certain number of granules and epithelial cells, derived from the secretions mixed with it, and the characteristic sperm cells and spermatozoa which form its essential constituents. The sperm cells are those occupying the tubuli semeniferi of the testicle. Several kinds of sperm cells are described which receive their name from the position they occupy with regard to the lumen of the tubule (fig. 43). The cells which are next to the wall of the tubule are called the outer or lining cells. They are more or less flattened in form, and are situated on a distinct basement membrane. Internal to this layer is another, consisting of round cells the

Generation.

The semen.

Formation of the spermatozoa.

nuclei of which are in a state of proliferation; this is the
intermediate layer. Between this and the lumen of the

Fig. 43.

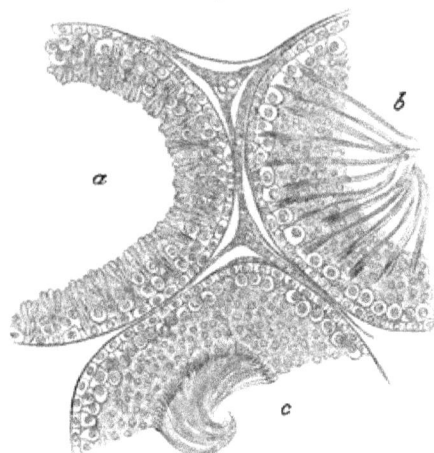

SECTION OF PARTS OF THREE SEMINIFEROUS TUBULES OF THE RAT.

a. With the spermatozoa least advanced in development. *b*. More advanced. *c*. Containing fully-developed spermatozoa. Between the tubules are seen strands of interstitial cells and lymph spaces. (From a preparation by Mr. A. Frazer.)

tubule are a number of cells, irregular in shape, amongst
which are imbedded the heads of the spermatozoa, the tails
of which project into the lumen. The spermatozoa are
thought to arise from this innermost layer in the following
manner: the nuclei of the sperm cells proliferate, and from
their subdivisions arise the heads of the spermatozoa, the
bodies of which originate from the protoplasm of the cells. By
the decomposition of the substance in which the heads of
the spermatozoa are imbedded the contained spermatozoa
become liberated, and move about freely in the seminal fluid.

Their microscopic characters. As seen under the microscope, the spermatozoa, which exist
in healthy semen in enormous numbers, present the appear-
ance of minute particles, not unlike a tadpole in shape.
The head is oval and flattened, measuring about $\frac{1}{10000}$ of an
inch in breadth, and attached to it is a delicate filamentous
expansion or tail, which tapers to a point so fine that its
termination cannot be seen by the highest powers of the
microscope. The whole spermatozoon measures from $\frac{1}{400}$ to

Their power of motion. $\frac{1}{600}$ of an inch in length. The spermatozoa are observed
to be in constant motion, sometimes very rapid, sometimes

more gentle, which is supposed to be the means by which they pass upwards through the female genital organs. They retain their vitality and power of movement for a considerable time after emission, provided the semen is kept at a temperature similar to that of the body. Under such circumstances they have been observed in active motion from forty-eight to seventy-two hours after ejaculation, and they have also been seen alive in the testicle as long as twenty-four hours after death. In all probability they continue active much longer within the generative organs, as many physiologists have observed them in full vitality in bitches and rabbits, seven or eight days after copulation. The recent experiments of Haussman, however, show that they lose their power of motion in the human vagina within twelve hours after coitus, although they doubtless retain it longer in the uterus and Fallopian tubes. Abundant leucorrheal discharges and acrid vaginal secretions destroy their movements, and may thus cause sterility in the female. On account of their mobility, the spermatozoa were long considered to be independent animalcules, a view which is by no means exploded, and has been maintained in modern times by Pouchet, Joulin, and other writers, while Coste, Robin, Kölliker, etc., liken their motion to that of ciliated epithelium. There can be no doubt that the fertilising power of the semen is due to the presence of the spermatozoa, although some of the older physiologists assigned it to the spermatic fluid. The former view, however, has been conclusively proved by the experiments of Prévost and Dumas, who found that on carefully removing the spermatozoa by filtration the semen lost its fecundating properties.

There has been great difference of opinion as to the part of the genital tract in which the spermatozoa and the ovule come into contact, and in which impregnation, therefore, occurs. Spermatozoa have been observed in all parts of the female genital organs in animals killed shortly after coitus, especially in the Fallopian tubes, and even on the surface of the ovary. The phenomena of ovarian gestation, and the fact that fecundation has been proved to occur in certain animals within the ovary, tend to support the idea that it may also occur in the human female before the rupture of the Graafian follicle. In order to do so, however, it is necessary

Sites of impregnation.

for the spermatozoa to penetrate the proper structure of the follicle and the epithelial covering of the ovary, and no one has actually seen them doing so. Most probably the contact of the spermatozoa and the ovule occurs very shortly after the rupture of the follicle, and in the outer part of the Fallopian tubes. Coste maintains that, unless the ovule is impregnated, it very rapidly degenerates after being expelled from the ovary, partly by inherent changes in the ovule itself, and partly because it then soon becomes invested by an albuminous covering which is impermeable to the spermatozoa. He believes, therefore, that impregnation can only occur either on the surface of the ovary, or just within the fimbriated extremity of the tube.

Mode in which the ascent of the semen is effected.

The semen is probably carried upwards chiefly by the inherent mobility of the spermatozoa. It is believed by some that this is assisted by other agencies: amongst them are mentioned the peristaltic action of the uterus and Fallopian tubes; a sort of capillary attraction affected when the walls of the uterus are in close contact, similar to the movement of fluid in minute tubes; and also the vibratile action of the cilia of the epithelium of the uterine mucous membrane. The action of the latter is extremely doubtful, for they are also supposed to effect the descent of the ovule, and they can hardly act in two opposite ways. The movement of the cilia being from within outwards it would certainly oppose rather than favour the progress of the spermatozoa. It must, therefore, be admitted that they ascend chiefly through their own powers of motion. They certainly have this power to a remarkable extent, for there are numerous cases on record in which impregnation has occurred without penetration, and even when the hymen was quite entire, and in which the semen has simply been deposited on the exterior of the vulva; in such cases, which are far from uncommon, the spermatozoa must have found their way through the whole length of the vagina. It is probable, however, that under

Fig. 44.

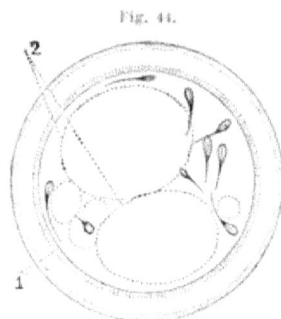

OVUM OF RABBIT CONTAINING SPERMATOZOA.

1. Zona pellucida. 2. The germs, consisting of two large cells, several smaller cells, and spermatozoa.

ordinary circumstances the passage of the spermatic fluid into the uterus is facilitated by changes which take place in the cervix during the sexual orgasm, in the course of which the os uteri is said to dilate and close again in a rhythmical manner.[1]

The precise method in which the spermatozoa effect impregnation was long a matter of doubt. It is now, however, certain that they actually penetrate the ovule, and reach its interior. This has been conclusively proved by the observations of Barry, Meissner, and others, who have seen the spermatozoa within the external membrane of the ovule in rabbits (fig. 44). In some of the invertebrata a canal or opening exists in the zona pellucida, through which the spermatozoa pass. No such aperture has yet been demonstrated in the ovules of mammals, but its existence is far from improbable. According to the observations of Newport, several spermatozoa enter the ovule, and the greater the number that do so the more certain fecundation becomes. After the spermatozoa penetrate the zona pellucida, they disintegrate and mingle with the yelk, having, while doing so, imparted to the ovule a power of vitality, and initiated its development into a new being.

Mode of impregnation.

The length of time which lapses before the fecundated ovule arrives in the cavity of the uterus has not yet been ascertained, and it probably varies under different circumstances. It is known that in the bitch it may remain eight or ten days in the Fallopian tube, in the guinea-pig three or four. In the human female the ovum has never been discovered in the cavity of the uterus before the tenth or twelfth day after impregnation.

Progress of the impregnated ovule towards the uterus.

The changes which occur in the human ovule immediately before and after impregnation, and during its progress through the Fallopian tube, are only known to us by analogy, as, of course, it is impossible to study them by actual observation. We are in possession, however, of accurate information of what has been made out in the lower animals, and it is reasonable to suppose that similar changes occur in man. Immediately after the ovule has passed into the Fallopian tube, it is found to be surrounded by a layer of granular cells, a portion of the lining membrane of the Graafian follicle, which was described as the discus proligerus.

Changes which the ovule undergoes immediately before and after impregnation.

[1] *How do the Spermatozoa enter the Uterus?* by J. Beck, M.D.

As it proceeds along the tube these surrounding cells disappear, partly, it is supposed, by friction on the walls of the tube, and partly by being absorbed to nourish the ovule. Be this as it may, before long they are no longer observed, and the zona pellucida forms the outermost layer of the ovule. When the ovule has advanced some distance along the tube, it becomes invested with a covering of albuminous material, which is deposited around it in successive layers, the thickness of which varies in different animals. It is very abundant in birds, in whom it forms the familiar white of the egg. In some animals it has not been detected, so that its presence in the human ovule is uncertain. Where it exists it doubtless contributes to the nourishment of the ovule. Coincident with these changes is the disappearance of the germinal vesicle. At the same time the yelk contracts and becomes more solid ; retiring from close contact with the zona pellucida, and thus forming a species of cavity, between the outer edge of the yelk and the vitelline membrane, which in some animals is filled with a transparent liquid. Coincident with the shrinking of the yelk, a small granular mass of a rounded form is extruded from the yelk into the clear space beneath the zona pellucida. At a later period another similar mass is extruded. These are the *polar globules* (fig. 45), the origin of which is thought to be in connection with the disappearance of the germinal vesicle and the germinal spot. These changes occur in all ovules, whether they are impregnated or not, but if the ovule is not fecundated, no further alterations occur. Supposing impregnation has taken place, a bright, clear vesicle, called the *vitelline nucleus*, very similar in appearance to a drop of oil, appears in the centre of the yelk. After this occurs the very peculiar phenomenon known as the cleavage of the yelk, which results in the formation of the layer of cells from which the fœtus is developed. The segmentation of the yelk (fig. 46)

Disappearance of the germinal vesicle.

Cleavage of the yelk.

Fig. 45.

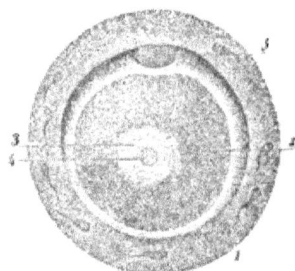

FORMATION OF THE 'POLAR GLOBULE.'
1. Zona pellucida, containing spermatozoa. 2. Yelk. 3 and 4. Germinal vesicle. 5. The polar globule.

occupies in mammals the whole of its substance. In birds the cleavage is confined to a small area of the yelk called the cicatricula or Blastoderm. Hence the term Holoblastic has been applied to the ova of mammals, Mesoblastic to those of birds. It divides at first into two halves, and at the same time the vitelline nucleus becomes constricted in its centre, and separates into two portions, one of which forms a centre for each of the halves into which the yelk has divided. Each of these immediately divides into two, as does its contained portion of the vitelline nucleus, and so on in rapid succession until the whole yelk is divided into a number of spheres, each of which consists of a clump of nucleated protoplasm.

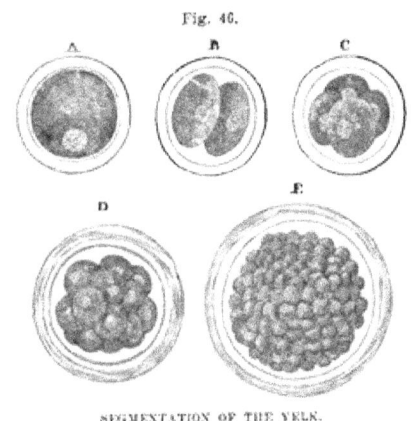

Fig. 46.

SEGMENTATION OF THE YELK.

A. Ovum with first embryo cell.
B. Division of embryo cell and cleavage of the yelk around it.
C, D, E. Further division of the yelk.

By these continuous dichotomous divisions the whole yelk is formed into a granular mass, which, from its supposed resemblance to a mulberry, has been named the *muriform body*. When the subdivision of the yelk is completed, its separate spheres become converted into cells, consisting of a fine membrane with granular contents. These cells unite by their edges to form a continuous membrane (fig. 47), which, through the expansion of the muriform body by fluid which forms in its interior, is distended until it forms a lining to the zona pellucida. This is the *blastodermic membrane*, from which the fœtus is developed. By this time the ovum has reached the uterus, and, before proceeding to consider the further changes which it undergoes, it will be well to study the alteration which the stimulus of impregnation has set on foot in the mucous membrane of the uterus, in order to prepare it for the reception and growth of its contents.

Even before the ovum reaches the uterus, the mucous membrane becomes thickened and vascular, so that its op-

(margin: Formation of the blastodermic membrane.)

(margin: Changes in the uterine)

mucous membrane consequent on pregnancy.

posing surfaces entirely fill the uterine cavity. These changes may be said to be the same in kind, although more

Fig. 47.

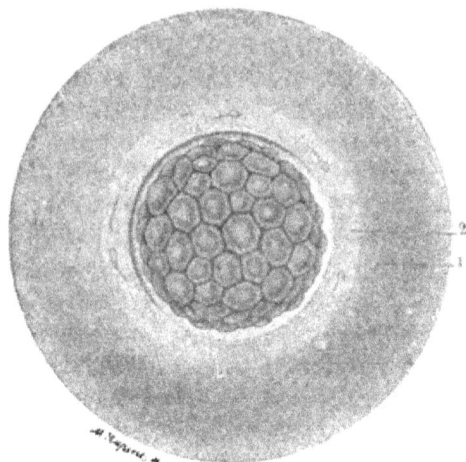

FORMATION OF THE BLASTODERMIC MEMBRANE FROM THE CELLS OF THE MURIFORM BODY. (After Joulin.)

1. Layer of albuminous material surrounding 2. The zona pellucida.

marked and extensive in degree, as the alterations which take place in the mucous membrane in connection with each menstrual period. The result is the formation of a distinct membrane, which affords the ovum a safe anchorage and protection, until its connections with the uterus are more fully developed. After delivery, this membrane, which is by that time quite altered in appearance, is at least partially thrown off with the ovum; on which account it has received the name of the *decidua* or *caduca*.

Divisions of the decidua.

The decidua consists of two principal portions, which, in early pregnancy, are separated from each other by a considerable interspace. One of these, called the *decidua vera*, lines the entire uterine cavity, and is, no doubt, the original mucous lining of the uterus greatly hypertrophied. The second, the *decidua reflexa*, is closely applied round the ovum; and it is probably formed by the sprouting of the decidua vera around the ovum at the point on which the latter rests, so that it eventually completely surrounds it. As the ovum enlarges, the decidua reflexa is necessarily stretched, until it comes everywhere into contact with the decidua vera, with which it firmly unites. After the third month of pregnancy true union has occurred, and the two layers of decidua are no longer separate. The *decidua sero-tina*, which is described as a third portion, is merely that part of the decidua vera on which the ovum rests, and where the placenta is eventually developed.

It is needless to refer to the various views which have been held by anatomists as to the structure and formation of the decidua. That taught by John Hunter was long believed to be correct, and down to a recent date it received the adherence of most physiologists. He believed the decidua to be an inflammatory exudation which, on account of the stimulus of pregnancy, was thrown out all over the cavity of the uterus, and soon formed a distinct lining membrane to it. When the ovum reached the uterine orifice of the Fallopian tube it found its entrance barred by this new membrane, which accordingly it pushed before it. This separated portion formed a covering to the ovum, and became the decidua reflexa, while a fresh exudation took place at that portion of the uterine wall which was thus laid bare, and this became the decidua vera. William Hunter had much more correct views of the decidua, the accuracy of which was at the time much contested, but which have recently received full recognition. He describes the decidua in his earlier writings as an hypertrophy of the uterine mucous membrane itself, a view which is now held by all physiologists.

When the decidua is first formed it is a hollow triangular sac lining the uterine cavity (fig. 48), and having three openings into it, those of the Fallopian tubes at its upper angles, and one, corresponding to the internal os uteri, below. If, as is generally the case, it is thick and pulpy, these openings are closed up, and can no longer be detected. In early pregnancy it is well developed, and continues to grow up to the third month of utero-gestation. After that time it commences to atrophy, its adhesion with the uterine walls lessens, it becomes thin and transparent, and is ready for expulsion when delivery is effected. When it is most developed, a careful examination of the decidua enables us to detect in it all the elements of the uterine mucous membrane greatly hypertrophied. Its substance chiefly consists of large round or oval nucleated cells and elongated fibres,

Views of William and John Hunter.

Structure of the decidua.

mixed with the tubular uterine gland ducts, which are much
elongated, and filled with cylindrical epithelium cells, and a

Fig. 48.

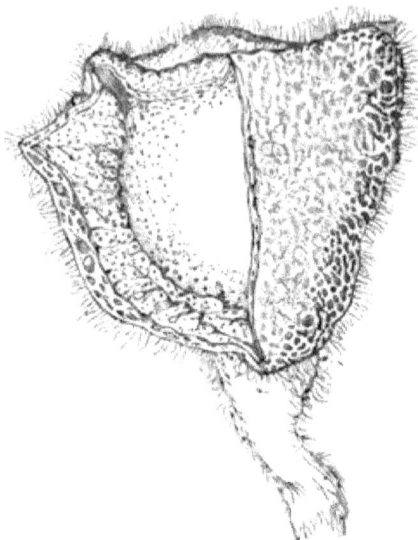

ABORTED OVUM OF ABOUT FORTY DAYS, SHOWING THE TRIANGULAR SHAPE OF
THE DECIDUA (WHICH IS LAID OPEN), AND THE APERTURE OF THE FALLOPIAN
TUBE. (After Coste.)

small quantity of milky fluid. According to Friedlander,
the decidua is divisible into two layers : the inner being
formed by a proliferation of the corpuscles of the sub-
epithelial connective tissue of the mucous membrane; the
deeper, in contact with the uterine walls, out of flattened or
compressed gland ducts.   In an early abortion the extremities of these ducts may be observed by a lens on the external
or uterine surface of the decidua, occupying the summit of
minute projections, separated from each other by depressions.
If these projections be bisected they will be found to contain
little cavities, filled with lactescent fluid, which were first
described by Montgomery of Dublin, and are known as
Montgomery's cups.   They are in fact the dilated canals of
the uterine tubular glands.   On the internal surface of such
an early decidua a number of shallow depresions may be
made out, which are the open mouths of these ducts.

Formation   When the ovum reaches the uterine cavity it soon be-
of the   comes imbedded in the folds of the hypertrophied mucous

membrane, which almost entirely fills the uterine cavity. decidua
As a rule it is attached to some point near the opening of reflexa.
a Fallopian tube, the swollen folds of mucous membrane
preventing its descent to the lower part of the uterus; in

Fig. 49.   Fig. 50.

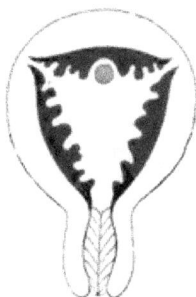

FORMATION OF DECIDUA.

(The decidua is coloured black, the
ovum is represented as engaged
between two projecting folds of
membrane.)

PROJECTING FOLDS OF MEMBRANE
GROWING UP AROUND THE OVUM.

(After Dalton.)

exceptional circumstances, however—as in women who have
borne many children, and have a more than usually dilated
uterine cavity—it may fix itself at
a point much nearer the internal os
uteri. According to the now generally accepted opinion of Coste, the
mucous membrane at the base of the
ovum soon begins to sprout around
it, and gradually extends until it
eventually covers the ovum (figs.
49-51), and forms the decidua
reflexa. Coste describes, under the
name of the *umbilicus*, a small depression at the most prominent part
of the ovum, which he considers to
be the indication of the point where
the closure of the decidua reflexa is
effected. There are some objections to this theory, for no
one has seen the decidua reflexa incomplete and in the
process of formation, and on examining its external surface,
that is, the one farthest from the ovum, its microscopical

Fig 51.

SHOWING OVUM COMPLETELY
SURROUNDED BY THE DE-
CIDUA REFLEXA.

appearance is identical with that of the inner surface of the decidua vera. To meet these difficulties, Weber and Goodsir, whose views have been adopted by Priestley, contended that the decidua reflexa is 'the primary lamina of the mucous membrane, which, when the ovum enters the uterus, separates in two thirds of its extent from the layers beneath it, to adhere to the ovum; the remaining third remains attached, and forms a centre of nutrition.' According to this view the decidua vera would be a subsequent growth over the separated portion, and the decidua serotina the portion of the primary lamina which remained

Fig. 52.

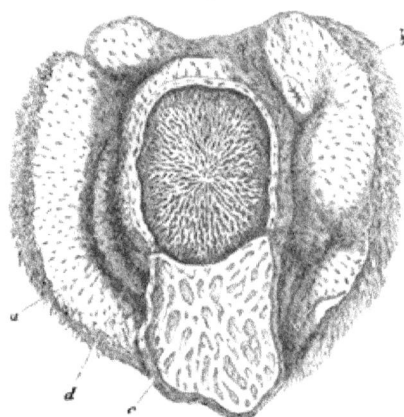

AN OVUM REMOVED FROM UTERUS, AND PART OF THE DECIDUA VERA CUT AWAY.
(After Coste.)

a. Decidua vera, showing the follicles opening on its inner surface. *b.* Inner extremity of Fallopian tube. *c.* Flap of decidua reflexa. *d.* Ovum.

attached. In this way the fact of the opposed surfaces of the decidua vera and reflexa being identical in structure would be accounted for. The difficulty which this theory is intended to meet does not seem so great as is supposed, for if, as is likely, it is only the epithelial or internal surface of the mucous membrane which sprouts over the ovum, and not its deeper layers, the facts of the case would be sufficiently met by Coste's view.

Up to the third month the decidua vera and

Up to the third month of pregnancy the decidua reflexa and vera are not in close contact, and there may even be a considerable interspace between them, which sometimes contains a small quantity of mucous fluid, called the *hydro-*

perione. This fact may account for the curious circum- reflexa are not in contact.
stance, of which many instances are on record, that a uterine
sound may be passed into a gravid uterus in the early months
of pregnancy without necessarily producing abortion, and
also for the occasional occurrence of menstruation after
conception (figs. 52 and 79).   Eventually, by the growth of
the ovum, the decidua reflexa comes closely into contact
with the vera, and the two become intimately blended and
inseparable.

As pregnancy advances the decidua alters in appearance Decidua at the end of pregnancy and after delivery.
and becomes fibrous and thin.   In the later months of utero-
gestation fatty degeneration of its structure commences, its
vessels and glands are obliterated, and its adhesion to the
uterine walls is lessened, so as to prepare it for separation.
As we shall subsequently see, this fatty degeneration was
assumed by Simpson to be the determining cause of labour at
term.   After the eighth month, thrombi form in the veins
lying underneath the decidua serotina, and at the end of
pregnancy they are described by Leopold[1] as having become,
to a great extent, obliterated.   This, he supposes, may have
some effect in inducing the contractions of the uterus in
labour.

It was long believed that the entire decidua was thrown
off after labour, leaving the muscular coat of the uterus bare
and denuded, and that a new mucous membrane was formed
during convalescence.   According to Robin,[2] whose views are Views of Robin.
corroborated by Priestley, no such denudation of the mus-
cular tissue of the uterus ever occurs, but a portion of the
decidua always remains attached after delivery.   After the
fourth month of pregnancy they believe that a new mucous
membrane is formed under the decidua, which remains in a
somewhat imperfect condition till after delivery, when it
rapidly develops and assumes the proper functions of the
mucous lining of the uterus.   Robin also believes that that
portion of the decidua which covers the placental site, the
so-called decidua serotina, is not thrown off with the membranes, like the decidua vera and reflexa, but remains
attached to the uterine walls, a thin layer of it only being
expelled with the placenta, on which it may be observed.

[1] *Arch. f. Gyn.* B. xi. H. 3.
[2] *Mémoires de l'Acad. Imp. de Méd.* 1861.

Duncan[1] entirely dissents from these views, and does not admit the formation of a new mucous membrane during the later months of utero-gestation. He believes that the greater portion of the decidua is thrown off, but that part remains, and from this the fresh mucous membrane is developed. This view is similar to that of Spiegelberg, who holds that the portion of the decidua that is expelled is the more superficial of the two layers described by Friedlander, composed chiefly of the epithelial elements, while the deeper or glandular layer remains attached to the walls of the uterus. From the epithelium of the glands a new epithelial layer is rapidly developed after delivery. Leopold[2] has shown that the uterine mucous membrane is completely re-formed within six weeks after delivery, and that its re-generation is sometimes completed as early as the end of the third week. This theory bears on the well-known analogy of the uterus after delivery to the stump of an amputated limb, an old simile, principally based on the erroneous theory that the whole muscular tissue of the uterus was laid bare. This, as we have seen, is not the case, but the simile so far holds good in that the mucous lining is deprived of its epithelial covering; and this fact, together with the existence of numerous open veins on the interior of the uterus, readily explains the extreme susceptibility to septic absorption which forms so peculiar a characteristic of the puerperal state.

Analogy of the interior of the uterus after delivery to the stump of an amputated limb.

Before we commenced the study of the decidua we had traced the impregnated ovum into the uterine cavity, and described the formation of the blastodermic membrane by the junction of the cells of the muriform body. We must now proceed to consider the further changes which result in the development of the fœtus, and of the membranes that surround it. It would be quite out of place in a work of this kind to enter into the subject of embryology at any length, and we must therefore be content with such details as are of importance from a practical point of view.

Changes in the ovum.

The blastodermic membrane, which forms a complete spherical lining to the ovum, between the yelk and the zona pellucida, soon divides into two layers, the most external,

Division of the blastodermic membrane into layers.

[1] *Researches in Obstetrics*, p. 186 *et seq.*
[2] *Arch. f. Gyn.* B. xii. H. 2.

called the *epiblast*, and an internal, the *hypoblast*, and
between them is subsequently developed a third, known as
the *mesoblast*. From these three layers are formed the
entire fœtus : the epiblast giving origin to the central
nervous system, to the superficial layer of the skin, and aiding in formation of the organs of special sense, and of the
amnion ; the hypoblast forming the epithelial lining membrane of the alimentary and respiratory tracts, and of the
tubes and glands in connection with them, and helping in
the development of the yelk sac ; the mesoblast giving rise
to the skeleton, the muscles, the connective tissues, the
vascular system, the genito-urinary organs, and taking part
in the formation of all the membranes.

Almost immediately after the separation of the blastodermic membrane into layers, one part of it becomes
thickened by the aggregation of cells and is called the *area* The area
germinativa. This is at first round and then oval in shape, germina-
and in its centre the first indication of the embryo may be tiva.
detected in the form of a narrow straight line, the *primitive
trace*. Surrounding it are some cells more translucent than
those of the rest of the area germinativa, and hence called
the *area pellucida* (fig. 53). In front of the primitive
trace two elevated ridges soon arise, the *laminæ dorsales*,
which include between them a groove, the medullary groove,
and gradually unite posteriorly to form a cavity within which

Fig. 53.

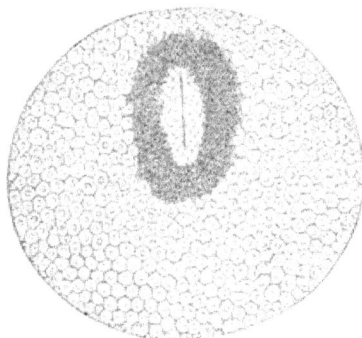

DIAGRAM OF AREA GERMINATIVA, SHOWING THE PRIMITIVE TRACE AND
AREA PELLUCIDA.

the cerebro-spinal axis is subsequently developed. The medullary groove as it grows backwards overlaps the primitive

trace, which disappears. The embryo is differentiated from the rest of the blastoderm by a fold anteriorly, which is called the cephalic or head fold. Another fold afterwards appears posteriorly, which is called the caudal or tail fold. Laterally folds also arise. These folds all tend to grow towards the centre of the under surface of what will be the embryo.

The mesoblastic layer of the blastoderm, except that part which forms the axis of the embryo, splits into an upper layer, the somato-pleure, which is beneath the epiblast, and a lower layer, the splancho-pleure, which lies upon the hypoblast. The space formed by this cleavage of the mesoblast is called the pleuro-peritoneal cavity. The somato-pleure is engaged in the formation of the body walls of the embryo. The splanchno-pleure forms the walls of the alimentary tract.

Formation of the amnion.

Processes arise from the somato-pleure anteriorly, posteriorly, and laterally, which gradually arch over the dorsal surface of the fœtus, until they meet each other and form a complete envelope to it. At the ventral surface these processes are separated by the whole length of the embryo, but they here also gradually approach each other, and eventually surround what is subsequently the umbilical cord, and blend with the integument of the fœtus at the point of its insertion. In this way is formed the *amnion* (fig. 54), consisting of two layers: the internal, derived from

Fig. 54.

DEVELOPMENT OF THE AMNION.
1. Vitelline membrane. 2. External layer of blastodermic membrane. 3. Internal layers forming the umbilical vesicle. 4. Umbilical vessels. 5. Projections forming amnion. 6. Embryo. 7. Allantois.

the epiblast, is formed of tesselated epithelial cells; the external, arising from the mesoblast, is formed of cells like those of young connective tissue. Before the folds of the amnion unite, the free edge of each is bent outwards and spreads around the ovum, immediately within the zona pellucida, forming a lining to it, termed by Turner the *sub-*

zonal membrane, which is connected with the development of the chorion. The amnion is the most internal of the membranes surrounding the fœtus, and will presently be studied more in detail. It soon becomes distended with fluid, the *liquor amnii*, and as this increases in amount it separates the amnion more and more from the fœtus.

During this time the innermost layer of the blastodermic membrane or hypoblast is also developing two projections at **Changes in the hypoblast.**

Fig. 55.

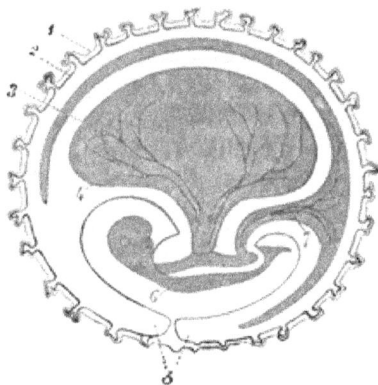

1. Exo-chorion.   2. External layer of blastodermic membrane.   3. Umbilical vesicle.
4. Its vessels.   5. Amnion.   6. Embryo.   7. Allantois increasing in size.

either extremity of the fœtus, and these gradually approach each other anteriorly. As the hypoblast is in contact with the yelk, when these meet they have the effect of dividing the yelk into two portions. One, and the smaller of the two, forms eventually the intestinal canal of the fœtus; the other, and much the larger, contains the greater portion of the yelk, and forms the ephemeral structure known as the *umbilical vesicle*, from which the fœtus derives most of its **The umbilical vesicle.** nourishment during the early stage of its existence. Its communication with the abdominal cavity of the fœtus is through the constricted portion at the point of division called the *vitelline duct* (fig. 55). An artery and vein, the *omphalo-mesenteric*, ramify on the vesicle and its duct.

As the amnion increases in size, it pushes back the umbilical vesicle towards the external membrane of the ovum, between which and the amnion it lies (fig. 56); and when the allantois is developed, it ceases to be of any use, and rapidly shrinks and dwindles away. In most mammals no trace of it can be found after the fourth month of utero-gestation; in some, including the human female, it is said to exist as a minute vesicle at the placental end of the umbilical cord at the full period of pregnancy. The umbilical vesicle is filled with a yellowish fluid, containing many oil and fat globules, similar to the yelk of an egg.

The allan-tois.

Somewhere about the twentieth day after conception a small vesicle is formed towards the caudal extremity of the fœtus, which is called the *allantois*. This membrane in mammals is impor-tant, as it forms the greater part of the fœtal placenta, a small portion of it remaining inside the body permanently as the bladder. It begins as a diverti-culum from the lower part of the intestinal canal. This, at first spherical, rapidly developes and becomes pyriform in

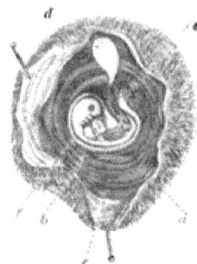

Fig. 56.

AN EMBRYO OF ABOUT TWENTY-FIVE DAYS LAID OPEN. (After Coste.)

a. Chorion. b. Amnion. c. Cavity of chorion. d. Umbilical vesicle e. Pedicle of allantois. f. Embryo.

Fig. 57.

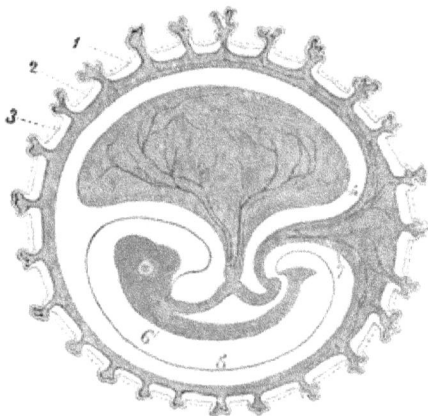

1. Exo-chorion. 2. External layer of the blastodermic membrane. 3. Allantois. 4. Umbilical vesicle. 5. Amnion. 6. Embryo. 7. Pedicle of allantois.

shape, while by a process of constriction, similar to that which occurs in the vitellus to form the umbilical vesicle, it becomes divided into two parts, communicating with each other, the

smaller of them being eventually developed into the urinary
bladder. The larger portion, leaving the abdominal cavity
along with the vitelline duct, rapidly grows until it comes
into contact with the most external ovular membrane, the
chorion, over the entire inner surface of which it spreads.
In this part vessels soon develop: namely, the two umbilical
arteries, derived from the abdominal aorta, and two umbilical
veins, one of which subsequently disappears; these, along
with the vitelline duct and the pedicle of the allantois, form
the umbilical cord. The main and very important function
of the allantois, therefore, is to carry the fœtal vessels up to
the inner surface of the sub-zonal membrane. Besides this
purpose, the allantois, at a very early period, may receive the
excretions of the fœtus, and serve as an excrementitious
organ. According to Cazeaux, scarcely a trace of the allantois
can be seen a few days after its formation. Its lower part or
pedicle, however, long remains distinct, and forms part of the
umbilical cord; and traces of it may be found even in adult
life in the form of the urachus, which is really the dwindled   The ura-
pedicle, and forms one of the ligaments of the bladder.   chus.

Between the chorion and amnion is often found a gelatinous fluid, with minute filamentous processes traversing it,
called by Velpeau the *corps reticulé*, which is not met with   The *corps*
until the allantois comes into contact with the chorion, and   *reticulé* or
which seems to be formed out of the tissues of that vesicle.   vitriform
It is analogous to the so-called Wharton's jelly found in the   body.
umbilical cord. When first formed it is highly vascular, but
the vessels entirely disappear after the placenta is formed,
and the remainder of the chorionic villi atrophy. Sometimes
it exists in considerable quantities, and should the chorion
rupture at the end of pregnancy, it may escape and give rise
to an erroneous impression that the liquor amnii has been
discharged.

Before proceeding to consider the fœtal envelopes more   Recapitu-
at length, it may be useful to recapitulate the structures   lation.
already alluded to as forming the ovum. In this we find :—

1. The *embryo* itself.

2. A fluid, the *liquor amnii*, in which it floats.

3. The *amnion*, a purely fœtal membrane surrounding
the embryo, and containing the liquor amnii.

4. The *umbilical vesicle*, containing the greater portion

of the yelk, serving as a source of nutrition to the early embryo through the vitelline duct, and on which ramify the omphalo-mesenteric vessels.

5. The *allantois*, a vesicle proceeding from the caudal extremity of the embryo, spreading itself over the interior of the ovum, and serving as a channel of vascular communication between the chorion and the fœtus, through the umbilical vessels.

6. An interspace between the outer layer of the ovum and the amnion, in which is contained the *umbilical vesicle* and *allantois*, and the *corps reticulé* of Velpeau.

7. The outer layer of the ovum, along with the sub-zonal membrane, forming the *chorion* and *placenta*.

The amnion.

The *amnion* is the most internal of the two membranes surrounding the fœtus ; its origin at an early period of fœtal life has already been described. It is a perfectly smooth, transparent, but tough membrane, continuous with the integument of the fœtus at the insertion of the umbilical cord, round which it forms a sheath. Soon after it is formed it becomes distended with a fluid, the *liquor amnii*, in which the fœtus is suspended and floats. This fluid increases gradually in quantity, distending the amnion as it does so, until this is brought into contact with the inner surface of the chorion, from which it was at first separated by a considerable interspace.

Structure of the amnion.

The internal surface of the amnion is smooth and glistening, and on microscopic examination it is found to consist of a layer of flattened cells, each containing a large nucleus. These rest on a stratum of fibrous tissue which gives to the membrane its toughness, and by which it is attached to the inner surface of the chorion. It is entirely destitute of

The liquor amnii.

vessels, nerves, and lymphatics. The quantity of the liquor amnii varies much at different periods of pregnancy. In the early months it is relatively greater in amount than the fœtus, which it outweighs. As pregnancy advances, the weight of the fœtus becomes four or five times greater than that of the liquor amnii, although the actual quantity of fluid increases during the whole period of gestation. The amount of fluid varies much in different pregnancies. Sometimes there is comparatively little ; while at others the quantity is immense, reaching several pounds in weight, greatly distending the uterus, and thus, it may be, producing difficulty in labour.

At first the liquid is clear and limpid. As pregnancy advances it becomes more turbid and dense, from the admixture of epithelial débris derived from the cutaneous surface of the foetus. In some cases, without actual disease, it may be dark green in colour, and thick and tenacious in consistency. It has a peculiar heavy odour, and it consists chemically of water containing albumen, with various salts, principally phosphates and chlorides. *Its quality.*

The source of the liquor amnii has been much disputed. Some maintain that it is derived chiefly from the foetus, a view sufficiently disproved by the fact that the liquor amnii continues to increase in amount after the death of the foetus. Burdach believed that it is secreted by the internal surface of the uterus, and arrives in the cavity of the amnion by transudation through the membrane. Priestley—and this seems the most probable hypothesis—thinks that it is secreted by the epithelial cells lining the membrane, which become distended with fluid, burst, and pour their contents into the amniotic cavity. *Its source.*

The most obvious use of the liquor amnii is to afford a fluid medium in which the foetus floats, and so is protected from the shocks and jars to which it would otherwise be subjected, and from undue pressure upon the uterine walls. By distending the uterus it saves the uterus from injury, which the movements of the foetus might otherwise inflict, and the foetus is thus also enabled to change its position freely. The facility with which version by external manipulation can be effected depends entirely on the mobility of the foetus in the fluid which surrounds it. Some have also supposed that it prevents the foetus, in the early months of pregnancy, from forming adhesions to the amnion. In labour it is of great service, by lubricating the passages, but chiefly by forming, with the membranes, a fluid wedge, which dilates the circle of the os uteri. *Functions and uses.*

The chorion is the more external of the truly foetal membranes, although external to it is the decidua, having a strictly maternal origin. It is a perfectly closed sac, its external surface, in contact with the decidua, being rough and shaggy from the development of villi (fig. 55), its *The chorion.*

internal smooth and shining. As the ovum passes along the
Fallopian tube it receives, as we have seen, an albuminous
coating, and this, with the zona pellucida, is developed into
The primi- tive cho- rion. a temporary structure, the *primitive chorion*. On its external surface villous prominences soon appear, which have
no ascertained structure, and which seem to supply the early
ovum with nutriment by endosmotic absorption from the
mucous membrane of the uterus. This primitive chorion,
however, has not been observed in the human subject,
although it may be readily seen in the ova of some of the
lower animals, such as the dog and the rabbit. Some **twelve**
days after conception, when the blastodermic membrane
The true chorion. is formed, the true chorion appears. This is, in fact, formed
by the epiblast layer of the blastodermic membrane, which
everywhere lines the zona pellucida or primitive chorion and,
by pressure, causes its absorption and disappearance. On
the surface of the true chorion thus formed, which is now the
external envelope of the ovum, villi soon appear.

Formation of the villi. These villi are hollow projections like the fingers of a
glove, which are raised up from the surface of the chorion
(the hollows looking into the chorionic cavity), and they
cover the whole external surface of the ovum, giving it the
peculiar shaggy appearance observed in early abortions.
They push themselves into the substance of the decidua,
with which they soon become so firmly united that they
Junction of the allantois with the chorion. cannot be separated without laceration. At first they are
absolutely non-vascular, but soon the allantois, previously
described, reaches the inner surface of the chorion, and
spreads itself over the whole of it. Each villus now receives
a separate artery and vein, the former having a branch to
each of the subdivisions into which the villus divides. These
vessels are encased in a fine sheath of the allantois, which
enters the villus along with them and forms a lining to it,
described by some as the *endo-chorion*; the external epithelial membrane of the villus, derived from the epiblast layer
of the blastodermic membrane, being called the *exo-chorion*.
The artery and vein lie side by side in the centre of the
villus, and anastomose at its extremity; each villus thus
having a separate circulation.

Growth and atro- As soon as the union of the allantois with the chorion
has been effected, the villi grow very rapidly, give off

branches, which, in their turn, give off secondary branches, and so form root-like processes of great complexity. In the early months of gestation they exist equally over the whole surface of the ovum. As pregnancy advances, however, those which are in contact with the decidua reflexa shrivel up, and by the end of the second month disappear, being no longer required for the nutrition of the ovum. The chorion and decidua thus come into close contact, being united together by fibrous shreds, which on microscopic examination are found to consist of atrophied villi. A certain number of villi, viz. those which are in contact with the decidua serotina, instead of dwindling away, increase greatly in size, and eventually develop into the organ by which the fœtus is nourished—the *placenta*.

phy of the villi.

This important organ serves the purpose of supplying nutriment to, and aerating the blood of, the fœtus, and on its integrity the existence of the fœtus depends. It is met with in all mammals, but is very different in form and arrangement in different classes. Thus, in the sow, mare, and in the cetacea, it is diffused over the whole interior of the uterus ; in the ruminants, it is divided into a number of separate small masses, scattered here and there over the uterine walls ; while in the carnivora and elephant, it forms a zone or belt round the uterine cavity. In the human race, as well as in rodentia, insectivora, etc., the placenta is in the form of a circular mass, attached generally to some part of the uterus near the orifices of one Fallopian tube ; but it may be situated anywhere in the uterine cavity, even over the internal os uteri. As it is expelled after delivery with the fœtal membranes attached to it, and as the aperture in these corresponds to the os uteri, we can generally determine pretty accurately the situation in which the placenta was placed by examining them after expulsion. The maternal surface of the placenta is somewhat convex, the fœtal concave. Its size varies greatly in different cases, and it is usually largest when the child is big, but not necessarily so. Its average diameter is from 6 to 8 inches, its weight from 18 to 24 ozs., but in exceptional cases it has been found to weigh several pounds. Abnormalities of form are not very rare. Thus, the placenta has been found to be divided into distinct parts, a form said by Professor Turner to be normal

Form of the placenta in animals.

Form of the placenta in man.

in certain genera of monkeys; or smaller supplementary placentæ (*placentæ succentariæ*) may exist round a central mass. These variations of shape are only of importance in consequence of a risk of part of the detached placenta being left in the uterus after delivery, and giving rise to septicæmia or secondary hæmorrhage.

Attachment of the membranes.
The fœtal membranes cover the whole fœtal surface of the placenta, being reflected from its edges so as to line the uterine cavity, and being expelled with it after delivery. They also leave it at the insertion of the cord, to which they form a sheath. The cord is generally attached near the centre of the placenta, and from its insertion the umbilical vessels may be seen dividing and radiating over the whole fœtal surface.

Its maternal surface.
The maternal surface is rough and divided by numerous sulci, which are best seen if the placenta is rendered convex, so as to resemble its condition when attached to the uterus. A careful examination shows that a delicate membrane covers the entire maternal surface, unites the sulci together, and dips down between them. This is, in fact, the cellular layer of the decidua serotina, which is separated and expelled with the placenta, the deeper layer remaining attached to the uterus. Numerous small openings may be seen on the surface, which are the apertures of the veins torn off from the uterus, as also those of some arteries, which, after taking several sharp turns, open suddenly into the substance of the organ.

Minute structure of the placenta.
As regards the minute structure of the placenta, it is certain that it consists essentially of two distinct portions— one *fœtal*, consisting of the greatly hypertrophied chorion villi, with their contained vessels, which carry the fœtal blood so as to bring it into intimate relation with the maternal blood, and thus admit of the necessary changes occurring in it connected with the nutrition of the fœtus; and the other *maternal*, formed out of the decidua serotina and the maternal bloodvessels. These two portions are in the human female so intimately blended as to form the single deciduous organ which is thrown off after delivery. These main facts are admitted by all, but considerable differences of opinion still exist among anatomists as to the precise arrangement of these parts. In the following sketch of the

subject I shall describe the views most generally entertained, merely briefly indicating the points which are contested by various authorities.

The fœtal portion of the placenta consists essentially of the ultimate ramifications of the chorion villi, which may be seen on microscopic examination in the form of club-shaped digitations, which are given off at every possible angle from the stem of a parent trunk, just like the branches of a

Fœtal portion of the placenta.

Fig. 58.

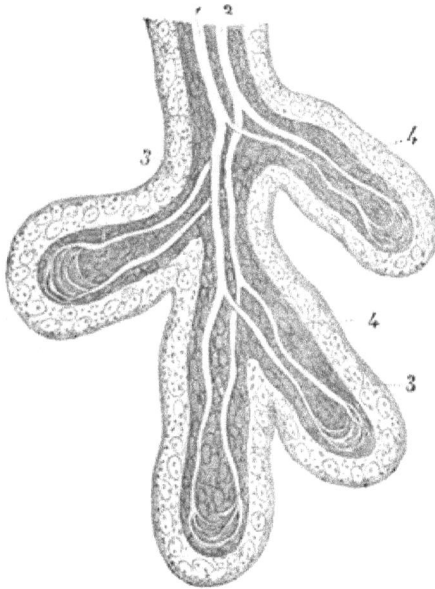

PLACENTAL VILLUS, GREATLY MAGNIFIED.   (After Joulin.)

1, 2. Placental vessels, forming terminal loops.   3. Chorion tissue, forming external walls of villus.   4. Tissue surrounding vessels.

plant. Within the transparent walls of the villi the capillary tubes of the contained vessels may be seen lying, distended with blood, and presenting an appearance not unlike loops of small intestine. The capillaries are the terminal ramifications of the umbilical arteries and veins, which, after reaching the site of the placenta, divide and subdivide until they at last form an immense number of minute capillary vessels, with their convexities looking towards the maternal portion of the placenta, each terminal loop being contained in one of the digitations of the chorion villi. Each arterial twig is accompanied by a corresponding venous branch, which unites with it to form the terminal arch or loop (fig. 58). The fœtal blood is carried through these arterial twigs to the villi, where it comes into intimate contact with the maternal blood, in consequence of the anatomical arrangements presently to be described ; but the two do not directly mix, as the older physiologists believed, for none of the maternal blood escapes when the umbilical cord is cut, nor can the minutest injections through the fœtal vessels be made to pass into the maternal vascular system, or vice versâ. In

Fig. 59.

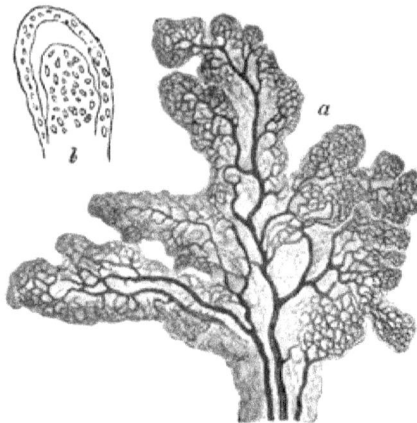

a. Terminal villus of fœtal tuft, minutely injected. b. Its nucleated non-vascular sheath. (After Farre.)

addition to the looped terminations of the umbilical vessels, Farre and Schroeder van der Kolk have described another set of capillary vessels in connection with each villus (fig. 59). This consists of a very fine network covering each villus, and very different in appearance from the convoluted vessels lying in its interior, which are the only ones which have been usually described. Dr. Farre believes that these vessels only exist in the early months of pregnancy, and that they disappear as pregnancy advances. Priestley[1] suggests that they may not be vessels at all, but lymphatics, which may possibly absorb nutrient material from the mother's blood, and throw it into the fœtal vascular system. The

[1] *The Gravid Uterus*, p. 52.

existence of lymphatics, or nerves, in the placenta, however, has never been demonstrated, and they are believed not to exist.

As generally described, the maternal portion of the placenta consists of large cavities, or of a single large cavity, which contain the maternal blood, and into which the villi of the chorion penetrate (fig. 60). Into this maternal part of the viscus the curling arteries of the uterus pour their

Maternal portion of the placenta.

Fig. 60.

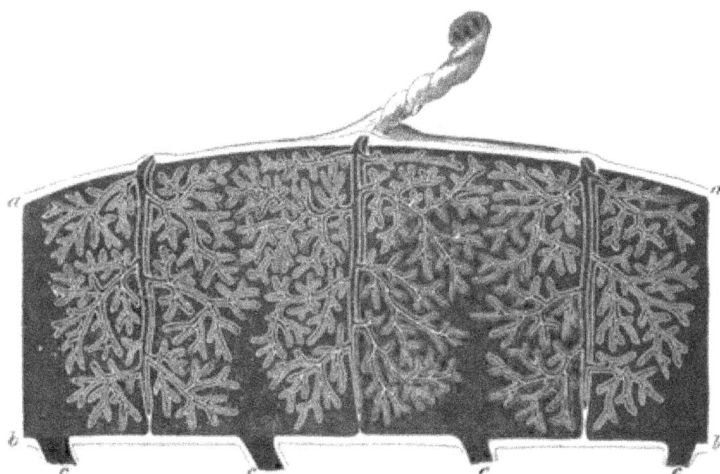

DIAGRAM REPRESENTING A VERTICAL SECTION OF THE PLACENTA. (After Dalton.)
a, a. Chorion. *b, b.* Decidua. *c, c, c, c.* Orifices of uterine sinuses.

blood, which is collected from it by the uterine sinuses. The villi of the chorion, therefore, are suspended in a sac filled with maternal blood, which penetrates freely between them, and with which they are brought into very intimate contact. Dr. John Reid believed that only the delicate internal lining of the maternal vessels entered the substance of the placenta, to form the sac just spoken of. Into this the villi project, pushing before them the membrane forming the limiting wall of the placental sinuses, each of them in this way receiving an investment, just as the fingers of a hand are covered by a glove (fig. 61).

Theory of Reid.

Schroeder van der Kolk and Goodsir (fig. 62) were of opinion that not only were the maternal bloodvessels continued into the substance of the placenta, but also the pro-

Theory of Goodsir.

cesses of the decidua, which accompanied the vessels and were prolonged over each villus, so as to separate it from the limiting membrane of the maternal sinuses. Each villus would thus be covered by two layers of fine tissue, one from the internal lining membrane of the maternal bloodvessels, the other from the epithelial cells of the decidua.

Theory of Turner.

Turner, whose valuable researches on the comparative anatomy of the placenta have thrown much light on its structure, points out that the placentæ of all animals are formed on the same fundamental type,[1] in which the *fœtal portion* consists of a smooth, plane-surfaced vascular mem-

Fig. 61.

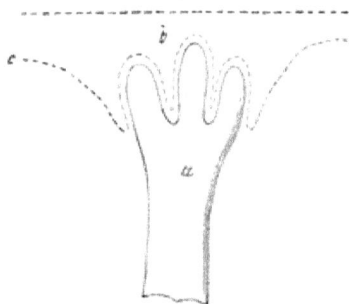

DIAGRAM ILLUSTRATING THE MODE IN WHICH A PLACENTAL VILLUS DERIVES A COVERING FROM THE VASCULAR SYSTEM OF THE MOTHER. (After Priestley.)

a. Villus having three terminal digitations projecting into *b.* Cavity of the mother's vessel. *c.* Dotted lines representing coat of vessel.

Fig. 62.

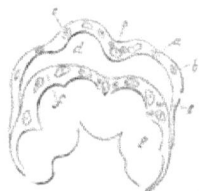

THE EXTREMITY OF A PLACENTAL VILLUS. (After Goodsir.)

a. External membrane of villus (the lining membrane of vascular system of Weber).
b. External cells of villus derived from decidua.
c, c. Nuclei of ditto.
d. The space between the maternal and fœtal portions of villus.
e. Its internal membrane.
f. Its internal cells.
g. The loop of umbilical vessels.

brane, covered with pavement epithelium, which is brought into contact with the *maternal portion*, consisting of a smooth, plane-surfaced vascular membrane covered with columnar epithelium. The fœtal capillaries are separated from the maternal capillaries only by two opposed layers of epithelium. In various animals the placentæ are more or less specialised from the generalised form, in some to a much greater extent than others. In the human placenta the maternal vessels have lost their normal cylindrical form, and are dilated into a system of freely intercommunicating placental sinuses, which are, in fact, maternal capillaries enormously enlarged, with their walls so expanded and thinned out that they cannot be recognised as a distinct layer limiting

---

[1] *Introduction to Human Anatomy*, Part 2.

the sinus.   Each fœtal chorion villus projecting into these sinuses is covered with a layer of cells distinct from those of the epithelial layer of the villus, and readily stripped from it.   These are maternal in their origin, and are derived from the decidua, which sends prolongations of its tissue into the placenta.   These cells, he believes, form a secreting epithelium which separates from the maternal blood a secretion for the nourishment of the fœtus, which is, in its turn, absorbed by the villi of the chorion.

A view not very dissimilar to this has been advanced by Professor Ercolani of Bologna, who maintains that the maternal portion of the placenta is a new formation, strictly glandular, and not vascular, in its structure.   It is formed, he thinks, by the submucous connective tissue of the decidua serotina, and it dips down into the placenta and forms a sheath to each of the chorion villi, which it separates from the maternal blood.   This new glandular structure he describes as secreting a fluid, termed the ' uterine milk,' which is absorbed by the villi of the chorion, just as the mother's milk is absorbed by the villi of the intestines, and it is with this fluid alone that the chorion villi are in direct contact. The sheath thus formed to each villus is doubtless analogous to the layer of cells which Goodsir described as encasing each villus, but is attributed to a new structure formed after conception.

Theory of Ercolani.

The existence of the maternal sinus system in the placenta is altogether denied by anatomists of eminence whose views are worthy of careful consideration.   Prominent amongst these is Braxton Hicks,[1] who has written an elaborate paper on the subject.   He holds that there is no evidence to prove that the maternal blood is poured out into a cavity in which the chorion villi float, and he believes that the curling arteries, instead of entering the so-called maternal portion of the placenta, terminate in the decidua serotina.   The hypertrophied chorion villi at the site of the placenta are firmly attached to the decidual surface, into which their tips are imbedded.   The line of junction between the decidua reflexa and serotina forms a circumferential margin to, and limits, the placenta.   The arrangement of the fœtal portion of the placenta on this view is very similar to that generally

Theory of Braxton Hicks.

[1] *Obst. Trans.* vol. xiv.

described, but the villi are not surrounded by maternal blood at all, and nothing exists between them, unless it be a small quantity of serous fluid. The change in the fœtal blood is effected by endosmosis, and Hicks suggests that the follicles of the decidua may secrete a fluid, which is poured into the intervillous spaces for absorption by the villi.

It will thus be seen that anatomists of repute are still undecided as to important points in the minute anatomy of the placenta, which further investigation will doubtless clear up. The main functions of the organ are, however, sufficiently clear. During the entire period of its existence it fills the important office of both stomach and lungs to the fœtus. Whatever view of the arrangement of the maternal bloodvessels be taken, it is certain that the fœtal blood is propelled by the pulsations of the fœtal heart into the numberless villi of the chorion, where it is brought into very intimate relation with the mother's blood, gives off its carbonic acid, absorbs oxygen, and passes back to the fœtus, through the umbilical vein, in a fit state for circulation. The mode of respiration, therefore, in the fœtus is analogous to that in fishes, the chorion villi representing the gills, the maternal blood the water in which they float. Nutrition is also effected in the organ, and, by absorption through the chorion villi, the pabulum for the nourishment of the fœtus is taken up. It also probably serves as an emunctory for the products of excretion in the fœtus. Picard found that the blood in the placenta contained an appreciably larger quantity of urea than that in other parts of the body, this urea probably being derived from the fœtus. Claude Bernard also attributed to it a glycogenic function,[1] supposing it to take the place of the fœtal liver until that organ was sufficiently developed.

Finally, we find that the temporary character of the placenta is indicated by certain degenerative changes, which take place in it previous to expulsion. These consist chiefly in the deposit of calcareous patches on its uterine surface, and in fatty degeneration of the villi, and of the decidual layer between the placenta and the uterus. If this degeneration be carried to excess, as is not unfrequently the case, the fœtus may perish from want of a sufficient number of

Functions of the placenta.

Degenerative changes previous to expulsion.

---

[1] *Acad. des Sciences*, April 1859.

healthy villi through which its respiration and nutrition may
be effected.

The *umbilical* cord is the channel of communication be-   Umbilical
tween the fœtus and placenta, being attached to the former   cord.
at the umbilicus, to the latter generally near its centre, but
sometimes, as in the battledore placenta, at its edge. It
varies much in length, measuring on an average from 18 to
24 inches, but in exceptional cases being found as long as 50
or 60, and as short as 5 or 6 inches.

When fully formed it consists of an external membranous layer formed of the amnion, two umbilical arteries, one
umbilical vein, and a considerable quantity of a transparent
gelatinous substance surrounding the vessels, called Wharton's
jelly, which is contained in a fine network of fibres, and is
formed out of the tissue of the allantois. At an early period
of pregnancy, in addition to these structures, the cord contains the pedicle of the umbilical vesicle, with the omphalo-
mesenteric vessels ramifying on it, and two umbilical veins,
one of which soon atrophies and disappears. No nerves or
lymphatics have been satisfactorily demonstrated in the cord,
although such have been described as existing. The vessels   Course of
of the cord are at first straight in their course, but shortly   the ves-
they become greatly twisted, the arteries being external to   sels.
the vein, and in nine cases out of ten the twist is from left
to right. Various explanations have been given of this
peculiarity, none of them entirely satisfactory. Tyler Smith
attributed it to the movements of the fœtus twisting the
cord, its attachment to the placenta being a fixed point ; this
would not, however, account for the frequency with which
the spiral turns occur in one direction. Mr. John Simpson
attributed it to the greater pressure of the blood through
the right hypogastric artery, on account of that vessel having
a more direct relation to the aorta than the left. The umbilical arteries give off no branches, and the vein contains no
valves, nor can any vasa vasorum be detected in their coats
after they have left the umbilicus. The umbilical arteries
increase in size after they leave the cord, to divide on the
surface of the placenta. This is the only example in the
body in which arteries are larger near their terminations than
their origin, and the object of this arrangement is probably
to effect a retardation of the current of the blood distributed

to the placenta. The tortuous course of the vein probably compensates for the absence of valves, and moderates the flow of blood through it. Distinct knots are not unfrequently observed in the cord, but they rarely have the effect of obstructing the circulation through it. They no doubt form when the fœtus is very small. They may sometimes also be produced in labour by the child being propelled through a coil of the cord lying circularly round the os uteri. The so-called false knots are merely accidental nodosities due to local enlargements of the vessels.

CHAPTER II.

It is obviously impossible to attempt anything like a full account of the development of the various fœtal structures, or of their growth during intra-uterine life. To do so would lead us far beyond the scope of this work, and would involve a study of complex details only suitable in a treatise on embryology. It is of importance, however, that the practitioner should have it in his power to determine approximately the age of the fœtus in abortions or premature labours, and for this purpose it is necessary to describe briefly the appearance of the fœtus at various stages of its growth.

1st Month.—The fœtus in the first month of gestation is a minute gelatinous and semi-transparent mass, of a greyish colour, in which no definite structure can be made out, and in which no head or extremities can be seen. It is rarely to be detected in abortions, being lost in surrounding blood-clots. In the few examples which have been carefully examined it did not measure more than a line in length. It is, however, already surrounded by the amnion, and the pedicle of the umbilical vesicle can be traced into the unclosed abdominal cavity.

Appearance of the fœtus at various stages of development.

2nd Month.—The embryo becomes more distinctly apparent, and is curved on itself, weighing about 62 grains, and measuring 6 to 8 lines in length. The head and extremities are distinctly visible—the latter in the form of rudimentary projections from the body. The eyes are to be seen as small black spots on the side of the head. The spinal column is divided into separate vertebræ. The independent circulatory system of the fœtus is now beginning to form, the heart consisting of only one ventricle and one auricle, from the former of which both the aorta and pulmonary arteries arise. On either side of the vertebral column, reaching from the heart to the pelvis, are two large glandular structures, the *corpora Wolffi-ania*, which consist of a series of convoluted tubes opening into an excretory duct, running along their external borders, and connected below with the common cloaca of the genito-urinary and digestive tracts. They seem to act as secreting glands, and fulfil the functions of the kidneys before they are formed. Towards the end of the second month they atrophy and disappear, and the only trace of them in the fœtus at term is to be found in the parovarium lying between the folds of the broad ligaments. At this stage of develop-ment there are met with in the human embryo, as in that of all mammals, four transverse fissures opening into the pharynx, which are analogous to the permanent branchiæ of fishes. Their vascular supply is also similar, as the aorta at this time gives off four branches on each side, each of which forms a branchial arch, and these afterwards unite to form the descending aorta. By the end of the sixth week these, as well as the transverse fissures to which they are distributed, dis-appear. By the end of the second month the kidneys and supra-renal capsules are forming, and the single ventricle is divided into two by the growth of the inter-ventricular septum. The umbilical cord is quite straight, and is inserted into the lower part of the abdomen. Centres of ossification are showing themselves in the inferior maxillary bones and the clavicle.

3rd Month.—The embryo weighs from 70 to 300 grains, and measures from 2½ to 3½ inches in length. The forearm is well formed, and the first traces of the fingers can be made out. The head is large in proportion to the rest of the body, and the eyes are prominent. The umbilical vesicle and allantois have disappeared, the greater portion of the chorion villi have atrophied, and the placenta is distinctly formed.

4th Month.—The weight is from 4 to 6 oz., and the length about 6 inches. The convolutions of the brain are beginning to develop. The sex of the child can now be ascertained on inspection. The muscles are sufficiently formed to produce distinct movements of the limbs. Ossifi-cation is extending, and can be traced in the occipital and frontal bones, and in the mastoid processes. The sexual organs are differentiated.

5th Month.—Weight about 10 oz. Length, 9 or 10 inches. Hair is observed covering the head, which forms about one-third of the length of the whole fœtus. The nails are beginning to form, and ossification has commenced in the ischium.

6th Month.—Weight about 1 lb. Length, 11 to 12½ inches. The hair is darker. The eyelids are closed, and the membrana pupillaris exists; eyelashes have now been formed. Some fat is deposited under the skin. The testicles are still in the abdominal cavity. The clitoris is prominent. The pubic bones have begun to ossify.

7th Month.—Weight, from 3 to 4 lbs. Length, 13 to 15 inches. The skin is covered with unctuous, sebaceous matter, and there is a more considerable deposit of subcutaneous fat. The eyelids are open. The testicles have descended into the scrotum.

8th Month.—Weight, from 4 to 5 lbs. Length, 16 to 18 inches, and the fœtus seems now to grow in thickness rather than in length. The nails are completely developed. The membrana pupillaris has disappeared.

At the completion of pregnancy the fœtus weighs on an average 6½ lbs., and measures about 20 inches in length. *Fœtus at term.* These averages are, however, liable to great variation. Remarkable histories are given by many writers of fœtuses of extraordinary weight, which have been probably greatly exaggerated. Out of 3,000 children delivered under the care of Cazeaux at various charities, one only weighed 10 lbs. There are, however, several carefully recorded instances of weight far exceeding this; but they are undoubtedly much more uncommon than is generally supposed. Dr. Ramsbottom mentions a fœtus weighing 16½ lbs.; Cazeaux tells us of one which he delivered by turning which weighed 18 lbs., and measured 2 feet 1½ inches, and the birth of one weighing 21 lbs. has been recently recorded.[1] Such overgrown children are almost invariably stillborn.[2]

[1] *Brit. Med. Journ.* Feb. 1, 1879.

[2] Probably the largest fœtus on record was that of Mrs. Captain Bates, the Nova Scotia giantess, a woman of 7ft. 9 in., whose husband is also of gigantic build, reaching 7 ft. 7 in. in height. This child, born in Ohio, was their second, and was lost in its birth, as no forceps could be procured of sufficient size to grasp the head. The fœtus weighed 23¾ lbs., and was 30 in. in length. Their first infant weighed 19 lbs. We have had children born in this city (Philadelphia) at maturity and live that weighed but one pound. The well-remembered 'Pincus baby' weighed a pound and an ounce.—Harris, note to 3rd American edition.

The average size of male children at birth, as in after-life, is somewhat greater than that of female. Thus Simpson [1] found that out of 100 cases the male children averaged 10 oz. more in weight than the female, and half an inch more in length. A new-born child at term is generally covered to a greater or less extent with a greasy, unctuous material, the **Vernix caseosa.** *vernix caseosa*, which is formed of epithelial scales and the secretion of the sebaceous glands, and which is said to be of use in labour by lubricating the surface of the child. The head is generally covered with long dark hair, which frequently falls off or changes in colour shortly after birth. Dr. Wiltshire [2] has called attention to an old observation, that the eyes of all new-born children are of a peculiar dark steel-grey colour, and that they do not acquire their permanent tint until some time after birth. The umbilical cord is generally inserted below the centre of the body.

Anatomy of the foetal head. The most important part of the foetus from an obstetrical point of view is the head, which requires a separate study, as it is the usual presenting part, and the facility of the labour depends on its accurate adaptation to the maternal passages.

The chief anatomical peculiarity of interest, in the head of the foetus at term, is that the bones of the skull, especially of its vertex—which, in the vast majority of cases, has to pass first through the pelvis—are not firmly ossified as in adult life, but are joined loosely together by membrane or cartilage. The result of this is, that the skull is capable of being moulded and altered in form to a very considerable extent by the pressure to which it is subjected, and thus its passage through the pelvis is very greatly facilitated. This, however, is chiefly the case with the cranium proper, the bones of the face and of the base of the skull being more firmly united. By this means the delicate structures at the base of the brain are protected from pressure while the change of form which the skull undergoes during labour implicates a portion of the skull where pressure on the cranial contents is least likely to be injurious.

The divisions between the bones of the cranium are further of obstetric importance in enabling us to detect the precise position of the head during labour, and an

[1] *Selected Obst. Works*, p. 327.   [2] *Lancet*, February 11, 1871.

accurate knowledge of them is therefore essential to the obstetrician.

We talk of them as *sutures* and *fontanelles*: the former being the lines of junction between the separate bones, which overlap each other to a greater or less extent during labour; The sutures and fonta-nelles. the latter membranous interspaces where the sutures join each other.

Fig. 63.

The principal sutures are: 1st. The *sagittal*, which separates the two parietal bones, and extends longitudinally backwards along the vertex of the head. 2nd. The *frontal*, which is a continuation of the sagittal, and divides the two halves of the frontal bone, at this time separate from each other. 3rd. The *coronal*, which separates the frontal from the parietal

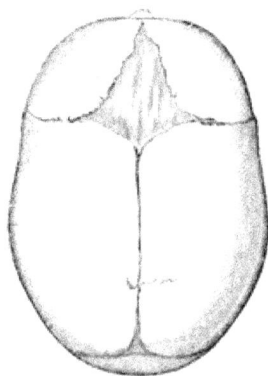

ANTERIOR AND POSTERIOR
FONTANELLES.

bones, and extends from the squamous portion of the temporal bone across the head to a corresponding point on the opposite side; and 4th, the *lambdoidal*, which receives its name from its resemblance to the Greek letter Λ, and separates the occipital from the parietal bones on either side. The fontanelles (fig. 63) are the membranous interspaces where the sutures join—the *anterior* and larger being lozenge-shaped, and formed by the junction of the frontal, sagittal, and two halves of the coronal sutures. It will be well to note that there are, therefore, four lines of sutures,

Fig. 64.

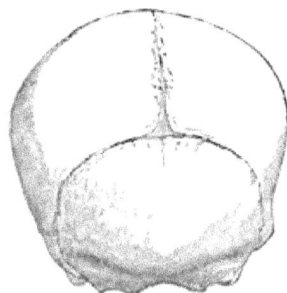

BI-PARIETAL DIAMETER, SAGITTAL AND
LAMBDOIDAL SUTURES, WITH POSTE-
RIOR FONTANELLE.

running into it, and four angles, of which the anterior, formed by the frontal suture, is most elongated and well marked. The *posterior* fontanelle (fig. 64) is formed by the junction of the sagittal suture with the two legs of the lambdoidal. It is, therefore, triangular in shape, with three lines of suture entering it in three angles, and is much smaller than the anterior fontanelle, forming merely a depression into which the tip of the finger can be placed, while the latter is a hollow as big as a shilling, or even larger. As it is the posterior fontanelle which is generally lowest, and the one most commonly felt during labour, it is important for the student to familiarise himself with it, and he should lose no opportunity of studying the sensations imparted to the finger by the sutures and fontanelles in the head of the child after birth.

The diameters of the foetal skull.
For the purpose of understanding the mechanism of labour, we must study the measurements of the foetal head in relation to the cavity through which it has to pass. They are taken from corresponding points opposite to each other, and are known as the diameters of the skull (fig. 65). Those of most importance are: 1st. The *occipito-mental*, from the occipital protuberance to the point of the chin, 5·25″ to 5·50″. 2nd. The *occipito-frontal*, from the occiput to

Fig. 65.

1 & 2. Occipito-frontal diameter.
3 & 4. Occipito-mental.
5 & 6. Cervico-bregmatic.
7 & 8. Fronto-mental.

the centre of the forehead, 4·50″ to 5″. 3rd. The *sub-occipito-bregmatic*, from a point midway between the occipital protuberance and the margin of the foramen magnum to the centre of the anterior fontanelle, 3·25″. 4th. The *cervico-bregmatic*, from the anterior margin of the foramen magnum to the centre of the anterior fontanelle, 3·75″. 5th. *Transverse*, or *bi-parietal*, between the parietal protuberances, 3·75″ to 4″. 6th. *Bi-temporal*, between the ears, 3·50″. 7th. *Fronto-mental*, from the apex of the forehead to the chin, 3·25″.

Alteration of diameters by compression and
The length of these respective diameters, as given by different writers, differs considerably—a fact to be explained by the measurements having been taken at different times; by some just after birth, when the head was altered in shape

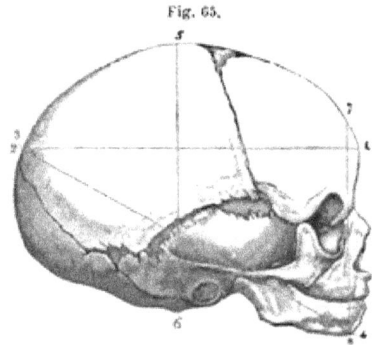

by the moulding it had undergone ; by others when this had
either been slight, or after the head had recovered its normal
shape. The above measurements may be taken as the average of those of the normally shaped head, and it is to be noted
that the first two are most apt to be modified during labour.
The amount of compression and moulding to which the head
may be subjected, without proving fatal to the fœtus, is not
certainly known, but it is doubtless very considerable. Some
interesting examples of the extent to which the head may be
altered in shape in difficult labours have been given by
Barnes,[1] who has shown by tracings of the shape of the
head taken immediately after delivery, that in protracted
labour the occipito-mental and occipito-frontal diameters
may be increased more than an inch in length, while
lateral compression may diminish the bi-parietal diameter
to the same length as the inter-auricular. The fœtal head
is movable on the vertical column to the extent of a quarter
of a circle ; and it seems probable that the laxity of the
ligaments admits with impunity a greater circular movement
than would be possible in the adult.

moulding during labour.

On taking the average of a large number of measurements, it is found that the heads of male children are larger
and more firmly ossified than those of females, the former
averaging about half an inch more in circumference. Sir
James Simpson attributed great importance to this fact, and
believed that it was sufficient to account for the larger proportion of still-births in male than in female children, as well
as for the greater difficulty of labour and the increased maternal mortality that are found to attend on male births. His
well-known paper on this subject, which has given rise to
much controversy, is full of the most elaborate details,
and so great did he believe the fœtal influence to be, that he
calculated that between the years 1834 and 1837 there were
lost in Great Britain, as a consequence of the slightly larger
size of the male than of the female head at birth, about
50,000 lives, including those of about 46,000 or 47,000
infants, and of between 3,000 and 4,000 mothers who died
in childbed.[2] It is probable that race and other conditions,
such as civilisation and intellectual culture, have considerable
influence on the size of the fœtal skull, but we are not

Influence of sex and race on the fœtal head.

[1] *Obst. Trans.*, vol. vii.   [2] *Selected Obst. Works*, p. 363.

in possession of sufficiently accurate data to justify any very positive opinion on these points.

In the very large majority of cases the fœtus lies in utero with head downwards, and is so placed as to be adapted in the most convenient way to the cavity in which it is placed. The uterine cavity is most roomy at the fundus, and narrowest at the cervix, and the greatest bulk of the fœtus is at the breech, so that the largest part of the child usually lies in the part of the uterus best adapted to contain it. The various parts of the child's body are further so placed, in regard to each other, as to take up the least possible amount of space. (See frontispiece.) The body is bent so that the spine is curved with its convexity outwards, this curvature existing from the earliest period of development ; the chin is flexed on the sternum ; the forearms are flexed on the arms, and lie close together on the front of the chest ; the legs are flexed on the thighs, and the thighs drawn up on the abdomen ; the feet are drawn up towards the legs ; the umbilical cord is generally placed out of reach of injurious pressure, in the space between the arms and the thighs. Variations from this attitude, however, are not uncommon, and are not, as a rule, of much consequence. Although the cranial presentations are much the most common, averaging 96 out of every 100 cases, other presentations are by no means rare, the next most frequent being either that of the breech, in which the long diameter of the child lies in the long diameter of the uterine cavity, or some variety of transverse presentation, in which the long diameter of the fœtus lies obliquely across the uterus, and no longer corresponds to its longitudinal axis.

It was long believed that the head presentation was only assumed towards the end of pregnancy, when it was supposed to be produced by a sudden movement on the part of the fœtus, known as the *culbute*. It is now well known that, in the large majority of cases, the head is lowest during all the latter part of pregnancy, although changes in position are more common than is generally believed to be the case, and presentation of parts other than the head is much more frequent in premature labour than in delivery at term. In evidence of the last statement, Churchill says that in labour at the seventh month the head presents only 83 times out of

100 when the child is living, and that as many as 53 per cent. of the presentations are preternatural when the child is still-born. The frequency with which the fœtus changes its position before delivery has been made the subject of investigation by various German obstetricians, and the fact can be readily ascertained by examination. Valenta[1] found that out of nearly 1,000 cases, carefully and frequently examined by him, in 57·6 per cent. the presentation underwent no change in the latter months of pregnancy, but in the remaining 42·4 per cent. a change could be readily detected. These alterations were found to be most frequent in multiparæ, and the tendency was for abnormal presentations to alter into normal ones. Thus it was common for transverse presentations to alter longitudinally, and but rare for breech presentations to change into head. The ease with which these changes are effected no doubt depends, in a considerable degree, on the laxity of the uterine parietes, and on the greater quantity of amniotic fluid, by both of which the free mobility of the fœtus is favoured.

The facility with which the position of the fœtus in utero can be ascertained by abdominal palpation has not been generally appreciated in obstetric works, and yet, by a little practice, it is easy to make it out. Much information of importance can be gained in this way, and it is quite possible, under favourable circumstances, to alter abnormal presentations before labour has begun. For the purpose of making this examination, the patient should lie at the edge of the bed, with her shoulders slightly raised, and the abdomen uncovered. The first observation to make is to see if the longitudinal axis of the uterine tumour corresponds with that of the mother's abdomen ; if it does, the presentation must be either a head or a breech. By spreading the hands over the uterus (fig. 66) a greater sense of resistance can be felt, in most cases, on one side than on the other, corresponding to the back of the child. By striking the tips of the fingers suddenly inwards at the fundus, the hard breech can generally be made out, or the head still more easily, if the breech be downwards. When the uterine walls are unusually lax, it is often possible to feel the limbs of the child. These observations can be generally corroborated by auscultation,

Detection of fœtal position by abdominal palpation.

---

[1] *Mon. f. Geburt.* 1866.

for in head presentations the fœtal heart can usually be heard below the umbilicus, and in breech cases above it. Transverse presentations can even more easily be made out

Fig. 66.

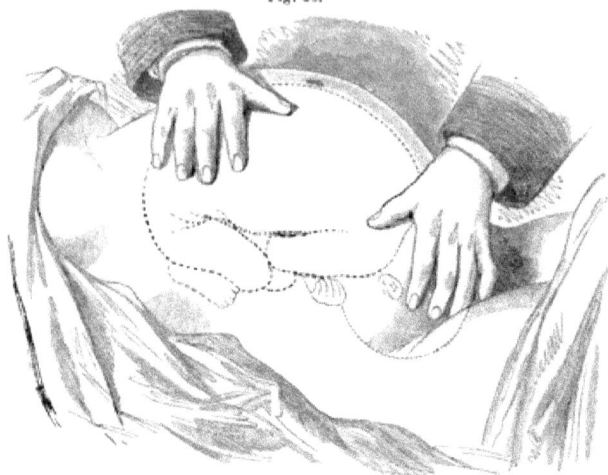

MODE OF ASCERTAINING THE POSITION OF THE FŒTUS BY PALPATION.

by abdominal palpation. Here the long axis of the uterine tumour does not correspond with the long axis of the mother's abdomen, but lies obliquely across it. By palpation the rounded mass of the head can be easily felt in one of the mother's flanks, and the breech in the other, while the fœtal heart is heard pulsating nearer to the side at which the head is detected.

Explanation of the position of the fœtus in utero.

The reason why the head presents so frequently has been made the subject of much discussion. The oldest theory was, that the head lay over the os uteri as the result of gravitation, and the influence of gravity, although contested by many obstetricians, prominent among whom were Dubois and Simpson, has been insisted upon as the chief cause by others, Dr. Duncan being one of the most strenuous advocates of this view. The objections urged against the gravitation theory were drawn partly from the result of experiments, and partly from the frequency with which abnormal presentations occur in premature labours, when the action of gravity cannot be supposed to be suspended. The experiments made by Dubois went to show, that when the fœtus was suspended in water gravitation caused the shoulders, and not the head, to fall lowest. He, therefore, advanced the hypothesis that the position of the fœtus was due to instinctive movements, which it made to adapt itself to the most comfortable position in which it could lie. It need only be remarked that there is not the slightest evidence of the fœtus possessing any such power. Simpson proposed a theory which was much more plausible. He assumed that the fœtal position was due to reflex movements produced by physical irritations to which the cutaneous surface of the fœtus is subjected from changes of the mother's position, uterine contractions, and the like. The absence of these movements, in the case of the death of the fœtus, would readily explain the frequency of mal-presentations under such circumstances. The obvious objection to this theory, complete as it seems to be, is the absence of any proof that such constant extensive reflex movements really do occur in utero. Dr. Duncan has very conclusively disposed of the principal objections which have been raised against the influence of gravitation, and, when an obvious explanation of so simple a kind exists, it seems useless to seek further for another. He has shown that Dubois's experiments did not accurately represent the state of the fœtus in utero, and that during

Fig. 67.

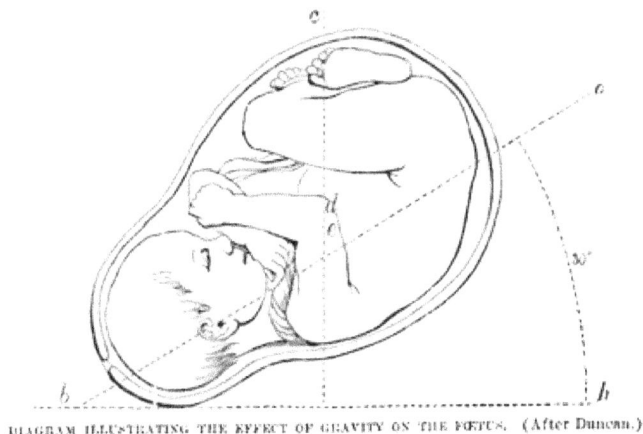

DIAGRAM ILLUSTRATING THE EFFECT OF GRAVITY ON THE FŒTUS. (After Duncan.)

a, b, is parallel to the axis of the pregnant uterus and pelvic brim. c, d, e, is a perpendicular line. e, the centre of gravity of the fœtus. d, the centre of flotation.

the greater part of the day, when the woman is upright, or lying on her back, the fœtus lies obliquely to the horizon at

an angle of about 30°.   The child thus lies, in the former
case, on an inclined plane, formed by the anterior uterine
wall and by the abdominal parietes, in the latter by the
posterior uterine wall and the vertebral column.   Down the
inclined plane so formed the force of gravity causes the
fœtus to slide, and it is only when the woman lies on her
side that the fœtus is placed horizontally, and is not subjected
in the same degree to the action of gravity (fig. 67).   The
frequency of mal-presentations in premature labours is explained by Dr. Duncan partly by the fact that the death of
the child (which so frequently precedes such cases) alters its
centre of gravity, and partly by the greater mobility of the

Fig. 68.

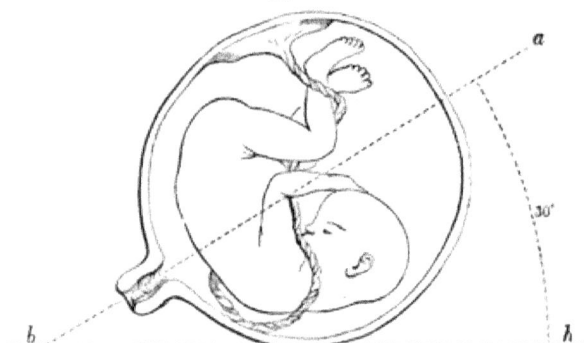

ILLUSTRATING THE GREATER MOBILITY OF THE FŒTUS AND THE LARGER RELATIVE
AMOUNT OF LIQUOR AMNII IN EARLY PREGNANCY.   (After Duncan.)

a, b. Axis of pregnant uterus.   *b, h.* A horizontal line.

child and the greater relative amount of liquor amnii (fig.
68).   The influence of gravitation is probably greatly assisted by the contractions of the uterus which are going on
during the greater part of pregnancy.   The influence of
these was pointed out by Dr. Tyler Smith, who distinctly
showed that the contractions of the uterus preceding delivery
exerted a moulding or adapting influence on the fœtus, and
prevented undue alterations of its position.   Dr. Hicks
proved [1] that these uterine contractions are of constant
occurrence from the earliest period of pregnancy, and there
can be little doubt that they must have an important
influence on the body contained within the uterus.   The

---

[1] *Obst. Trans.* vol. xiii. p. 216.

whole subject has been recently considered by Pinard,[1] who shows that many factors are in action to produce and maintain the usual position of the fœtus in utero, which may be either of an active or a passive character : the former being chiefly the active movements of the fœtus and the contractions of the uterus and the abdominal muscles; the latter, the form of the uterus and the fœtus, the slippery surface of the amnion, pressure of the amniotic fluid, etc. When any of these factors are at fault, mal-presentation is apt to occur.

The functions of the fœtus are in the main the same, with differences depending on the situation in which it is placed, as those of the separate being. It breathes, it is nourished, it forms secretions, and its nervous system acts. The mode in which some of these functions are carried on in intra-uterine life requires separate consideration.

Functions of the fœtus.

During the early part of pregnancy, and before the formation of the umbilical vesicle and the allantois, it is certain that nutritive material must be supplied to the ovum by endosmosis through its external envelope. The precise source, however, from which this is obtained is not positively known. By some it is believed to be derived from the granulations of the discus proligerus which surround it as it escapes from the Graafian follicle, and subsequently from the layer of albuminous matter which surrounds the ovum before it reaches the uterus ; while others think it probable that it may come from a special liquid secreted by the interior of the Fallopian tube as the ovum passes along it. As soon as the ovum has reached the uterus, there is every reason to believe that the umbilical vesicle is the chief source of nourishment to the embryo, through the channel of the omphalo-mesenteric vessels, which convey matters absorbed from the interior of the vesicle to the intestinal canal of the fœtus. At this time the exterior of the ovum is covered by the numerous fine villosities of the primitive chorion, which are imbedded in the mucous membrane of the uterus, and it is thought that they may absorb materials from the maternal system, which may be either directly absorbed by the embryo, or which may serve the purpose of replacing the nutritive matter which has been removed from the

Nutrition.

[1] *Annal. de Gyn.* May and July 1878.

umbilical vesicle by the omphalo-mesenteric vessels. This
point it is, of course, impossible to decide. Joulin, however,
thinks that these villi probably have no direct influence on
the nourishment of the fœtus, which is at this time solely
effected by the umbilical vesicle, but that they absorb fluid
from the maternal system, which passes through the amnion
and forms the liquor amnii. As soon as the allantois is
developed, vascular communication between the fœtus and
the maternal structures is established, and the temporary
function of the umbilical vesicle is over; that structure,
therefore, rapidly atrophies and disappears, and the nutrition
of the fœtus is now solely carried on by means of the chorion
villi, lined as they now are by the vascular endo-chorion, and
chiefly by those which go to form the substance of the
placenta.

This statement is opposed to the views of many physiologists, who believe that a certain amount of nutritive
material is conveyed to the fœtus through the channel of the
liquor amnii, itself derived from the maternal system which
is supposed either to be absorbed through the cutaneous
surface of the fœtus, or carried to the intestinal canal by
deglutition. The reasons for assigning to the liquor amnii a
nutritive function are, however, so slight, that it is difficult
to believe that it has any appreciable action in this way.
They are based on some questionable observations, such as
those of Weydlich, who kept a calf alive for fifteen days by
feeding it solely on liquor amnii, and the experiments of
Burdach, who found the cutaneous lymphatics engorged in a
fœtus removed from the amniotic cavity, while those of the
intestine were empty. The deglutition of the liquor amnii
for the purposes of nutrition has been assumed from its
occasional detection in the stomach of the fœtus, the presence of which may, however, be readily explained by spasmodic efforts at respiration, which the fœtus undoubtedly
often makes before birth, especially when the placental circulation is in any way interfered with, and during which
a certain quantity of fluid would necessarily be swallowed.
The quantity of nutritive material, however, in the liquor
amnii is so small—not more than 6 to 9 parts of albumen in
1,000—that it is impossible to conceive how it could have

any appreciable influence in nutrition, even if its absorption, either by the skin or stomach, were susceptible of proof.

That the nutrition of the fœtus is effected through the placenta is proved by the common observation that whenever the placental circulation is arrested, as by disease of its structure, the fœtus atrophies and dies. The precise mode, however, in which nutritive materials are absorbed from the maternal blood is still a matter of doubt, and must remain so until the mooted points as to the minute anatomy of the placenta are settled. The various theories entertained on this subject by the upholders of the Hunterian doctrine of placental anatomy, and by those who deny the existence of a sinus system, have already been referred to in the chapter on the Anatomy of the Placenta, to which the reader is referred (pp. 103-10).

One of the chief functions of the placenta, besides that of nutrition, is the supply of oxygenated blood to the fœtus. That this is essential to the vitality of the fœtus, and that the placenta is the site of oxygenation, is shown by the fact that whenever the placenta is separated, or the access of fœtal blood to it arrested by compression of the cord, instinctive attempts at inspiration are made, and if aërial respiration cannot be performed, the fœtus is expelled asphyxiated. Like the other functions of the fœtus during intra-uterine life, that of respiration has been made the subject of numerous more or less ingenious hypotheses. Thus many have believed that the fœtus absorbed gaseous material from the liquor amnii, which served the purpose of oxygenating its blood, St. Hilaire thinking that this was effected by minute openings in its skin, Beclard and others through the bronchi, to which they believed the liquor amnii gained access. Independently of the entire want of evidence of the absorption of gaseous materials by these channels, the theory is disproved by the fact that the liquor amnii contains no air which is capable of respiration. Serres attributed a similar function to some of the chorion villi, which he believed penetrated the utricular glands of the decidua reflexa, and absorbed gas from the hydroperione, or fluid situated between it and the decidua vera, and in this manner he thought the fœtal blood was oxygenated until the fifth

Respiration.

month of intra-uterine life, when the placenta was fully
formed.

This hypothesis, however, rests on no accurate foundation, for it is certain that the chorion villi do not penetrate
the utricular glands in the manner assumed; or, even if
they did, the mode in which the oxygen thus absorbed by
the chorion villi reaches the fœtus, which is separated from
them by the amnion and its contents, would still remain unexplained.

The mode in which the oxygenation of the fœtal blood is
effected before the formation of the placenta remains, therefore, as yet unknown. After the development of that organ,
however, it is less difficult to understand, for the fœtal blood
is everywhere brought into such close contact with the
maternal, in the numerous minute ramifications of the umbilical vessels, that the interchange of gases can readily be
effected. The activity of respiration is doubtless much less
than in extra-uterine life, for the waste of tissue in the fœtus
is necessarily comparatively small, from the fact of its being
suspended in a fluid medium of its own temperature, and
from the absence of the processes of digestion and of respiratory movements. The quantity of carbonic acid formed
would, therefore, be much less than after birth, and there
would be a correspondingly small call for oxygenation of
venous circulation.

Circula-
tion.

The functions of the lungs being in abeyance, it is
necessary that all the fœtal blood should be carried to the
placenta to receive oxygen and nutritive materials. To
understand the mode in which this is effected we must bear
in mind certain peculiarities in the circulatory system which
disappear after birth.

Anatomi-
cal pecu-
liarities of
the fœtal
circula-
tion.

1. The two sides of the fœtal heart are not separate, as
in the adult. The right ventricle in the adult sends all the
venous blood to the lungs, through the pulmonary arteries,
to be aërated by contact with the atmosphere. In the fœtus,
however, only sufficient blood is passed through the pulmonary arteries to insure their being pervious and ready to
carry blood to the lungs immediately after birth.

An aperture of communication, the *foramen ovale*, exists
between the two auricles, which is arranged so as to permit
the blood reaching the right auricle to pass freely into the

left, but not *vice versâ*. By this means a large portion of the blood reaching the heart through the venæ cavæ, instead of passing, as in the adult, into the right ventricle, is directed into the **left auricle**.

2. Even with this arrangement, however, a larger portion of blood would pass into the pulmonary arteries than is required for transmission to the lungs, and a further provision is made to prevent its going to them by means of a fœtal vessel, the *ductus arteriosus* (fig. 69), which arises from the point of bifurcation of the pulmonary arteries, and opens into the arch of the aorta. In consequence of this arrangement only a very small portion of the blood reaches the lungs at all.

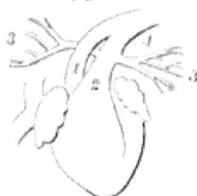

Fig. 69.

DIAGRAM OF FŒTAL HEART.
(After Dalton.)

1. Aorta.
2. Pulmonary artery.
3, 3. Pulmonary branches.
4. Ductus arteriosus.

3. The fœtal hypogastric arteries are continued into **large** arterial trunks, which, **passing into the cord, form** the *umbilical arteries*, and carry the impure fœtal **blood into** the placenta.

4. The purified blood is collected into the single *umbilical vein*, through which it is carried to the **under surface** of the liver, from which point it is conducted, **by means of** another special **fœtal vessel, the** *ductus venosus,* **into the** ascending vena cava, and the right auricle.

In order to understand the **course of the fœtal blood,** it may be most conveniently traced from the point where it reaches the under surface of the liver through the umbilical **vein. Part of it is** distributed to the liver itself, but the **greater** quantity is carried directly into the inferior vena cava, through the ductus venosus. The inferior vena cava also receives the blood from the fœtal veins of the lower **extremities,** and that portion of the blood of the umbilical vein which has passed through the liver. This mixed blood is carried up to the right auricle, from which by **far the greater** part of it is immediately directed into the left auricle, through the foramen **ovale.** From thence it passes into the left ventricle, which sends the greater part of it into the head and upper extremities through **the** aorta, a comparatively small quantity being transmitted to the inferior extremities.

Course of the fœtal circulation.

The blood which is thus sent to the upper part of the body is collected into the vena cava superior, by which it is thrown into the right auricle. Here the mass of it is probably directed into the right ventricle, which expels it into the pulmonary arteries, and from thence, through the ductus arteriosus, into the descending aorta. By this arrangement it will be seen that the descending aorta conveys to the lower part of the body the comparatively impure blood which has already circulated through the head, neck, and upper extremities. From the descending aorta a small quantity of blood is conveyed to the lower extremities, the greater part of it being carried for purification to the placenta through the umbilical arteries.

Establishment of independent circulation. As soon as the child is born it generally cries loudly, and inflates its lungs, and, in consequence, the pulmonary arteries are dilated, and the greater portion of the blood of the right ventricle is at once sent to the lungs, from whence, after being arterialised, it is returned to the left auricle, through the pulmonary veins. The left auricle, therefore, receives more blood than before, the right less, and the placental circulation being arrested, no more passes through the umbilical vein. In consequence of this, the pressure of the blood in the two auricles is equalised, the mass of the blood in the right auricle no longer passes into the left (the valve of the foramen ovale being closed by the equal pressure on both sides), but directly into the right ventricle and from thence into the pulmonary arteries, and the ductus arteriosus soon collapses and becomes impervious. The mass of blood in the descending aorta no longer finds its way into the hypogastric arteries, but passes into the lower extremities, and the adult circulation is established.

Changes in foetal circulation after birth. The changes which take place in the temporary vascular arrangements of the foetus, prior to their complete disappearance, are of some practical interest. The ductus arteriosus, as has been said, collapses, chiefly because the mass of blood is drawn to the lungs, and partly, perhaps, by its own inherent contractility. Its walls are found to be thickened, and its canal closes, first in the centre, and subsequently at its extremities, its aortic end remaining longer pervious on account of the greater pressure of blood from the left side of the heart (fig. 70). Practical closure occurs

within a few days after birth, although Flourens states that
it is not completely obliterated until eighteen months or two
years have elapsed.[1]   According to Schroeder, its walls unite
without the formation of any thrombus.   The foramen ovale
is soon closed by its valve, which contracts adhesion with the
edges of the aperture, so as effectually to occlude it.   Sometimes, however, a small canal of
communication between the two
auricles may remain pervious for
many months, or even a year and
more, without, however, any admixture of blood occurring.   A
permanently patulous condition
of this aperture, however, sometimes exists, giving rise to the
disease known as cyanosis.

The umbilical arteries and
veins and the ductus venosus
soon also become impermeable, in
consequence of concentric hyper-
trophy of their tissue and collapse of their walls.   The
closure of the former is aided by the formation of coagula
in the interior.   According to Robin, a longer time than is
usually supposed elapses before they become completely
closed, the vein remaining pervious until the twentieth or
thirtieth day after delivery, the arteries for a month or six
weeks.   He has also described [2] a remarkable contraction of
the umbilical vessels within their sheaths, at the point
where they leave the abdominal walls, which takes place
within three or four days after birth, and seems to prevent
hæmorrhage taking place when the cord is detached.

The liver, from its proportionately large size, apparently
plays an important part in the fœtal economy.   It is not
until about the fifth month of utero-gestation that it assumes
its characteristic structure, and forms bile, previous to that
time its texture being soft and undeveloped.   According to
Claude Bernard, after this period one of its most important
offices is the formation of sugar, which is found in much
larger amount in the fœtus than after birth.   Sugar is,
however, found in the fœtal structures long before the

Function
of the
liver.

Fig. 70.

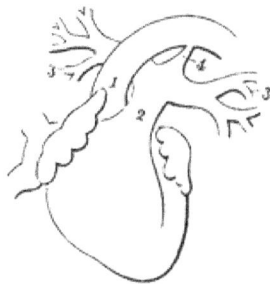

DIAGRAM OF HEART OF INFANT.
(After Dalton.)
1. Aorta.   2. Pulmonary artery.
3, 3. Pulmonary branches.
4. Ductus arteriosus becoming obliterated.

development of the liver, especially in the mucous and
cutaneous tissues, and it seems probable that these, as well
as the placenta itself, then fulfil the glycogenic function,
afterwards chiefly performed by the liver. The bile is
secreted after the fifth month of pregnancy, and passes into
the intestinal canal, and is subsequently collected in the
gall-bladder. By some physiologists it has been supposed
that the liver, during intra-uterine life, was the chief seat
of depuration of the carbonic acid contained in the venous
blood of the fœtus. It is, however, more generally believed
that this is accomplished solely in the placenta. The bile,
mixed with the mucous secretion of the intestinal tract,
forms the *meconium* which is contained in the intestines
of the fœtus, and which collects in them during the whole
period of intra-uterine life. It is a thick, tenacious, greenish
substance, which is voided soon after birth in considerable
quantity.

The meconium.

Urine is certainly formed during intra-uterine life, as
is proved by the fact familiar to all accoucheurs, that the
bladder is constantly emptied instantly after birth. It has
generally been supposed that the fœtus voids its urine into
the cavity of the amnion, and the existence of traces of urea
in the liquor amnii, as well as some cases of imperforate
urethra, in which the bladder was found to be enormously distended, and some cases of congenital hydro-nephrosis associated with impervious ureters, have been supposed to corroborate this assumption. The question has been very fully
studied by Joulin, who has collected together a large number
of instances in which there was imperforate urethra without
any undue distension of the bladder. He holds, also, that the
amount of urea found in the liquor amnii is far too minute
to justify the conclusion that the urine of the fœtus was
habitually passed into it, although a small quantity may, he
thinks, escape into it from time to time; and he, therefore,
believes that the urine of the fœtus is only secreted regularly
and abundantly after birth, and that during intra-uterine
life its retention is not likely to give rise to any functional
disturbance.[1]

The urine.

There is no doubt that the nervous system acts to a considerable extent during intra-uterine life, and some authors

Function
of the
nervous
system.

[1] *Acad. des Sciences*, p. 308.

have even supposed that the fœtus was endowed with the power of making instinctive or voluntary movements for the purpose of adapting itself to the form of the uterine cavity. Most probably, however, the movements the fœtus performs are purely reflex.   That it responds to a stimulus applied to the cutaneous nerves is proved by the experiments of Tyler Smith, who laid bare the amnion in pregnant rabbits, and found that the fœtus moved its limbs when these were irritated through it.   Pressure on the mother's abdomen, cold applications, and similar stimuli will also produce energetic fœtal movements.   The grey matter of the brain in the new-born child is, however, quite rudimentary in its structure, and there is no evidence of intelligent action of the nervous system until some time after birth, and *a fortiori* during pregnancy.

As soon as conception has taken place a series of remarkable changes commence in the uterus, which progress until the termination of pregnancy, and are well worthy of careful study. They produce those marvellous modifications which effect the transformation of the small undeveloped uterus of the non-pregnant state, into the large and fully developed uterus of pregnancy, and have no parallel in the whole animal economy.

A knowledge of them is essential for the proper comprehension of the phenomena of labour, and for the diagnosis of pregnancy which the practitioner is so frequently called upon to make. Excluding the varieties of abnormal pregnancy, which will be noticed in another place, we shall here limit ourselves to the consideration of the modifications of the maternal organism which result from simple and natural gestation.

Changes
in the
uterus.

The unimpregnated uterus measures $2\frac{1}{2}$ inches in length, and weighs about 1 oz., while at the full term of pregnancy it has so immensely grown as to weigh 24 oz. and measure 12 inches. The growth commences as soon as the ovum reaches the uterus, and continues uninterruptedly until delivery. In the early months the uterus is contained entirely in the cavity of the pelvis, and the increase of size is only apparent on vaginal examination, and that with difficulty. Before the third month the enlargement is chiefly in the lateral direction, so that the whole body of the uterus assumes more of a spherical shape than in the non-pregnant state. If an opportunity of examining the gravid uterus *post mortem* should occur at this time, it will be found to have the form of a sphere flattened somewhat posteriorly, and bulging anteriorly.

After the ascent of the organ into the abdomen, it develops more in the vertical direction, so that at term it has the form of an ovoid, with its large extremity above and its narrow end at the cervix uteri, and its longitudinal axis corresponds to the long diameter of the mother's abdomen, provided the presentation be either of the head or breech. The anterior surface is now even more distinctly

Fig. 71.

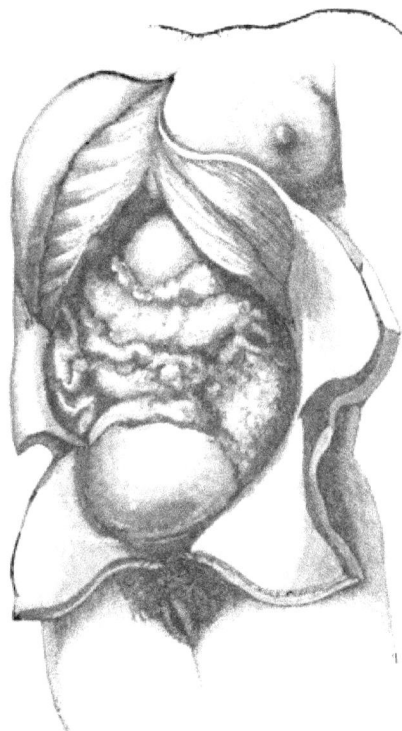

RELATIONS OF THE PREGNANT UTERUS AT SIXTH MONTH TO THE SURROUNDING PARTS.
(After Martin.)

projecting than before—a fact which is explained by the proximity of the posterior surface to the rigid spinal column behind, while the anterior is in relation with the lax abdominal parietes, which yield readily to pressure, and so allow of the more marked prominence of the anterior uterine wall.

Before the gravid uterus has risen out of the pelvis no

Change in situation.

appreciable increase in the size of the abdomen is perceptible. On the contrary, it is an old observation that at this early stage of pregnancy the abdomen is flatter than usual, on account of the partial descent of the uterus in the pelvic cavity as a result of its increased weight. As the growth of the organ advances it soon becomes too large to be contained any longer within the pelvis, and about the middle of the third or the beginning of the fourth month the fundus rises above the pelvic brim—not suddenly, as is often erroneously thought, but slowly and gradually—when it may be felt as a smooth rounded swelling.

It is about this time that the movements of the fœtus first become appreciable to the mother, when '*quickening*' is said to have taken place.

Size of uterine tumour at various periods of pregnancy.

Towards the end of the fourth month the uterus reaches to about three fingers' breadth above the symphysis pubis. About the fifth month it occupies the hypogastric region, to which it imparts a marked projection, and the alteration in the figure is now distinctly perceptible to visual examination.

Fig. 72.

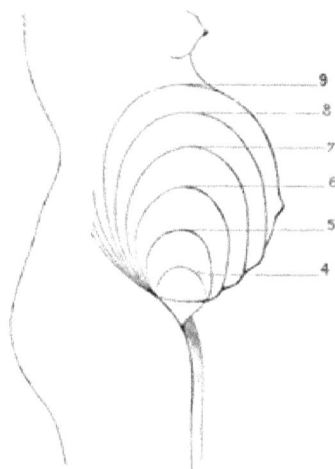

SIZE OF UTERUS AT VARIOUS PERIODS OF PREGNANCY.

About the sixth month it is on a level with, or a little above, the umbilicus. About the seventh month it is about two inches above the umbilicus, which is now projecting and prominent, instead of depressed, as in the non-pregnant state. During the eighth and ninth months it continues to increase until the summit of the fundus is immediately below the ensiform cartilage (fig. 72). A knowledge of the size of the uterine tumour at various periods of pregnancy, as thus indicated, is of considerable practical importance, as forming the only guide by which we can estimate the probable period of delivery in certain cases in which the usual data for calculation are absent, as, for example, when the patient has conceived during lactation.

For about a week or more before labour the uterus gene- *The uterus sinks before delivery.*
rally sinks somewhat into the pelvic cavity, in consequence
of the relaxation of the soft parts which precedes delivery,
and the patient now feels herself smaller and lighter than
before. This change is familiar to all childbearing women,
to whom it is known as 'the lightening before labour.'

While the uterus remains in the pelvis its longitudinal *The direction of the uterus.*
axis varies in direction, much in the same way as that of the
non-pregnant uterus, sometimes being more or less vertical,
at others in a state of anteversion or partial retroversion.
These variations are probably dependent on the distension or
emptiness of the bladder, as its state must necessarily affect
the position of the movable organ poised behind it. After
the uterus has risen into the abdomen its tendency is to
project forwards against the abdominal wall, which forms its
chief support in front. In the erect position the long axis of
the uterine tumour corresponds with the axis of the pelvic
brim, forming an angle of about 30° with the horizon. In
the semi-recumbent position, on the other hand, as Duncan[1]
has pointed out, its direction becomes much more nearly
vertical. In women who have borne many children, the
abdominal parietes no longer afford an efficient support, and
the uterus is displaced anteriorly, the fundus in extreme
cases even hanging downwards.

In addition to this anterior obliquity, on account of the *Lateral obliquity of the uterus.*
projection of the spinal column, the uterus is very generally
also displaced laterally, and sometimes to a very marked
degree, so that it may be felt entirely in one flank, instead of
in the centre of the abdomen. In a large proportion of cases
this lateral deviation is to the right side, and many hypotheses have been brought forward to explain this fact, none
of them being satisfactory. Thus, it has been supposed to
depend on the greater frequency with which women lie on
their right side during sleep, on the greater use of the right
leg during walking, on the supposed comparative shortness
of the right round ligament, which drags the tumour to that
side, or on the frequent distension of the rectum on the left
side, which prevents the uterus being displaced in that direction. Of these the last is the cause which seems most constantly in operation, and most likely to produce the effect.

[1] *Researches in Obstetrics*, p. 10.

Changes in the direction of the cervix.

The cervix must obviously adapt itself to the situation of the body of the uterus. We find, therefore, that in the early months, when the uterus lies low in the pelvis, it is more readily within reach. After the ascent of the uterus, it is drawn up, and frequently so much so as to be reached with difficulty. When the uterus is much anteverted, as is so often the case, the os is displaced backwards, so that it cannot be felt at all by the examining finger.

Relation of the uterus to the surrounding parts.

Towards the end of pregnancy the greater part of the anterior surface of the uterus is in contact with the abdominal wall, its lower portion resting on the posterior surface of the symphysis pubis. The posterior surface rests on the spinal column, while the small intestines are pushed to either side, the large intestines surrounding the uterus like an arch.

Changes in the uterine parietes.

The great distension of the uterus during pregnancy was formerly supposed to be mainly due to the mechanical pressure of the enlarging ovum within it. If this were so, then the uterine walls would be necessarily much thinner than in the non-pregnant state. This is well known not to be the

Thickness.

case, and the immense increase in the size of the uterine cavity is to be explained by the hypertrophy of its walls. At the full period of pregnancy the thickness of the uterine parietes is generally about the same as that of the non-pregnant uterus, rather more at the placental site, and less in the neighbourhood of the cervix. Their thickness, however, varies in different places, and in some women they are so thin as to admit of the fœtal limbs being very readily made

Density.

out by palpation. Their density is, however, always much diminished, and, instead of being hard and inelastic, they become soft and yielding to pressure. This change coincides with the commencement of pregnancy, of which it forms, as recognisable in the cervix, one of the earliest diagnostic marks. At a more advanced period it is of value as admitting a certain amount of yielding of the uterine walls to movements of the fœtus, thus lessening the chance of their being injured.

Changes in the cervix during pregnancy.

Very erroneous views have long been taught, in most of our standard works on midwifery, as to the changes which occur in the cervix uteri during pregnancy. It is generally stated that, as pregnancy advances, the cervical cavity is greatly diminished in length, in consequence of its being gradually drawn up so as to form part of the general cavity of the uterus, so that in the latter months it no longer exists. In almost all midwifery works accurate diagrams are given of this progressive shortening of the cervix (figs. 73 to 76).

Figs. 73. 74, 75, 76.

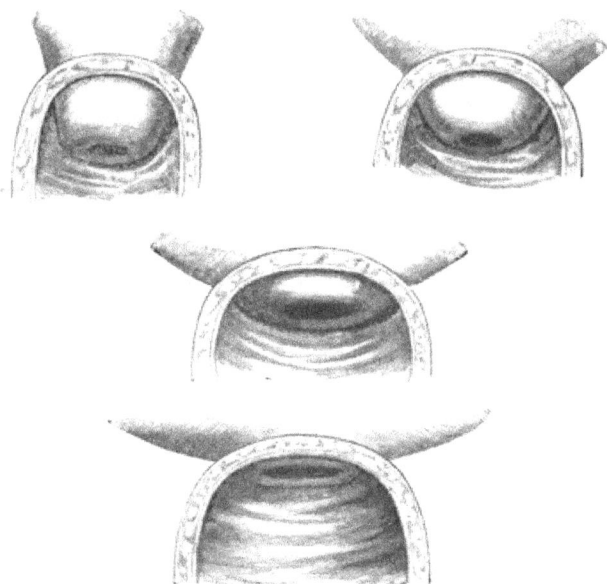

SUPPOSED SHORTENING OF THE CERVIX AT THE THIRD, SIXTH, EIGHTH, AND NINTH MONTHS OF PREGNANCY, AS FIGURED IN OBSTETRIC WORKS.

The cervix is generally described as having lost one half of its length at the sixth month, two thirds at the seventh, and to be entirely obliterated in the eighth and ninth. The correctness of these views was first called in question in recent times by Stoltz, in 1826, but Dr. Duncan,[1] in an elaborate historical paper on the subject, has shown that Stoltz was anticipated by Weitbrech in 1750, and, to a less degree, by Roederer and other writers. This opinion is now pretty generally admitted to be correct, and is upheld by Cazeaux, Arthur Farre, Duncan, and most modern obstetricians. Indeed, various post-mortem examinations in advanced pregnancy have shown that the cavity of the cervix remains in reality of its normal length of one inch, and it can often be measured

[1] *Researches in Obstetrics.*

during life by the examining finger, on account of its
patulous state (fig. 77). During the fortnight immediately

Fig. 77.

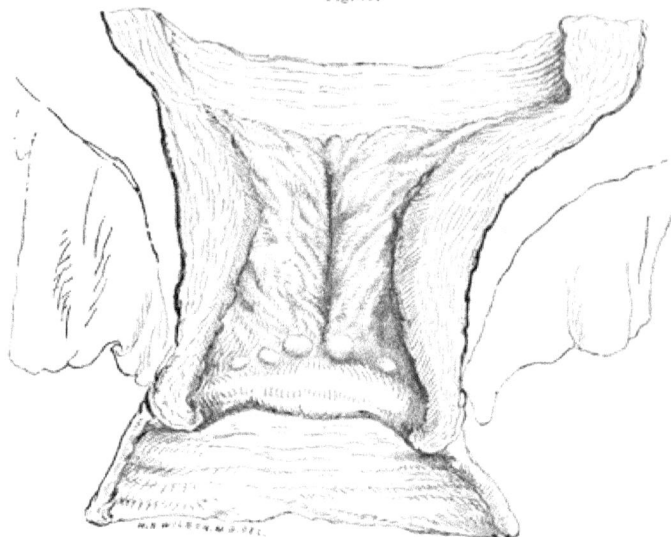

CERVIX FROM A WOMAN DYING IN THE EIGHTH MONTH OF PREGNANCY. (After Duncan.)

preceding delivery, however, a real shortening or obliteration
of the cervical cavity takes place; but this, as Duncan has
pointed out, seems to be due to the incipient uterine contractions which prepare the cervix for labour.

An appa-
rent short-
ening is
always
present.

There is, no doubt, an apparent shortening of the cervix
always to be detected during pregnancy, but this is a fallacious and deceptive feeling, due to the softness of the tissue
of the cervix, which is exceedingly characteristic of pregnancy,
and which to an experienced finger affords one of its best
diagnostic marks.

Softening
of the
cervix.

In the non-pregnant state the tissue of the cervix is hard,
firm, and inelastic. When conception occurs, softening begins at the external os, and proceeds gradually and slowly
upwards until it involves the whole of the cervix. By the
end of the fourth month both lips of the os are thick,
softened, and velvety to the touch, giving a sensation, likened
by Cazeaux to that produced by pressing on a table through
a thick, soft cover. By the sixth month at least one half of
the cervix is thus altered, and by the eighth the whole of it,

and so much so that at this time those unaccustomed to
vaginal examination experience some difficulty in distinguishing it from the vaginal walls. It is this softening, then,
which gives rise to the apparent shortening of the cervix so
generally described, and it is an invariable concomitant of
pregnancy, except in some rare cases in which there has been
antecedent morbid induration and hypertrophic elongation
of the cervix. If, therefore, on examining a woman supposed
to be advanced in pregnancy, we find the cervix to be hard
and projecting into the vaginal canal, we may safely conclude
that pregnancy does not exist. The existence of softening,
however, it must be remembered, will not itself justify an
opposite conclusion, as it may be produced, to a very considerable extent, by various pathological conditions of the
uterus.

Value of softening as a sign of pregnancy.

At the same time that the tissue of the cervix is softened,
its cavity is widened, and the external os becomes patulous.
This change varies considerably in primiparæ and multiparæ.
In the former the external os often remains closed until the
end of pregnancy ; but even in them it generally becomes
more or less patulous after the seventh month, and admits
the tip of the examining finger.   In women who have borne
children this change is much more marked.   The lips of the
external os are in them generally fissured and irregular, from
slight lacerations of its tissue in former labours.   It is also
sufficiently open to admit the tip of the finger, so that in the
latter months of pregnancy it is often quite possible to touch
the membrane, and through them to feel the presenting
part of the child.

The os uteri is generally patulous.

The remarkable increase in size of the uterus during
pregnancy is, as we have seen, chiefly to be explained by the
growth of its structures, all of which are modified during
gestation. The peritoneal covering is considerably increased,
so as still to form a complete covering to the uterus when at
its largest size. William Hunter supposed that its extension was effected rather by the unfolding of the layers of the
broad ligament than by growth. That the layers of the
broad ligament do unfold during gestation, especially in the
early months, is probable ; but this is not sufficient to
account for the complete investment of the uterus, and it
is certain that the peritoneum grows *pari passu* with the

Changes in the texture of the uterine tissues. The peritoneal coat.

enlargement of the uterus. In addition, there is a new
formation of fibrous tissue between the peritoneal and the
muscular coats, which affords strength, and diminishes the
risk of laceration during labour.

The mus-
cular coat.

The hypertrophy of the muscular tissue of the uterus is,
however, the most remarkable of the changes produced by

The
muscular
fibres.

pregnancy. Not only do the previously existing rudimentary
fibre-cells become enormously increased in size—so as to
measure, according to Kölliker, from seven to eleven times
their former length, and from two to five times their former
breadth—but new unstriped fibres are largely developed,
especially in the inner layers. These new cells are chiefly
found in the first months of pregnancy, and their growth
seems to be completed by the sixth month. The connective
tissue between the muscular layers is also largely increased
in amount. The weight of the muscular tissue of the gravid
uterus is, therefore, much increased, and it has been estimated by Heschl that it weighs at term from 1 to 1·5 lbs.,
that is, about sixteen times more than in the unimpregnated
state. This great development of the muscular tissue admits
of its dissection in a way which is quite impossible in the
unimpregnated state, and the researches of Hélie (p. 38)
enable us to understand much better than before how the
muscles forming the walls of the gravid uterus act during
the expulsion of the child.

The
mucous
coat.

The changes in the mucous coat of the uterus, which
result in the formation of the decidua, have already been discussed at length elsewhere (p. 91).

Circula-
tory appa-
ratus.

The circulatory apparatus of the uterus during pregnancy
has been described when the anatomy of the placenta was
under consideration (p. 104).

Lym-
phatics.

The lymphatics are much increased in size; and recent
theories, on the production of certain puerperal diseases
attribute to them a more important action than has been
commonly assigned to them.

Nerves.

The question of the growth of the nerves has been hotly
discussed. Robert Lee took the foremost place among those
who maintained that the nerves of the uterus share the general growth of its other constituent parts. Dr. Snow Beck,
however, believed that they remain of the same size as in the
unimpregnated state, and this view is supported by Hirch-

feld, Robin, and other recent writers. Robin thought that
there is an apparent increase in the size of the nerve-tubes,
which, however, is really due to increase in the neurilemma.
Kilian describes the nerves as increasing in length but not
in thickness, while Schroeder states that they participate
equally with the lymphatics in the enlargement the latter
undergo. Whichever of these views may ultimately be
found to be correct, it is certain that analogy would lead us
to expect an increase of nervous as well as of vascular supply.

It is not in the uterus alone that pregnancy is found to
produce modifications of importance. There are few of the
more important functions of the body which are not, to a
greater or less extent, affected; to some of these it is necessary briefly to direct attention, inasmuch as, when carried to
excess, they produce those disorders which often complicate
gestation, and which prove so distressing and even dangerous
to the patients. Such of them as are apparent and may aid
us in diagnosis are discussed in the chapter which treats of
the signs and symptoms of pregnancy; in this place it is
only necessary to refer to those which do not properly fall
into that category.

General modifications in the body produced by pregnancy.

Amongst those which are most constant and important
are the alterations in the composition of the blood. The
opinion of the profession on this subject has, of late years,
undergone a remarkable change. Formerly it was universally believed that pregnancy was, as the rule, associated with
a condition analogous to plethora, and that this explained
many characteristic phenomena of common occurrence, such
as headache, palpitation, singing in the ears, shortness of
breath, and the like. As a consequence it was the habitual
custom, not yet by any means entirely abandoned, to treat
pregnant women on an antiphlogistic system; to place them
on low diet, to administer lowering remedies, and very often
to practise venesection, sometimes to a surprising extent.
Thus it was by no means rare for women to be bled six
or eight times during the latter months, even when no definite symptoms of disease existed; and many of the older
authors record cases where depletion was practised every
fortnight, as a matter of routine, and, when the symptoms
were well marked, even from fifty to ninety times in the
course of a single pregnancy.

Changes in the blood.

Numerous careful analyses have conclusively proved that the composition of the blood during pregnancy is very generally—perhaps it would not be too much to say always—profoundly altered. Thus it is found to be more watery, its serum is deficient in albumen, and the amount of coloured globules is materially diminished, averaging, according to the analysis of Becquerel and Rodier, 111·8 against 127·2 in the non-gravid state. At the same time the amount of fibrine and of extractive matter is considerably increased. The latter observation is of peculiar importance, as it goes far to explain the frequency of certain thrombotic affections observed in connection with pregnancy and delivery; this hyperinosis of the blood is also considerably increased after labour by the quantity of effete material thrown into the mother's system at that time, to be got rid of by her emunctories. The truth is, that the blood of the pregnant woman is generally in a state much more nearly approaching the condition of anæmia than of plethora, and it is certain that most of the phenomena attributed to plethora may be explained equally well and better on this view. These changes are much more strongly marked at the latter end of pregnancy than at its commencement, and it is interesting to observe that it is then that the concomitant phenomena alluded to are most frequently met with. Cazeaux, to whom we are chiefly indebted for insisting on the practical bearing of these views, contends that the pregnant state is essentially analogous to chlorosis, and that it should be so treated. More recently the accurate observations of Willcocks[1] have shown that the blood of pregnancy differs from that of chlorosis in the fact that while in both the amount of hæmoglobin is lessened, in pregnancy the individual blood-cells are not impoverished as they are in chlorosis, but simply lessened in comparative number, owing to an increase in the water of the plasma, due to the progressive enlargement of the vascular area during gestation. Objection has not unnaturally been taken to Cazeaux's theory, as implying that a healthy and normal function is associated with a morbid state, and it has been suggested that this deteriorated state of the blood may be a wise provision of nature instituted for a purpose

[1] 'Comparative Observations on the Blood in Chlorosis and Pregnancy.' By Fred. Willcocks, M.D., The Lancet, December 3, 1881.

we are not as yet able to understand. It may certainly be
admitted that pregnancy, in a perfectly healthy state of the
system, should not be associated with phenomena in themselves in any degree morbid. It must not be forgotten,
however, that our patients are seldom—we might safely say
never—in a state that is physiologically healthy. The influence of civilisation, climate, occupation, diet, and a thousand
other disturbing causes that, to a greater or less degree, are
always to be met with, must not be left out of consideration.
Making every allowance, therefore, for the undoubted fact
that pregnancy *ought* to be a perfectly healthy condition, it
must be conceded, I think, that in the vast majority of
cases coming under our notice it is not entirely so; and the
deductions drawn by Cazeaux, from the numerous analyses of
the blood of pregnant women, seem to point strongly to the
conclusion that the general blood-state is tending to poverty
and anæmia, and that a depressing and antiphlogistic treatment is distinctly contra-indicated.

Closely connected with the altered condition of the blood
is the physiological hypertrophy of the heart, which is now
well known to occur during pregnancy. This was first pointed
out by Larcher in 1828, and it has been since verified by
numerous observers. It seems to be constant and considerable, and to be a purely physiological alteration intended
to meet the increased exigencies of the circulation, which
the complex vascular arrangements of the gravid uterus produce. The hypertrophy is limited to the left ventricle; the
right ventricle, as well as both auricles, being unaffected. Blot
estimates that the whole weight of the heart increases one-
fifth during gestation. The more recent researches of
Löhlein[1] render it probable that the hypertrophy is less than
those authors have supposed. According to Duroziez[2] the
heart remains enlarged during lactation, but diminishes in
size immediately after delivery in women who do not suckle,
while in women who have borne many children it remains
permanently somewhat larger than in nulliparæ. Similar
increase in the size of other organs has been pointed out by
various writers, as, for example, in the lymphatics, the spleen,
and the liver. Tarnier states that in women who have died
after delivery, the organs always show signs of fatty de-

*Modifica-
tions in
certain
viscera—
In the
heart;*

*In the
liver, lym-
phatics,
and
spleen.*

[1] *Zeitschrift für Geburtshülfe, &c.*, 1876.　　[2] *Gaz. des Hôpit.* 1868.

generation. According to Gassner the whole body increases
in weight during the latter months of pregnancy, and this
increase is somewhat beyond that which can be explained by
the size of the womb and its contents.

Formation
of osteo-
phytes.

Irregular bony deposits between the skull and the dura
mater, in some cases so largely developed as to line the whole
cranium, have been so frequently detected in women who
have died during parturition that they are believed by some
to be a normal production connected with pregnancy. Ducrest
found these osteophytes in more than one-third of the cases
in which he performed post-mortem examinations during the
puerperal period. Rokitansky, who corroborated the observation, believed this peculiar deposit of bony matter to be
a physiological, and not a pathological, condition connected
with pregnancy; but whether it be so, or how it is produced,
has not yet been satisfactorily determined.

Changes
in the
nervous
system.

More or less marked changes connected with the nervous
system are generally observed in pregnancy, and sometimes
to a very great extent. When carried to excess they produce
some of the most troublesome disorders which complicate
gestation, such as alterations in the intellectual functions,
changes in the disposition and character, morbid cravings,
dizziness, neuralgia, syncope, and many others. They are
purely functional in their character, and disappear rapidly
after delivery, and may be best described in connection with
the disorders of pregnancy.

Changes
in the re-
spiratory
organs.

Respiration is often interfered with, from the mechanical
results of the pressure of the enlarged uterus. The longitudinal dimensions of the thorax are lessened by the upward
displacement of the diaphragm, and this necessarily leads to
some embarrassment of the respiration, which is, however,
compensated, to a great extent, by an increase in breadth of
the base of the thoracic cavity.

Changes
in the
urine.

Certain changes, which are of very constant occurrence,
in the urine of pregnant women have attracted much attention, and have been considered by many writers to be
pathognomonic. They consist in the presence of a peculiar
deposit, formed when the urine has been allowed to stand for
some time, which has received the name of *kiestein*. Its
presence was known to the ancients, and it was particularly
mentioned by Savonarola in the fifteenth century, but it has

more especially been studied within the last thirty years by
Eguisier, Golding Bird, and others.   If the urine of a pregnant woman be allowed to stand in a cylindrical vessel,
exposed to light and air, but protected from dust, in a period,
varying from two to seven days, a peculiar flocculent sediment,
like fine cotton wool, makes its appearance in the centre of
the fluid, and soon afterwards rises to the surface and forms a
pellicle, which has been compared to the fat on cold mutton-
broth.   In the course of a few days the scum breaks up and
falls to the bottom of the vessels.   On microscopic examination it is found to be composed of fat particles, with crystals
of ammoniaco-magnesium phosphates and phosphate of lime,
and a large quantity of vibriones.   These appearances are
generally to be detected after the second month of pregnancy,
and up to the seventh or eighth month, after which they are
rarely produced.   Regnauld explains their absence during
the latter months of gestation by the presence in the urine,
at that time, of free lactic acid, which increases its acidity,
and prevents the decomposition of the urea into carbonate
of ammonia.   He believes that kiestein is produced by the
action of free carbonate of ammonia on the phosphate of
lime contained in the urine, and that this reaction is prevented by the excess of acid.

Golding Bird believed kiestein to be analogous to casein,
to the presence of which he referred it, and he states that
he has found it in twenty-seven out of thirty cases.   Braxton
Hicks so far corroborates his view, and states that the deposit
of kiestein can be much more abundantly produced if one or
two teaspoonfuls of rennet be added to the urine, since that
substance has the property of coagulating casein.   Much less
importance, however, is now attached to the presence of
kiestein than formerly, since a precisely similar substance is
sometimes found in the urine of the non-pregnant, especially
in anæmic women, and even in the urine of men.   Parkes
states that it is not of uniform composition, that it is produced by the decomposition of urea, and consists of the free
phosphates, bladder mucus, infusoria, and vaginal discharges.
Neugebauer and Vogel give a similar account of it, and hold
that it is of no diagnostic value.   That it is of interest, as
indicating the changes going on in connection with pregnancy,
is certain ; but inasmuch as it is not of invariable occurrence,

and may even exist quite independently of gestation, it is obviously quite undeserving of the extreme importance that has been attached to it.

Glyco-
suria in
preg-
nancy.

Towards the end of pregnancy sugar may sometimes be detected in the urine, and after delivery and during lactation it exists in considerable abundance ; thus out of thirty-five cases tested in the Simpson Memorial Hospital in Edinburgh during the puerperium, it was found in all, the amount varying from 1 to 8 per cent.[1] Kaltenbach has shown that this temporary glycosuria is due to the presence of milk-sugar in the urine, and that it ceases with the disappearance of milk from the breasts.[2] This physiological glycosuria must be carefully distinguished from true diabetes, which is a grave complication of pregnancy.

[1] *Edin. Med. Journ.* Aug. 1881.   [2] *Zeit. f. Gynk.* September 13, 1879.

CHAPTER IV.

SIGNS AND SYMPTOMS OF PREGNANCY.

In attempting to ascertain the presence or absence of preg- nancy, the practitioner has before him a problem which is often beset with great difficulties, and on the proper solution of which the moral character of his patient, as well as his own professional reputation, may depend. The patient and her friends can hardly be expected to appreciate the fact that it is often far from easy to give a positive opinion on the point; and it is always advisable to use much caution in the examination, and not to commit ourselves to a positive opinion, except on the most certain grounds. This is all the more important because it is just in those cases in which our opinion is most frequently asked that the statements of the patient are of least value, as she is either anxious to conceal the existence of pregnancy, or, if desirous of an affirmative diagnosis, unconsciously colours her statements, so as to bias the judgment of the examiner.

Importance of the subject.

Constant attempts have been made to classify the signs of pregnancy: thus some divide them into the *natural* and *sensible* signs, others into the *presumptive*, the *probable*, and the *certain*. The latter classification, which is that adopted by Montgomery in his classical work on the 'Signs and Symptoms of Pregnancy,' is no doubt the better of the two, if any be required. The simplest way of studying the subject, however, is the one, now generally adopted, of considering the signs of pregnancy in the order in which they occur, and attaching to each an estimate of its diagnostic value.

Classification.

From the earliest ages authors have thought that the occurrence of conception might be ascertained by certain obscure signs, such as a peculiar appearance of the eyes, swelling of the neck, or by unusual sensations connected

Signs of a fruitful conception.

with a fruitful intercourse. All of these, it need hardly be said, are far too uncertain to be of the slightest value. The last is a symptom on which many married women profess themselves able to depend, and one to which Cazeaux is inclined to attach some importance.

Cessation of menstruation.

① The first appreciable indication of pregnancy on which any dependence can be placed is the cessation of the customary menstrual discharge, and it is of great importance, as forming the only reliable guide for calculating the probable period of delivery. In women who have been previously perfectly regular, in whom there is no morbid cause which is likely to have produced suppression, the non-appearance of the catamenia may be taken as strong presumptive evidence of the existence of pregnancy ; but it can never be more than this, unless verified and strengthened by other signs, inasmuch as there are many conditions besides pregnancy which may lead to its non-appearance. Thus exposure to cold, mental emotion, general debility, especially when connected with incipient phthisis, may all have this effect. Mental impressions are peculiarly liable to mislead in this respect. It is far from uncommon in newly married women to find that menstruation ceases for one or more periods, either from the general disturbance of the system connected with the married life, or from a desire on the part of the patient to find herself pregnant. Also in unmarried women, who have subjected themselves to the risk of impregnation, mental emotion and alarm often produce the same result.

Menstruation is often arrested independently of pregnancy.

Menstruation during pregnancy.

A further source of uncertainty exists in the fact, that in certain cases menstruation may go on for one or more periods after conception, or even during the whole pregnancy. The latter occurrence is certainly of extreme rarity, but one or two instances are recorded by Perfect, Churchill, and other writers of authority, and therefore its possibility must be admitted. The former is much less uncommon, and instances of it have probably come under the observation of most practitioners. The explanation is now well understood. During the early months of gestation, when the ovum is not yet sufficiently advanced in growth to fill the whole uterine cavity, there is a considerable space between the decidua reflexa which surrounds it and the decidua vera lining the uterine cavity. It is from this free surface of the decidua

Its explanation.

vera that the periodical discharge comes, and there is not
only ample surface for it to come from, but a free channel
for its escape through the os uteri.   After the third month
the decidua reflexa and the decidua vera blend together, and
the space between them disappears.   Menstruation after this
time is, therefore, much more difficult to account for.   It is
probable that, in many supposed cases, occasional losses of
blood from other sources, such as placenta prævia, an abraded
cervix uteri, or a small polypus, have been mistaken for true
menstruation.   If the discharge really occurs periodically
after the third month, it can only come from the canal of the
cervix.   The occurrence, however, is so rare that if a woman
is menstruating regularly and normally who believes herself
to be more than four months advanced in pregnancy, we are
justified *ipso facto* in negativing her supposition.   In an unmarried woman all statements as to regularity of menstruation are absolutely valueless, for in such cases nothing is
more common than for the patient to make false statements
for the express purpose of deception.

Conception may unquestionably occur when menstruation **Pregnancy**
is normally absent.   This is far from uncommon in women **sometimes occurs**
during lactation, when the function is in abeyance, and who **when men-**
therefore have no reliable data for calculating the true period **struation is nor-**
of their delivery.   Authentic cases are also recorded in which **mally absent.**
young girls have conceived before menstruation is estab-
lished, and in which pregnancy has occurred after the change
of life.

Taking all these facts into account, we can only look **Estimate**
upon the cessation of menstruation as a fairly presumptive **of its diagnostic**
sign of pregnancy in women in whom there is no clear reason **value.**
to account for it, but one which is undoubtedly of great value
in assisting our diagnosis.

Shortly after conception various sympathetic disturbances **Sympa-**
of the system occur, and it is only very exceptionally that **thetic dis-**
these are not established.   They are generally most developed **turbances.**
in women of highly nervous temperament ; and they are,
therefore, most marked in patients in the upper classes of
society, in whom this class of organisation is most common.

Amongst the most frequent of these are various disorders **Morning**
of the gastro-intestinal canal.   Nausea or vomiting is very **sickness.**
common ; and as it is generally felt on first rising from the

recumbent position, it is probably known amongst women as
the 'morning sickness.' It sometimes commences almost
immediately after conception, but more frequently not until
the second month, and it rarely lasts after the fourth month.
Generally there is nausea rather than actual vomiting. The
woman feels sick and unable to eat her breakfast, and often
brings up some glairy fluid. In other cases she actually
vomits; and sometimes the sickness is so excessive as to
resist all treatment, seriously to affect the patient's health,
and even imperil her life. These grave forms of the affection
will require separate consideration.

Cause of the sickness.

Very different opinions have been held as to the cause of
morning sickness. Dr. Henry Bennet believes that, when at
all severe, it is always associated with congestion and inflammation of the cervix uteri. Dr. Graily Hewitt maintains
that it depends entirely on flexion of the uterus, producing
irritation of the uterine nerves at the seat of the flexion,
and consequent sympathetic vomiting. This theory, when
broached at the Obstetrical Society, was received with little
favour; it seems to me to be sufficiently disproved by the
fact, which I believe to be certain, that more or less nausea
is a normal and nearly constant phenomenon in pregnancy,
for it is difficult to believe that nearly every pregnant woman
has a flexed uterus. The generally received explanation is,
probably, the correct one, viz. that nausea, as well as other
forms of sympathetic disturbance, depend on the stretching
of the uterine fibres by the growing ovum, and consequent
irritation of the uterine nerves. It is, therefore, one, and
only one, of the numerous reflex phenomena naturally accompanying pregnancy. It is an old observation that when the
sickness of pregnancy is entirely absent, other, and generally
more distressing, sympathetic derangements are often met
with, such as a tendency to syncope. Dr. Bedford [1] has laid
especial stress on this point, and maintains that under such
circumstances women are peculiarly apt to miscarry.

Other derange-ments of the diges-tive func-tions.

Other derangements of the digestive functions, depending
on the same cause, are not uncommon, such as excessive or
depraved appetite, the patient showing a craving for strange
and even disgusting articles of diet. These cravings may be
altogether irresistible, and are popularly known as 'longings.'

[1] *Diseases of Women and Children*, p. 551.

Of a similar character is the disturbed condition of the bowels frequently observed, leading to constipation, diarrhœa, and excessive flatulence.

Certain glandular sympathies may be developed, one of the most common being an excessive secretion from the salivary glands. A tendency to syncope is not infrequent, rarely proceeding to actual fainting, but rather to that sort of partial syncope, unattended with complete loss of consciousness, which the older authors used to call 'lypothemia.' This often occurs in women who show no such tendency at other times, and, when developed to any extent, it forms a very distressing accompaniment of pregnancy. Toothache is common, and is not rarely associated with actual caries of the teeth. When any of these phenomena are carried to excess it is more than probable that some morbid condition of the uterus exists, which increases the local irritation producing them.

Other sympathetic phenomena.

Mental phenomena are very general. An undue degree of despondency, utterly beyond the patient's control, is far from uncommon; or a change which renders the bright and good-tempered woman fractious and irritable; or even the more fortunate, but less common, change, by which a disagreeable disposition becomes altered for the better.

Mental peculiarities.

All these phenomena of exalted nervous susceptibility are but of slight diagnostic value. They may be taken as corroborating more certain signs, but nothing more; and they are chiefly interesting from their tendency to be carried to excess and to produce serious disorders.

The diagnostic value of these sympathetic disturbances is small.

Certain changes in the mammæ are of early occurrence, dependent, no doubt, on the intimate sympathetic relations at all times existing between them and the uterine organs, but chiefly required for the purpose of preparing for the important function of lactation, which, on the termination of pregnancy, they have to perform.

Mammary changes.

Generally about the second month of pregnancy the breasts become increased in size and tender. As pregnancy advances they become much larger and firmer, and blue veins may be seen coursing over them. The most characteristic changes are about the nipples and areolæ. The nipples become turgid, and are frequently covered with minute branny scales, formed by the desiccation of sero-lactescent

Changes in the areolæ.

fluid oozing from them.   The areolæ become greatly enlarged
and darkened from the deposit of pigment (fig. 78).   The

Fig. 78.

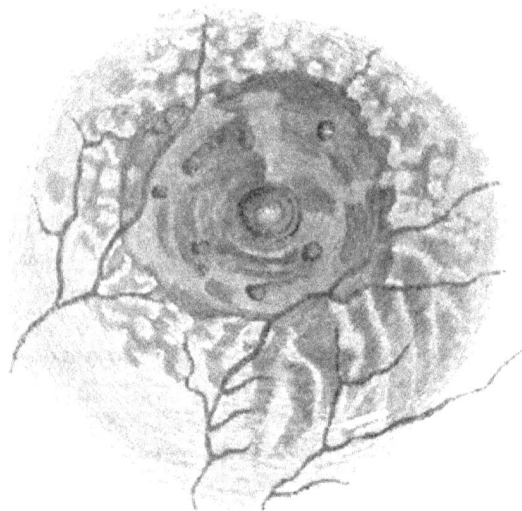

APPEARANCE OF THE AREOLA IN PREGNANCY.

extent and degree of this discoloration vary much in different women.   In fair women it may be so slight as to be
hardly appreciable ; while in dark women it is generally
exceedingly characteristic, sometimes forming a nearly black
circle extending over a great part of the breast.   The areola
becomes moist as well as dark in appearance, and is somewhat
swollen, and a number of small tubercles are developed upon
it, forming a circle of projections round the nipple.   These
tubercles are described by Montgomery as being intimately
connected with the lactiferous ducts, some of which may
occasionally be traced into them and seem to open on their
summits.   As pregnancy advances they increase in size and
number.   During the latter months what has been called
'the secondary areola' is produced, and when well marked

The se-
condary
areola.

presents a very characteristic appearance.   It consists of a
number of minute discoloured spots all round the outer
margin of the areola where the pigmentation is fainter, and
which are generally described as resembling spots from which
the colour has been discharged by a shower of water-drops.

This change, like the darkening of the primary areola, is more marked in brunettes. At this period, especially in women whose skin is of fine texture, whitish silvery streaks are often seen on the breasts. They are produced by the stretching of the cutis vera, and are permanent.

By pressure on the breasts a small drop of serous-looking fluid can very generally be forced out from the nipple, often as early as the third month, and on microscopic examination milk and cholostrum globules can be seen in it.

The diagnostic value of these mammary changes has been variously estimated. When well marked they are considered by Montgomery to be certain signs of pregnancy. To this statement, however, some important limitations must be made. In women who have never borne children they, no doubt, are so; for, although various uterine and ovarian diseases produce some darkening of the areola, they certainly never produce the well-marked changes above described. In multiparæ, however, the areolæ often remain permanently darkened, and in them these signs are much less reliable. In first pregnancies the presence of milk in the breasts may be considered an almost certain sign, and it is one which I have rarely failed to detect even from a comparatively early period. It is true that there are authenticated instances of non-pregnant women having an abundant secretion of milk established from mammary irritation. Thus Baudelocque presented to the Academy of Surgery of Paris a young girl, eight years of age, who had nursed her little brother for more than a month. Dr. Tanner states—I do not know on what authority—that 'it is not uncommon in Western Africa for young girls who have never been pregnant to regularly employ themselves in nursing the children of others, the mammæ being excited to action by the application of the juice of one of the euphorbiaceæ.' Lacteal secretion has even been noticed in the male breast. But these exceptions to the general rule are so uncommon as merely to deserve mention as curiosities; and I have hardly ever been deceived in diagnosing a first pregnancy from the presence of even the minutest quantity of lacteal secretion in the breasts, although even then other corroborative signs should always be sought for. In multiparæ the presence of milk is by no means so valuable, for it is common for milk to remain

Diagnostic value of mammary changes.

in the mammae long after the cessation of lactation, even for several years. Tyler Smith correctly says that ' suppression of the milk in persons who are nursing and liable to impregnation, is a more valuable sign of pregnancy than the converse condition.' This is an observation I have frequently corroborated.

They are of most value in first pregnancies.
As a diagnostic sign, therefore, the mammary appearances are of great importance in primiparæ, and when well marked they are seldom likely to deceive. They are specially important when we suspect pregnancy in the unmarried, as we can easily make an excuse to look at the breast without explaining to the patient the reason; and a single glance, especially if the patient be dark-complexioned, may so far strengthen our suspicion as to justify a more thorough examination. In married multiparæ they are less to be depended upon.

Other pigmentary changes.
In connection with this subject may be mentioned various irregular deposits of pigment which are frequently observed. The most common is a dark-brownish or yellowish line starting from the pubes and running up to the centre of the abdomen, sometimes as far as the umbilicus only, at others forming an irregular ring round the umbilicus, and reaching to the epigastrium. It is, however, of very uncertain occurrence, being well marked in some women, while in others it is entirely absent. Patches of darkened skin are often observed about the face, chiefly on the forehead, and this bronzing sometimes gives a very peculiar appearance. Joulin states that it only occurs on parts of the face exposed to the sun, and that it is therefore most frequently observed in women of the lower order who are freely exposed to atmospheric influences. These pigmentary changes are of small diagnostic value, and may continue for a considerable time after delivery.

Enlargement of the abdomen.
The progressive enlargement of the abdomen, and the size of the gravid uterus at various periods of pregnancy, as well as the method of examination by means of abdominal palpation, have already been described (pp. 121 and 136).

We will now consider the well-known phenomena produced by the movements of the fœtus in utero, which are so familiar to all pregnant women. These, no doubt, take place from the earliest period of fœtal life at which the mus-

cular tissue of the fœtus is sufficiently developed to admit of
contraction, but they are not felt by the mother until somewhere about the sixteenth week of utero-gestation, the
precise period at which they are perceived varying considerably in different cases.   The error of the law on this subject,
which supposes the child not to be alive, or ' quick,' until
the mother feels its movements, is well known, and has frequently been protested against by the medical profession.
The so-called *quickening*—which certainly is felt very suddenly by some women—is believed to depend on the rising
of the uterine tumour sufficiently high to permit of the impulse
of the fœtus being transmitted to the abdominal walls of the
mother, through the sensory nerves of which its movements
become appreciable.   The sensation is generally described as
being a feeble fluttering, which, when first felt, not unfrequently causes unpleasant nervous sensations.   As the uterus
enlarges, the movements become more and more distinct, and
generally consist of a series of sharp blows or kicks, sometimes quite appreciable to the naked eye, and causing distinct
projections of the abdominal walls.   Their force and frequency
will also vary during pregnancy according to circumstances.
At times they are very frequent and distressing ; at others,
the fœtus seems to be comparatively quiet, and they may
even not be felt for several days in succession, and thus unnecessary fears as to death of the fœtus often arise.   The
state of the mother's health has an undoubted influence upon
them.   They are said to increase in force after a prolonged
abstinence from food, or in certain positions of the body.   It
is certain that causes interfering with the vitality of the
fœtus often produce very irregular and tumultuous movements.   They can be very readily felt by the accoucheur on
palpating the abdomen, and sometimes, in the latter months,
so distinctly as to leave no doubt as to the existence of pregnancy.   They can also generally be induced by placing one
hand on each side of the abdomen and applying gentle pressure, which will induce fœtal motion, that can be easily
appreciated.

   As a diagnostic sign the existence of fœtal movements
has always held a high place, but care should be taken in relying on it.   It is certain that women are themselves very often
in error, and fancy they feel the movements of a fœtus when

Quicken-
ing.

The dia-
gnostic
value of
fœtal
move-
ments.

none exists, being probably deceived by irregular contractions of the abdominal muscles, or flatus within the bowels. They may even involuntarily produce such intra-abdominal movements as may readily deceive the practitioner. Of course, in advanced pregnancy, when the fœtal movements are so marked as to be seen as well as felt, a mistake is hardly possible, and they then constitute a certain sign. But in such cases there is an abundance of other indications and little room for doubt. In questionable cases, and at an early period of pregnancy, the fact that movements are not felt must not be taken as a proof of the non-existence of pregnancy, for they may be so feeble as not to be perceptible, or they may be absent for a considerable period.

Inter- mittent uterine contrac- tions.

Braxton Hicks[1] has directed attention to the value, from a diagnostic point of view, of intermittent contractions of the uterus during pregnancy. After the uterus is sufficiently large to be felt by palpation, if the hand be placed over it, and it be grasped for a time without using any friction or pressure, it will be observed to distinctly harden in a manner that is quite characteristic. This intermittent contraction occurs every five or ten minutes, sometimes oftener, rarely at longer intervals. The fact that the uterus did contract in this way had been previously described, more especially by Tyler Smith, who ascribed it to peristaltic action. But it is certain that no one, before Dr. Hicks, had pointed out the fact that such contractions are constant and normal concomitants of pregnancy, continuing during the whole period of utero- gestation, and forming a ready and reliable means of distin- guishing the uterine tumour from other abdominal enlarge-

Value of this sign.

ments. Since reading Dr. Hicks's paper I have paid considerable attention to this sign, which I have never failed to detect, even in the retroverted gravid uterus contained entirely in the pelvic cavity, and I am disposed entirely to agree with him as to its great value in diagnosis. If the hand be kept steadily on the uterus, its alternate hardening and relaxation can be appreciated with the greatest ease. The advantages which this sign has over the fœtal movements are that it is constant, that it is not liable to be simulated by anything else, and that it is independent of the life of the child, being equally appreciable when the uterus contains a degenerated

[1] *Obst. Trans.* v. 13.

ovum or dead fœtus. The only condition likely to give rise to error is an enlargement of the uterus in consequence of contents other than the results of conception, such as retained menses, or a polypus. The history of such cases—which are moreover of extreme rarity—would easily prevent any mistake. As a corroborative sign of pregnancy, therefore, I should give these intermittent contractions a high place.

The vaginal signs of pregnancy are of considerable importance in diagnosis. They are chiefly the changes which may be detected in the cervix, and the so-called *ballottement*, which depends on the mobility of the fœtus in the liquor amnii. *Vaginal signs of pregnancy.*

The alterations in the density and apparent length of the cervix have been already described (p. 139). When pregnancy has advanced beyond the fifth month the peculiar velvety softness of the cervix is very characteristic, and affords a strong corroborative sign, but one which it would be unsafe to rely on by itself, inasmuch as very similar alterations may be produced by various causes. When, however, in a supposed case of pregnancy advanced beyond the period indicated, the cervix is found to be elongated, dense, and projecting into the vaginal canal, the non-existence of pregnancy may be safely inferred. Therefore the negative value of this sign is of more importance than the positive. *Softening of the cervix.*

Ballottement, when distinctly made out, is a very valuable indication of pregnancy. It consists in the displacement, by the examining finger, of the fœtus, which floats up in the liquor amnii, and falls back again on the tip of the finger with a slight tap which is exceedingly characteristic. *Ballottement.*

In order to practise it most easily, the patient is placed on a couch or bed in a position midway between sitting and lying, by which the vertical diameter of the uterine cavity is brought into correspondence with that of the pelvis. Two fingers of the right hand are then passed high up into the vagina in front of the cervix. The uterus being now steadied from without by the left hand, the intravaginal fingers press the uterine wall suddenly upwards, when, if pregnancy exist, the fœtus is displaced, and in a moment falls back again, imparting a distinct impulse to the fingers. When easily appreciable it may be considered as a certain sign, for although an ante-flexed fundus, or a calculus in the bladder, may *Method of examination.*

give rise to somewhat similar sensations, the absence of other indications of pregnancy would readily prevent error. Ballottement is practised between the fourth and seventh months. Before the former time the fœtus is too small, while at a later period it is relatively too large, and can no longer be easily made to rise upwards in the surrounding liquor amnii. The absence of ballottement must not be taken as proving the non-existence of pregnancy, for it may be inappreciable from a variety of causes, such as abnormal presentations, or the implantation of the placenta upon the cervix uteri.

There are also some other vaginal signs of pregnancy of secondary consequence. Amongst these is the vaginal pulsation, pointed out by Osiander, resulting from the enlargement of the vaginal arteries, which may sometimes be felt beating at an early period. Often this pulsation is very distinct, at other times it cannot be felt at all, and it is altogether unreliable, as a similar pulsation may be felt in various uterine diseases.

Vaginal pulsation.

Dr. Rasch has drawn attention to a previously undescribed sign which he believes to be of importance in the diagnosis of early pregnancy.[1] It consists in the detection of fluctuation, through the anterior uterine wall, depending on the presence of the liquor amnii. In order to make this out, two fingers of the right hand must be used, as in ballottement, while the uterus is steadied through the abdomen. Dr. Rasch states that by this means the enlarged uterus in pregnancy can easily be distinguished from the enlargement depending on other causes, and that fluctuation can always be felt as early as the second month. If it is associated with suppressed menstruation and darkened areolæ, he considers it a certain sign. In order to detect it, however, considerable experience in making vaginal examinations is essential, and it can hardly be depended on for general use.

Uterine fluctuation.

A peculiar deep violet hue of the vaginal mucous membrane was relied on by Jacquemier and Klüge as affording a readily observed indication of pregnancy. In most cases it is well marked; sometimes, indeed, the change of colour is very intense, and it evidently depends on the congestion

Alteration in colour of the vagina.

[1] *Brit. Med. Jour.* vol. ii. 1873.

produced by pressure of the enlarged uterus. The same effect, however, is constantly seen where similar pressure is effected by large fibroid tumours of the uterus, and, therefore, for diagnostic purposes it is valueless.

By far the most important signs are those which can be detected by abdominal auscultation, and one of these—the hearing of the fœtal heart-sounds—forms the only sign which *per se*, and in the absence of all others, is perfectly reliable. *Auscultatory signs of pregnancy.*

The fact that the sounds of the fœtal heart are audible during advanced pregnancy was first pointed out by Mayor of Geneva in 1818, and the main facts in connection with fœtal auscultation were subsequently worked out by Kergaradec, Naegele, Evory Kennedy, and other observers. The pulsations first become audible, as a rule, in the course of the fifth month, or about the middle of the fourth month. In exceptional circumstances, and by practised observers, they have been heard earlier. Depaul believes that he detected them as early as the eleventh week, and Routh has also detected them at an earlier period by vaginal stethoscopy, which, however, for obvious reasons, cannot be ordinarily employed. Naegele never heard them before the eighteenth week, more generally at the end of the twentieth, and for practical purposes the pregnancy must be advanced to the fifth month before we can reasonably expect to detect them. From this period up to term they can almost always be heard, if not at the first attempt, at least afterwards, to a certainty, if we have the opportunity of making repeated examinations. Accidental circumstances, such as the presence of an unusual amount of flatus in the intestines, may deaden the sounds for a time, but not permanently. Depaul only failed to hear them in 8 cases out of 906 examined during the last three months of pregnancy; and out of 180 cases which Dr. Anderson, of Glasgow, carefully examined, he only failed in 12, and in each of these the child was stillborn. They, therefore, form not only a most certain indication of pregnancy, but of the life of the fœtus also.

Discovery of fœtal auscultation.

Period at which the fœtal heart-sounds are audible.

The sound has always been likened to the double tic-tac of a watch heard through a pillow, which it closely resembles. It consists of two beats, separated by a short interval, the first being the loudest and most distinct, the second being

Description of the sound.

sometimes inaudible. The rapidity of the foetal pulsations
forms an important means of distinguishing them from transmitted maternal pulsations with which they might be confounded. Their average number is stated by Slater, who
made numerous observations on this point, to be 132, but
sometimes they reach as high as 140, and sometimes as low
as 120. It will thus be seen that the pulsations are always
much more rapid than those of the mother's heart, unless,
indeed, the latter be unduly accelerated by transient mental
emotion or disease. To avoid mistakes, whenever the foetal
heart is heard its rate of pulsation should be carefully
counted, and compared with that of the mother's pulse; if
the rate differ, we may be sure that no error has been made.
The rapidity of the foetal pulsations remains, as a rule, the
same during the whole period of pregnancy, while their
intensity gradually increases. They may, however, be temporarily increased or diminished in frequency by disturbing
causes, such as the pressure of the stethoscope, which,
exciting tumultuous movements of the foetus, may induce
greatly increased frequency of its heart-beats. So also
during labour, after the escape of the liquor amnii, when the
contractions of the uterus have a very distinct influence on

Irregulari-
ties of the
foetal
heart-
sounds ;
their dia-
gnostic
value.

the foetus, they may be greatly modified. An acceleration or
irregularity of the pulsations, made out in the course of a
prolonged labour, may thus be of great practical importance,
by indicating the necessity for prompt interference. Similar
alterations, associated with tumultuous and unusual foetal
movements felt by the mother towards the end of pregnancy,
may point to danger to the life of the foetus during the latter
months, and may even justify the induction of premature
labour. This is especially the case in women who have previously given birth to a succession of dead children owing
to disease of the placenta, and, in them, careful and frequently repeated auscultations may warn us of the impending
danger.

Supposed
difference
of rapidity
according
to the sex
of the
foetus.

The rapidity of the foetal heart has been supposed by
some to afford a means of determining the sex of the child
before birth. Frankenhauser, who first directed attention to
this point, is of opinion that the average rate of pulsations of
the heart is considerably less in male than in female children, averaging 124 in the minute in the former, as against

144 in the latter. Steinbach makes the difference somewhat less, viz. 131 for males, and 138 for females. He predicted the sex correctly by this means in 45 out of 57 cases, while Frankenhauser was correct in the whole 50 cases which he specially examined with reference to the point. Dr. Hutton, of New York,[1] was also correct in 7 cases he fixed on for trial. Devilliers found the difference in the sexes to be the same as Steinbach; he attributes it, however, to the size and weight rather than to the sex of the child, and believes the pulsations to be least numerous in large and well-developed children. As male children are usually larger than female, he thus explains the relatively less frequent pulsations of their hearts. Dr. Cumming, of Edinburgh, also believes that the weight of the child has considerable influence on the frequency of its cardiac pulsations, so that a large female child may have a slower pulse than a small male.[2] The point, however, is more curious than practical, and the rapidity of the pulsations certainly would not justify any positive prediction on the subject. Circumstances influencing the maternal circulation seem to have no influence on that of the fœtus.

The fœtal heart-sounds are generally propagated best by the back of the child, and are, therefore, most easily audible when this is in contact with the anterior wall of the uterus, as is the case in the large majority of pregnancies. When the child is placed in the dorso-posterior position, the sounds have to traverse a larger amount of the liquor amnii, and are further modified by the interposition of the fœtal limbs. They are, therefore, less easily heard in such cases, but even in them they can almost always be made out. As the fœtus most frequently lies with the occiput over the brim of the pelvis, and the back of the child towards the left side of the mother, the heart-sounds are usually most distinctly audible at a point midway between the umbilicus and the left anterior superior spine of the ilium. In the next most common position, in which the back of the child lies to the right lumbar region of the mother, they are generally heard at a corresponding point at the right side, but in this case they are frequently more readily made out in the right flank, being then transmitted through the thorax of the child,

Site at which the sounds are heard.

[1] *New York Med. Jour.* July 1872.   [2] *Edin. Med. Jour.* 1875.

which is in contact with the side of the uterus. In breech
cases, on the other hand, the heart-sounds are generally heard
most distinctly *above* the umbilicus, and either to the right
or left, according to the side towards which the back of
the child is placed. It will thus be seen that the place
at which the fœtal heart-sounds are heard varies with the
position of the fœtus; and this, when combined with the
information derived from palpation, affords a ready means of
ascertaining the presentation of the child before labour.
The sounds are only audible over a limited space, about two
or three inches in diameter; therefore, if we fail to detect
them in one place, a careful exploration of the whole uterine
tumour is necessary before we are satisfied that they cannot
be heard.

Sources of
fallacy.

The only mistake that is likely to be made is taking the
maternal pulsations, transmitted through the uterine tumour,
for those of the fœtal heart. A little care will easily prevent
this error, and the frequency of the mother's pulse should
always be ascertained before counting the supposed fœtal
pulsations. If these are found to be 120 or more, while
the mother's pulse is only 70 or 80, no mistake is possible.
If the latter is abnormally quickened greater care may be
necessary, but even then the rate of pulsation of each will
be dissimilar. Braxton Hicks[1] has pointed out that in
tedious labour, when the muscular powers of the mother are
exhausted, the muscular subsurrus may produce a sound
closely resembling the fœtal pulsation; but error from this
source is obviously very improbable.

Mode of
practising
ausculta-
tion.

In listening for the fœtal heart-sound the patient
should be placed on her back, with the shoulders elevated
and the knees flexed. The surface of the abdomen should
be uncovered, and an ordinary stethoscope employed, the
end of which must be pressed firmly on the tumour, so as
to depress the abdominal walls. The most absolute still-
ness is necessary, as it is often far from easy to hear the
sounds. Sometimes, after failing with the ordinary stethoscope, I have succeeded with the bin-aural, which remarkably
intensifies them. When once heard they are most easily
counted during a space of five seconds, as, on account of

[1] *Obst. Trans.* vol. xv.

their frequency, it is not always possible to follow them over a longer period.

When the fœtal heart-sounds are heard distinctly, preg- *Value of this sign of preg- nancy.* nancy may be absolutely and certainly diagnosed. The fact that we do not hear them does not, however, preclude the possibility of gestation, for the fœtus may be dead, or the sounds temporarily inaudible.

There are some other sounds heard in auscultation which *Other sounds heard in preg- nancy. Umbilical souffle.* are of very secondary diagnostic value. One of these is the so-called *umbilical* or *funic souffle*, which was first pointed out by Evory Kennedy. It consists of a single blowing murmur, synchronous with the fœtal heart-sounds, and most distinctly heard in the immediate vicinity of the point where these are most audible. Most authors believe it to be pro- duced by pressure on the cord, either when it is placed between a hard part of the fœtus and the uterine walls, or is twisted round the child's neck. Schroeder and Hecker detected it in fourteen or fifteen per cent. of all cases, and the latter believed it to be caused by flexure of the first portion of the cord near the umbilicus. For practical purposes it is quite valueless, and need only be mentioned as a phenomenon which an experienced auscultator may occasionally detect.

The *uterine souffle* is a peculiar single whizzing murmur *The ute- rine souffle.* which is almost always audible on auscultation. It varies very remarkably in character and position. Sometimes it is a gentle blowing or even musical murmur; at others it is loud, harsh, and scraping; sometimes continuous, sometimes intermittent. It may also be heard at any point of the uterus, but most frequently low down, and to one or other side; more rarely above the umbilicus, or towards the fundus; and it often changes its position so as to be heard at a sub- sequent auscultation at a point where it was previously inaudible. It may be heard over a space of an inch or two only, or in some cases over the whole uterine tumour; or again it may sometimes be detected simultaneously over two entire distinct portions of the uterus. It is generally to be heard earlier than the fœtal heart-sounds, often as soon as the uterus rises above the brim of the pelvis, and it can almost always be detected after the commencement of the fourth month. The sound becomes curiously modified by the uterine contractions during labour, becoming louder and

more intense before the pain comes on, disappearing during
its acme, and again being heard as it goes off. Hicks attributes to a similar cause, viz. the uterine contractions during
pregnancy, the frequent variations in the sound which are
characteristic of it.[1] The uterine souffle is also audible after
the death of the fœtus, and it is believed by some to be
modified and to become more continuously harsh when that
event has taken place.

Theories as to its cause. Very various explanations have been given of the causes
of this sound. For long it was supposed to be formed in the
vessels of the placenta, and hence the name '*placental
souffle*,' by which it is often talked of; or if not in the placenta, in the uterine vessels in its immediate neighbourhood.
The non-placental origin of the sound is sufficiently demonstrated by the fact that it may be heard for a considerable
time after the expulsion of the placenta. Some have supposed that it is not formed in the uterus at all, but in the
maternal vessels, especially the aorta and the iliac arteries,
owing to the pressure to which they are subjected by the
gravid uterus. The extreme irregularity of the sound, its
occasional disappearance, and its variable site, seem to be conclusive against this view. The theory which refers the sound
to the uterine vessels is that which has received most adher-
ents, and which best meets the facts of the case; but it is
by no means easy, or even possible, to account for the exact
mode of its production in them. Each of the explanations
which have been given is open to some objection. It is
far from unlikely that the intermittent contractions of the
uterine fibres, which are known to occur during the whole
course of pregnancy, may have much to do with it, by modifying, at intervals, the rapidity of the circulation in the vessels. Its production in this manner may also be favoured by
the chlorotic state of the blood, to which Cazeaux and Scanzoni are inclined to attribute an important influence, likening
it to the anæmic murmur so frequently heard in the vessels
in weakly women.

Its diagnostic value. From a diagnostic point of view the uterine souffle is of
very secondary importance, because a similar sound is very
generally audible in large fibroid tumours of the uterus, and

[1] *Op. cit.* p. 223.

even in some few ovarian tumours; it is, therefore, of little or no value in assisting us to decide the character of the abdominal enlargement. The supposed dependence of the sound on the placental circulation has caused its site to be often identified with that of the placenta. It is, however, most frequently heard at the lower part of the uterus, while the placenta is generally attached near the fundus, so that its position cannot be taken as any safe guide in determining the situation of that organ. *It is no reliable indication of the site of the placenta.*

Occasionally, in practising auscultation, irregular sounds of brief duration may be heard, which are not susceptible of accurate description, and which doubtless depend on the sudden movement of the fœtus in the liquor amnii, or on the impact of its limbs on the uterine walls. When heard distinctly they are characteristic of pregnancy; and they may be sometimes heard when the other sounds cannot be detected. They are, however, so irregular, and so often entirely absent, that they can hardly be looked upon in any other light than as occasional phenomena. *Sounds produced by the movements of the fœtus.*

Two other sounds have been described as being sometimes audible, which may be mentioned as matters of interest, but which are of no diagnostic value. One is a rustling sound, said by Stoltz to be audible in cases in which the fœtus is dead, and which he refers to gaseous decomposition of the liquor amnii; its existence is, however, extremely problematical. The other is a sound heard after the birth of the child, and referred by Caillant to the separation of the placental adhesions. He describes it as a series of rapid short scratching sounds, similar to those produced by drawing the nails across the seat of a horse-hair sofa. Simpson [1] admitted the existence of the sound, but believed that it is produced by the mere physical crushing of the placenta, and artificially imitated it out of the body by forcing the placenta through an aperture the size of the os uteri. *Sounds referred to decomposition of the liquor amnii, and to separation of the placenta.*

It will be seen, then, that although there are numerous signs and symptoms accompanying pregnancy, many of them are unreliable by themselves, and apt to mislead. Those which may be confidently depended on are the pulsations of the fœtal heart, which, however, fail us in cases of dead chil- *Relative value of the signs and symptoms of pregnancy.*

---

[1] *Selected Obstet. Works*, p. 151.

dren : the fœtal movements when distinctly made out ; ballottement ; the intermittent contractions of the uterus ; and
to these we may safely add the presence of milk in the
breasts, provided we have to do with a first pregnancy.

The remainder are of importance in leading us to suspect
pregnancy, and in corroborating and strengthening other
symptoms, but they do not, of themselves, justify a positive
diagnosis.

CHAPTER V.

THE DIFFERENTIAL DIAGNOSIS OF PREGNANCY. SPURIOUS PREG-
NANCY. THE DURATION OF PREGNANCY. SIGNS OF RECENT
PREGNANCY.

THE differential diagnosis of pregnancy has of late years Import-
assumed much importance on account of the advance of ab- ance of the subject.
dominal surgery. The cases are so numerous in which even
the most experienced practitioners have fallen into error, and
in which the abdomen has been laid open in ignorance of the
fact that pregnancy existed, that the subject becomes one of
the greatest consequence. Fortunately it is less so from an
obstetrical than from a gynæcological point of view, inasmuch
as the converse error, of mistaking some other condition for
pregnancy, is of far less consequence, as it is one which time
will always rectify. But even in this way carelessness may
lead to very serious injury to the character, if not to the
health, of the patient; and it will be well to refer briefly to
some of the conditions most liable to be mistaken for pregnancy, and to the mode of distinguishing them.

Adipose enlargement of the abdomen may obscure the Adipose
diagnosis by preventing the detection of the uterus; and if, ment of
as is not uncommon with women of great obesity, it is asso- the abdo-
ciated with irregular menstruation, the increased size of the men.
abdomen might be supposed to depend on pregnancy. The
absence of corroborative signs, such as auscultatory phenomena, mammary changes, and the hardness of the cervix as
felt per vaginam, make it easy to avoid this error.

Distension of the uterus by retained menstrual fluid, or Distension
watery secretion, is an occurrence of rarity that could seldom uterus by
give rise to error. Still it occasionally happens that the retained
uterus becomes enlarged in this way, sometimes reaching menses,
even to the level of the umbilicus, and that the physical metra, &c

character of the tumour is not unlike that of the gravid uterus. The best safeguard against mistakes will be the previous history of the case, which will always be different from that of ordinary pregnancy. Retention of the menses almost always occurs from some physical obstruction to the exit of the fluid, such as imperforate hymen; or if it occur in women who have already menstruated, we may usually trace a history of some cause, such as inflammation following an antecedent labour, which has produced occlusion of some part of the genital tract. The existence of a pelvic tumour in a girl who has never menstruated will of itself give rise to suspicion, as pregnancy under such circumstances is of extreme rarity. It will also be found that general symptoms have existed for a period of time considerably longer than the supposed duration of pregnancy, as judged of by the size of the tumour. The most characteristic of them are periodic attacks of pain due to the addition, at each monthly period, to the quantity of retained menstrual fluid. Whenever, from any of these reasons, suspicion of the true character of the case has, arisen, a careful vaginal examination will generally clear it up. In most cases the obstruction will be in the vagina, and is at once detected, the vaginal canal above it, as felt per rectum, being greatly distended by fluid; and we may also find the bulging and imperforate hymen protruding through the vulva. The absence of mammary changes, and of ballottement, will materially aid us in forming a diagnosis.

Conges-
tive hyper-
trophy of
the uterus.
          The engorged and enlarged uterus, frequently met with in women suffering from uterine disease, might readily be mistaken for an early pregnancy, if it happened to be associated with amenorrhœa. A little time would, of course, soon clear up the point, by showing that progressive increase in size, as in pregnancy, does not take place. This mistake could only be made at an early stage of pregnancy, when a positive diagnosis is never possible. The accompanying symptoms—pain, inability to walk, and tenderness of the uterus on pressure—would prevent such an error.

Ascitic
distension
of the
abdomen.
          Ascites, *per se*, could hardly be mistaken for pregnancy; for the uniform distension and evident fluctuation, the absence of any definite tumour, the site of resonance on percussion changing in accordance with alteration of the position

of the woman, and the unchanged cervix and uterus, should
be sufficient to clear up any doubt. Pregnancy may, however, exist with ascites, and this combination may be difficult
to detect, and might readily be mistaken for ovarian disease,
associated with ascites. The existence of mammary changes,
the presence of the softened cervix, ballottement, and auscultation—provided the sounds were not masked by the surrounding fluid—would afford the best means of diagnosing such a
case.

One of the most frequent sources of difficulty is the dif- Uterine
and ova-
rian
tumours.
ferential diagnosis of large abdominal tumours, either fibroid
or ovarian, or of some enlargements due to malignant disease
of the peritoneum or abdominal viscera. The most experienced have been occasionally deceived under such circumstances. As a rule, the presence of menstruation will prevent
error, as this generally continues in ovarian disease, while in
fibroids it is often excessive. The character of the tumour
—the fluctuation in ovarian disease, the hard nodular masses
in fibroid—and the history of the case—especially the length
of time the tumour has existed—will aid in diagnosis, while
the absence of cervical softening and of auscultatory phenomena will further be of material value in forming a conclusion. Some of the most difficult cases to diagnose are those
in which pregnancy complicates ovarian or fibroid disease.
Then the tumour may more or less completely obscure the
physical signs of pregnancy. The usual shape of the abdomen will generally be altered considerably, and we may be
able to distinguish the gravid uterus, separated from the
ovarian tumour by a distinct sulcus, or with the fibroid masses
cropping out from its surface. Our chief reliance must then
be placed in the alteration of the cervix, and in the auscultatory signs of pregnancy.

The condition most likely to give rise to errors is that Spurious
preg-
nancy.
very interesting and peculiar state known as *spurious preg-*
nancy. In this most of the usual phenomena of pregnancy
are so strangely simulated that accurate diagnosis is often
far from easy. There are hardly any of the more apparent
symptoms of pregnancy which may not be present in marked
cases of this kind. The abdomen may become prominent,
the areolæ altered, menstruation arrested, and apparent
fœtal motions felt ; and, unless suspicion is aroused, and a

careful physical examination made, both the patient and the practitioner may easily be deceived.

Cases in which spurious pregnancy occurs. There is no period of the childbearing life in which spurious pregnancy may not be met with ; but it is most likely to occur in elderly women about the climacteric period, when it is generally associated with ovarian irritation connected with the change of life ; or in younger women, who are either very desirous of finding themselves pregnant, or who, being unmarried, have subjected themselves to the chance of being so. In all cases the mental faculties have much to do with its production, and there is generally either very marked hysteria, or even a condition closely allied to insanity. Spurious pregnancy is by no means confined to the human race. It is well known to occur in many of the lower animals. Harvey related instances in bitches, either after unsuccessful intercourse, or in connection with their being in heat, even when no intercourse had occurred. In such cases the abdomen swelled, and milk appeared in the mammæ. Similar phenomena are also occasionally met with in the cow. In these instances, as in the human female, there is probably some morbid irritation of the ovarian system.

Its signs and symptoms. The physical phenomena are often very well marked. The apparent enlargement is sometimes very great, and it seems to be produced by a projection forward of the abdominal contents due to depression of the diaphragm, together with rigidity of the abdominal muscles, and may even closely simulate the uterine tumour on palpation. After the climacteric it is frequently associated, as Gooch pointed out, with an undue deposit of fat in the abdominal walls and omentum, so that there may be even some dulness on percussion, instead of resonance of the intestines. The fœtal movements are curiously and exactly simulated, either by involuntary contractions of the abdominal walls, or by the movement of flatus in the intestines. The patient also generally fancies that she suffers from the usual sympathetic disorders of pregnancy, and thus her account of her symptoms will still further tend to mislead.

Sometimes followed by spurious labour. Not only may the supposed pregnancy continue, but, at what would be the natural term of delivery, all the phenomena of labour may supervene. Many authentic cases are on record in which regular pains came on, and continued to increase in force and frequency until the actual condition was diagnosed. Such mistakes, however, are only likely to happen when the statements of the patient have been received without further inquiry. When once an accurate examination has been made, error is no longer possible.

We shall generally find that some of the phenomena of pregnancy are absent. Possibly menstruation, more or less irregular, may have continued. Examination per vaginam will at once clear up the case, by showing that the uterus is not enlarged, and that the cervix is unaltered. It may then be very difficult to convince the patient or her friends that her symptoms have misled her, and for this purpose the inhalation of chloroform is of great value. As consciousness is abolished, the semi-voluntary projection of the abdominal muscles is prevented, the large apparent tumour vanishes, and the bystanders can be readily convinced that none exists. As the patient recovers the tumour again appears.

Methods of diagnosis.

The duration of pregnancy in the human female has always formed a fruitful theme for discussion amongst obstetricians. The reasons which render the point difficult of decision are obvious. As the large majority of cases occur in married women, in whom intercourse occurs frequently, there is no means of knowing the precise period at which conception took place. The only datum which exists for the calculation of the probable date of delivery is the cessation of menstruation. It is quite possible, however, and indeed probable, that conception occurred, in a considerable number of instances, not immediately after the last period, but immediately before the proper epoch for the occurrence of the next. Hence, as the interval between the end of one menstruation and the commencement of the next averages 25 days, an error to that extent is always possible. Another source of fallacy is the fact, which has generally been overlooked, that even a single coitus does not fix the date of conception, but only that of insemination. It is well known that in many of the lower animals the fertilisation of the ovule does not take place until several days after copulation, the spermatozoa remaining in the interval in a state of active vitality within the genital tract. It has been shown by Marion Sims that living spermatozoa exist in the cervical canal in the human female some days after intercourse. It

Duration of pregnancy. Sources of fallacy in calculation.

Conception may occur at any point of the menstrual interval.

Insemination and conception do not necessarily coincide.

'is very probable, therefore, that in the human female, as in the lower animals, a considerable but unknown interval occurs between insemination and actual impregnation, which may render calculations as to the precise duration of pregnancy altogether unreliable.

Average time between cessation of menstruation and delivery.

A large mass of statistical observations exist respecting the average duration of gestation, which have been drawn up and collated from numerous sources. It would serve no practical purpose to reprint the voluminous tables on this subject that are contained in obstetrical works. They are based on two principal methods of calculation. First, we have the length of time between the cessation of menstruation and delivery. This is found to vary very considerably, but the largest percentage of deliveries occurs between the 274th and 280th day after the cessation of menstruation, the average day being the 278th; but, in individual instances, very considerable variations both above and below these limits are found to exist. Next we have a series of cases, from various sources, in which only one coitus was believed to have taken place. These are naturally always open to some doubt, but, on the whole, they may be taken as affording tolerably fair grounds for calculation. Here, as in the other mode of calculation, there are marked variations, the average length of time, as estimated from a considerable collection of cases, being 275 days after the single intercourse. It may, therefore, be taken as certain that there is no definite time which we can calculate on as being the proper duration of pregnancy, and, consequently, no method of estimating the probable date of delivery on which we can absolutely rely.

Average time between a single coitus and delivery.

No precise date for delivery can be fixed.

Methods of predicting the probable date.

The prediction of the time at which the confinement may be expected is, however, a point of considerable practical importance, and one on which the medical attendant is always consulted. Various methods of making the calculation have been recommended. It has been customary in this country, according to the recommendation of Montgomery, to fix upon ten lunar months, or 280 days, as the probable period of gestation, and, as conception is supposed to occur shortly after the cessation of menstruation, to add this number of days to any day within the first week after the last menstrual period as the most probable period of delivery.

As, however, 278 days is found to be the average duration
of gestation after the cessation of menstruation, and as this
method makes the calculation vary from **281** to **287** days, it
is evidently liable to fix too late a date. Naegele's method
was to count 7 days from the first appearance of the last
menstrual period, and then reckon backwards three months
as the probable date. Thus, if a patient last commenced to
menstruate on August 10, counting in this way from August
17 would give May 17 as the probable date of the delivery.

Matthews Duncan has paid more attention than anyone
else to the prediction of the date of delivery. His method
of calculating is based on the fact of 278 days being the
average time between the cessation of menstruation and
parturition; and he claims to have had a greater average
of success in his predictions than on any other plan. His
rule is as follows: 'Find the day on which the female
ceased to menstruate, or the first day of being what she calls
"well." Take that day nine months forward as 275, unless
February is included, in which case it is taken as 273 days.
To this add three days in the former case, or five if February
is in the count, to make up the 278. This 278th day should
then be fixed on as the middle of the week, or, to make the
prediction the more accurate, of the fortnight in which the
confinement is likely to occur, by which means allowance is
made for the average variation of either excess or deficiency.'

Various periodoscopes and tables for facilitating the calculation have been made. The periodoscope of Dr. Tyler
Smith (sold by Messrs. John Smith & Co., 52 Long Acre) is
very useful for reference in the consulting room, giving at a
glance a variety of information, such as the probable period of
quickening, the dates for the induction of premature labour,
etc. The following table, prepared by Dr. Protheroe Smith,
is also easily read, and is very serviceable :—

TABLE FOR CALCULATING THE PERIOD OF UTERO-GESTATION.[1]

		Nine Calendar Months.			Ten Lunar Months.	
From		To		Days.	To		Days.
January	1	September	30	273	October	7	280
February	1	October	31	273	November	7	280
March	1	November	30	275	December	5	280
April	1	December	31	275	January	5	280
May	1	January	31	276	February	4	280
June	1	February	28	273	March	7	280
July	1	March	31	274	April	6	280
August	1	April	30	273	May	7	280
September	1	May	31	273	June	7	280
October	1	June	30	273	July	7	280
November	1	July	31	273	August	7	280
December	1	August	31	274	September	6	280

Quickening a fallacious guide in estimating date of delivery.

The date at which the quickening has been perceived is relied on by many practitioners, and still more by patients, in calculating the probable date of delivery, as it is generally supposed to occur at the middle of pregnancy. The great variations, however, at the time at which this phenomena is first perceived, and the difficulty which is so often experienced of ascertaining its presence with any certainty, render it a very fallacious guide. The only times at which the perception of quickening is likely to prove of any real value, are when impregnation has occurred during lactation (when menstruation is normally absent), or when menstruation is so uncertain and irregular that the date of its last appearance cannot be ascertained. As quickening is most commonly felt during the fourth month, more frequently in its first than in its last fortnight, it may thus afford the only guide we can obtain, and that an uncertain one, for predicting the date of delivery.

Is protraction of gestation possible ?

From a medico-legal point of view the question of the possible protraction of pregnancy beyond the average time, and of the limits within which such protraction can be admitted,

[1] The above obstetric 'Ready Reckoner' consists of two columns, one of calendar, the other of lunar months, and may be read as follows : A patient has ceased to menstruate on July 1 : her confinement may be expected at soonest about March 31 (*the end of nine calendar months*) ; or at latest on April 6 (*the end of ten lunar months*). Another has ceased to menstruate on January 20 : her confinement may be expected on September 30, plus 20 days (*the end of nine calendar months*), at soonest ; or on October 7, plus 20 days (*the end of ten lunar months*), at latest.

is of very great importance. The law on this point varies considerably in different countries. Thus in France it is laid down that legitimacy cannot be contested until 300 days have elapsed from the death of the husband, or the latest possible opportunity for sexual intercourse. This limit is also adopted by Austria, while in Prussia it is fixed at 302 days. In England and America no fixed date is admitted, but while 280 days is admitted as the 'legitimum tempus pariendi,' each case in which legitimacy is questioned is to be decided on its own merits. At the early part of the century the question was much discussed by the leading obstetricians in connection with the celebrated Gardner peerage case, and a considerable difference of opinion existed among them. Since that time many apparently perfectly reliable cases have been recorded, in which the duration of gestation was obviously much beyond the average, and in which all sources of fallacy were carefully excluded.

Not to burden these pages with a number of cases, it may suffice to refer, as examples of protraction, to four well-known instances recorded by Simpson,[1] in which the pregnancy extended respectively to 336, 332, 319, and 324 days after the cessation of the last menstrual period. In these, as in all cases of protracted gestation, there is the possible source of error that impregnation may have occurred just before the expected advent of the next period. Making an allowance of 23 days in each instance for this, we even then have a number of days much above the average, viz. 313, 309, 296, and 301. Numerous instances as curious may be found scattered through obstetric literature. Indeed, the experience of most accoucheurs will parallel such cases, which may be more common than is generally supposed, inasmuch as they are only likely to attract attention when the husband has been separated from the wife beyond the average and expected duration of the pregnancy.

Reliable cases of protraction.

Possibly they are more common than is generally believed.

The evidence in favour of the possible prolongation of gestation is greatly strengthened by what is known to occur in the lower animals. In some of these, as in the cow and the mare, the precise period of insemination is known to a certainty, as only a single coitus is permitted. Many tables of this kind have been constructed, and it has been shown

Protraction common in the lower animals.

---

[1] *Obstet. Memoirs*, p. 84.

that there is in them a very considerable variation. In some cases in the cow it has been found that delivery took place 45 days, and in the mare 43 days, after the calculated date. Analogy would go strongly to show that what is known to a certainty to occur in the lower animals may also take place in the human female. The fact, indeed, is now very generally admitted ; but we are still unable to fix, with any degree of precision, on the extreme limit to which protraction is possible. Some practitioners have given cases in which, on data which they believe to be satisfactory, pregnancy has been extremely protracted ; thus Meigs and Adler record instances which they believed to have been prolonged to over a year in one case, and over fourteen months in the other. These are, however, so problematical that little weight can be attached to them. On the whole, it would hardly be safe to conclude that pregnancy can go more than three or four weeks beyond the average time. This conclusion is justified by the cases we possess in which pregnancy followed a single coitus, the longest of which was 295 days.

Evidence from size of child.

Dr. Duncan [1] is inclined to refuse credence to every case of supposed protraction unless the size and weight of the child are above the average, believing that lengthened gestation must of necessity cause increased growth of the child. This point requires further investigation, and it cannot be taken as proved that the fœtus necessarily must be large because it has been retained longer than usual in utero ; or, even if this be admitted, it may have been originally small, and so, at the end of the protracted gestation, be little above the average weight. There are, however, many cases which certainly prove that a prolonged pregnancy is at least often associated with an unusually developed fœtus. Dr. Duncan himself cites several, and a very interesting one is mentioned by Leishman in which delivery took place 295 days after a single coitus, the child weighing 12 lbs. 3 oz.

In some cases labour may commence and be arrested.

It seems possible that, in some cases of protracted pregnancy, labour actually came on at the average time, but, on account of faulty positions of the uterus or other obstructing cause, the pains were ineffective and ultimately died away, not recurring for a considerable time. Joulin relates some instances of this kind. In one of them the labour was

[1] *Fecundity and Fertility*, p. 348.

expected from the 20th to the 25th of October. He was summoned on the 23rd, and found the pains regular and active, but ineffective ; after lasting the whole of the 24th and 25th they died away, and delivery did not take place until November 25th, after the lapse of a month. In this instance the apparent cause of difficulty was extreme anterior obliquity of the uterus. A precisely similar case came under my own observation. The lady ceased to menstruate on March 16th, 1870. On December 12th, that is, on the 273rd day, strong labour pains came on, the os dilated to the size of a florin, and the membranes became tense and prominent with each pain. After lasting all night they gradually died away, and did not recur until January 12th, 304 days from the cessation of the last period. Here there was no assignable cause of obstruction, and the labour, when it did come on, was natural and easy.

The curious fact that, in both these cases, as in others of the same kind that are recorded, labour came on exactly a month after the previous ineffectual attempt at its establishment, affords, so far as it goes, an argument in favour of the view maintained by many that labour is apt to come on at what would have been a menstrual period.

From a forensic point of view it often becomes of importance to be able to give a reliable opinion as to the fact of delivery having occurred, and a few words may be here said as to the signs of recent delivery. Our opinion is only likely to be sought in cases in which the fact of delivery is denied, and in which we must, therefore, entirely rely on the results of a physical examination. If this be undertaken within the first fortnight after labour, a positive conclusion can be readily arrived at.

Signs of recent delivery.

At this time the abdominal walls will still be found loose and flaccid, and bearing very evident marks of extreme distension in the cracks and fissures of the cutis vera. These remain permanent for the rest of the patient's life, and may be safely assumed to be signs of an antecedent pregnancy, provided we can be certain that no other cause of extreme abdominal distension has existed, such as ascites, or ovarian tumour.

Within the first few days after delivery, the hard round ball formed by the contracted and empty uterus can easily

be felt by abdominal palpation, and more certainly by combined external and internal examination. The process of
involution, however, by which the uterus is reduced to its
normal size, is so rapid that after the first week it can no
longer be made out above the brim of the pelvis. In cases
in which an accurate diagnosis is of importance, the increased
length of the uterus can be ascertained by the uterine sound,
and its cavity will measure more than the normal $2\frac{1}{2}$ inches
for at least a month after delivery. It should not be forgotten that the uterine parietes are now undergoing fatty
degeneration, and that they are more than usually soft and
friable, so that the sound should be used with great caution,
and only when a positive opinion is essential. The state of
the cervix and of the vagina may afford useful information.
Immediately after delivery the cervix hangs loose and patulous in the vagina, but it rapidly contracts, and the internal
os is generally entirely closed after the eighth or tenth day.
The remainder of the cervix is longer in returning to its
normal shape and consistency. It is generally permanently
altered after delivery, the external os remaining fissured and
transverse, instead of circular with smooth margins, as in
virgins. The vagina is at first lax, swollen, and dilated, but
these signs rapidly disappear, and cannot be satisfactorily
made out after the first few days. The absence of the fourchette may be recognised, and is a persistent sign.

The presence of the lochia affords a valuable sign of recent
delivery. For the first few days they are sanguineous, and
contain numerous blood-corpuscles, epithelial scales, and the
débris of the decidua. After the fifth day they generally
change in colour, and become pale and greenish, and from
the eighth or ninth day till about a month after delivery
they have the appearance of thick opalescent mucus. They
have, however, a peculiar, heavy, sickening odour, which
should prevent their being mistaken for either menstruation
or leucorrhœal discharge.

The appearance of the breasts will also aid the decision,
for it is impossible for the patient to conceal the turgid
swollen condition of the mammæ, with the darkened areolæ,
and, above all, the presence of milk. If, on microscopic
examination, the milk is found to contain colostrum corpuscles, the fact of very recent delivery is certain. In women

who do not nurse it should be remembered that the secretion
of milk often rapidly disappears, so that its absence cannot
be taken as a sign that delivery has not taken place. On the
whole, there should be no difficulty in deciding that a woman
has been delivered, as some of the signs are persistent for
the rest of her life ; but it is not so easy, unless we see the
case within the first eight or ten days, to say how long it is
since labour took place.

CHAPTER VI.

ABNORMAL PREGNANCY, INCLUDING MULTIPLE PREGNANCY, SUPER-
FŒTATION, EXTRA-UTERINE FŒTATION, AND MISSED LABOUR.

Plural
births an
abnormal
variety of
preg-
nancy.

THE occurrence of more than one fœtus in utero is far from
uncommon, but there are circumstances connected with it
which justify the conclusion that plural births must not be
classified as natural forms of pregnancy. The reasons for
this statement have been well collected by Dr. Arthur
Mitchell,[1] who conclusively shows that not only is there a
direct increase of risk both to the mother and her offspring,
but that many abnormalities, such as idiocy, imbecility, and
bodily deformity, occur with much greater frequency in twins
than in single-born children. He concludes that ' the whole
history of twin births is exceptional, indicates imperfect development and feeble organisation in the product, and leads
us to regard twinning in the human species as a departure
from the physiological rule, and therefore injurious to all
concerned.'

Frequency
of multiple
births.

The frequency of multiple births varies considerably
under different circumstances. Taking the average of a
large number of cases collected by authors in various countries, we find that twin pregnancies occur about once in 87
labours; triplets once in 7,679. A certain number of quad-
ruple pregnancies, and some cases of early abortion in which
there were five fœtuses, are recorded, so that there can be no
doubt of the possibility of such occurrences; but they are so
extremely uncommon that they may be looked upon as rare
exceptions, the relative frequency of which can hardly be
determined.

The frequency of multiple pregnancy varies remarkably

[1] *Med. Times and Gaz.*, Nov. 1862.

in different races and countries. The following table [1] will show this at a glance :—

RELATIVE FREQUENCY OF MULTIPLE PREGNANCIES IN EUROPE.

Countries	Proportion of Twin to Single Births	Proportion of Triplets	Proportion of Quadruplets
England	1 : 116	1 : 6,720	—
Austria	1 : 94	...	—
Grand Duchy of Baden .	1 : 89	1 : 6,575	—
Scotland . . .	1 : 95	...	...
France	1 : 99	1 : 8,256	1 : 2,074,366
Ireland . . .	1 : 64	1 : 4,395	1 : 167,226
Mecklenburg-Schwerin .	1 : 68·9	1 : 6,436	1 : 183,236
Norway	1 : 81·62	1 : 5,442	—
Prussia	1 : 89	1 : 7,820	1 : 394,690
Russia	1 : 50·05	1 : 4,054	—
Saxony	1 : 79	1 : 1,000	1 : 400,000
Switzerland . . .	1 : 102	—	—
Würtemberg . . .	1 : 862	1 : 6,464	1 : 110,991

It will be seen that the largest proportion of multiple births occurs in Russia, and that the number of triple births is greatest where twin pregnancies are most frequent. Puech concludes that the number of multiple pregnancies is in direct proportion to the general fecundity of the inhabitants.

Dr. Duncan has deduced some interesting laws, with regard to the production of twins, from a large number of statistical observations ; [2] especially that the tendency to the production of twins increases as the age of the woman advances, and is greater in each succeeding pregnancy, exception being made for the first pregnancy, in which it is greater than in any other. Newly married women appear more likely to have twins the older they are. There can be no doubt that there is often a strong hereditary tendency in individual families to multiple births. A remarkable instance of this kind is recorded by Mr. Curgenven,[3] in which a woman had four twin pregnancies, her mother and aunt each one, and her grandmother two. Simpson mentions a case of quadruplets, consisting of three males and one female, who all survived, the female subsequently giving birth to triplets.[4]

[1] Puech, *Des Naissances Multiples.*

[2] *On Fecundity, Fertility, and Sterility*, p. 99.

[3] *Obst. Trans.* vol. xi.

[4] *Obst. Works*, p. 830.

Sex of children.   In the largest number of cases of twins the children are of opposite sexes, next most frequently there are two females, and twin males are the most uncommon. Thus, out of 59,178 labours, Simpson calculates that twin male and female occurred once in 199 labours, twin females once in 226, and twin males once in 258. The proportion of male to female births is also notably less in twin than in single pregnancies.

Size of fœtuses.   Twins, and *à fortiori* triplets, are almost always smaller and less perfectly developed than single children. Hence the chances of their survival are much less, and Clarke calculates the mortality amongst twin children as one out of thirteen. Of triplets, indeed, it is comparatively rare that all survive; while in quadruplets, premature labour and the death of the fœtuses are almost certain. It is a common observation that twins are often unequally developed at birth. By some this difference is attributed to one of them being of a different age to the other. It is probable, however, that in most of these cases the full development of one fœtus has been interfered with by pressure of the other. This is far from uncommonly carried to the extent of destroying one of the twins, which is expelled at term, mummified and flattened between the living child and the uterine wall. In other cases, when one fœtus dies it may be expelled without terminating the pregnancy, the other being retained in utero and born at term; and those who disbelieve in the possibility of superfœtation explain in this way the cases in which it is believed to have occurred.

Causes.   Multiple pregnancies depend on various causes. The most common is probably the simultaneous, or nearly simultaneous, maturation and rupture of two Graafian follicles, the ovules becoming impregnated at or about the same time. It by no means necessarily follows, even if more than one follicle should rupture at once, that both ovules should be impregnated. This is proved by the occurrence of cases in which there are two corpora lutea with only one fœtus. There are numerous facts to prove that ovules thrown off within a short time of each other may become separately impregnated, as in cases in which negro women have given birth to twins, one of which was pure negro, the other half-caste.

It may happen, however, that a single Graafian follicle contains more than one ovule, as has actually been observed before its rupture ; or, as is not uncommon in the egg of the fowl, an ovule may contain a double germ, each of which may give rise to a separate fœtus.

The various modes in which twins may originate explain satisfactorily the variations which are met with in the arrangement of the fœtal membranes, and in the form and connections of the placentæ. In a large proportion of cases, there are two distinct bags of membranes, the septum between them being composed of four layers, viz., the chorion and amnion of each ovum. The placentæ are also entirely separate. Here it is obvious that each twin is developed from a distinct ovum, having its own chorion and amnion. On arriving in the uterus it is probable that each ovum becomes fixed independently in the mucous membrane, and is surrounded by its own decidua reflexa. As growth advances the decidua reflexa generally atrophies from pressure, as it is not usual to find more than four layers of membrane in the septum separating the ova. In other cases there is only one chorion, within which are two distinct amnions, the septum then consisting of two layers only. Then the placentæ are generally in close apposition, and become fused into a single mass ; the cords, separately attached to each fœtus, not infrequently uniting shortly before reaching the placental mass, their vessels anastomosing freely. In other more rare instances both fœtuses are contained in a common amniotic sac ; but as the amnion is a purely fœtal membrane, it is probable that, when this arrangement is met with, the originally existing septum between the amniotic sacs has been destroyed. In both these latter cases the twins must have been developed from a single ovule containing a double germ, and Schroeder states that they are then always of the same sex. Dr. Brunton [1] has started a precisely opposite theory, and has tried to prove that twins of the same sex are contained in separate bags of membrane, while twins of opposite sexes have a common sac. He says that out of twenty-five cases coming under his observation, in fifteen the children contained in different sacs were of the same sex, but in the remaining ten, in which there was only one sac, they were of opposite

Arrangement of the fœtal membranes and placentæ.

[1] Obst. Trans. vol. x.

sexes. It is difficult to believe that there is not an error in these observations, since twins contained in a single amniotic sac do not occur nearly as often as ten times out of twenty-five cases, and no distinction is made between a common chorion with two amnions and a single chorion and amnion.

Double monsters. The facts of double monstrosity also disprove this view, since conjoined twins must of necessity arise from a single ovule with a double germ, and there is no instance on record in which they were of opposite sexes.

Membranes and placentæ in triplets. In triplets the membranes and placentæ may be all separate, or, as is commonly the case, there is one complete bag of membranes, and a second having a common chorion, with a double amnion. It is probable, therefore, that triplets are generally developed from two ovules, one of which contained a double germ.

Diagnosis of multiple pregnancy. It is comparatively seldom that twin pregnancy can be diagnosed before the birth of the first child, and, even when suspicion has arisen, its indications are very defective. There is generally an unusual size and an irregularity of shape of the uterus, sometimes even a distinct depression or sulcus between the two fœtuses. When such a sulcus exists it may be possible to make out parts of each fœtus by palpation on either side of the uterus. The only sign, however, on which the least reliance can be placed is the detection of two fœtal hearts. If two distinct pulsations are heard at different parts of the uterus; if, on carrying the stethoscope from one point to another, there is an interspace where pulsations are no longer audible, or when they become feeble, and again increase in clearness as the second point is reached; and, above all, if we are able to make out a difference in frequency between them, the diagnosis is tolerably safe. It must be remembered, however, that the sounds of a single heart may be heard over a larger space than usual, and hence a possible source of error. Twin pregnancy, moreover, may readily exist without the most careful auscultation enabling us to detect a double pulsation, especially if one child lie in the dorso-posterior position, when the body of the other may prevent the transmission of its heart's beat. The so-called placental souffle is generally too diffuse and irregular to be of any use in diagnosis, even when it is distinctly heard at separate parts of the uterus.

Closely connected with the subject of multiple pregnancies are the conditions known as *superfecundation* and *superfœtation*, regarding which there has been much controversy and difference of opinion.

By the former is meant the fecundation, at or near the same period of time, of two separate ovules before the decidua lining the uterus has been formed, which by many is supposed to form an insuperable obstacle to subsequent impregnation. The possibility of this occurrence has been incontestably proved by the class of cases already referred to, in which the same woman has given birth to twins bearing evident traces of being the offspring of fathers of different races.

By *superfœtation* is meant the impregnation of a second ovule when the uterus already contains an ovum which has arrived at a considerable degree of development. The cases which are supposed to prove the possibility of this occurrence are very numerous. They are those in which a woman is delivered simultaneously of fœtuses of very different ages, one bearing all the marks of having arrived at term, the other of prematurity; or of those in which a woman is delivered of an apparently mature child, and, after the lapse of a few months, of another equally mature. The possibility of superfœtation is strongly denied by many practitioners of eminence, and explanations are given, which doubtless seem to account satisfactorily for a large proportion of the supposed examples. In the former class of cases it is supposed, with much probability, that there is an ordinary twin pregnancy, the development of one fœtus being retarded by the presence in utero of another. That this is not an uncommon occurrence is certain, and the fact has already been alluded to in treating of twin pregnancy. In cases of the latter kind it is possible that some of them may be due to separate impregnation in a bilobed uterus, the contents of one division being thrown off a considerable time before those of the other. Numerous authentic examples of this occurrence are recorded, but by far the most remarkable is that related by Dr. Ross, of Brighton, which has been already referred to (p. 45). In this case the patient had previously given birth to many children without any suspicion of her abnormal formation having arisen, and, had it not been detected by Dr. Ross, the case might fairly enough have been claimed as an indubitable example of superfœtation.

Some cases seem inexplicable except on the hypothesis of superfœtation. Making every allowance for these explanations, there remain a considerable number of cases which it is very difficult to account for, except on the supposition that the second child has been conceived a considerable time after the first. Those interested in the subject will find a large number of examples collected in a valuable paper by Dr. Bonnar, of Cupar.[1] He has adopted the ingenious plan of consulting the records of the British peerage, where the exact date of the birth of successive children of peers is given, without, of course, any reasonable possibility of error, and he has collected numerous examples of births rapidly succeeding each other which are apparently inexplicable on any other theory. In one case he cites, a child was born September 12, 1849, and the mother gave birth to another on January 24, 1850, after an interval of only 127 days. Subtracting from that 14 days, which Dr. Bonnar assumes to be the earliest possible period at which a fresh impregnation can occur after delivery, we reduce the gestation to 113 days, that is, to less than four calendar months. As both these children survived, the second child could not possibly have been the result of a fresh impregnation after the birth of the first; nor could the first child have been a twin prematurely delivered, for if so, it must have only reached rather more than the fifth month, at which time its survival would have been impossible.

Besides the numerous examples of cases of this kind recorded in most obstetric works, there are one or two of miscarriage in the early months, in which, in addition to a fœtus of four or five months' growth, a perfectly fresh ovum of not more than a month's development was thrown off. One such case was shown at the Obstetrical Society in 1862, which was reported on by Drs. Harley and Tanner, who stated that in their opinion it was an example of superfœtation. A still more conclusive case is recorded by Tyler Smith.[2] 'A young married woman, pregnant for the first time, miscarried at the end of the fifth month, and some hours afterwards a small clot was discharged, inclosing a perfectly healthy ovum of about one month. There were no signs of a double uterus

---

[1] *Edin. Med. Journ.* 1864-65.
[2] *Manual of Obstetrics*, p. 112.

in this case. The patient had menstruated regularly during
the time she had been pregnant.' This case is of special
interest from the fact of the patient having menstruated
during pregnancy—a circumstance only explicable on the
same anatomical grounds which render superfœtation possible. So far as I know, it is the only instance in which the
coincidence of superfœtation and menstruation during early
pregnancy has been observed.

The objections to the possibility of superfœtation are
based on the assumptions that the decidua so completely fills
up the uterine cavity that the passage of the spermatozoa
is impossible; that their passage is prevented by the mucous
plug which blocks up the cervix; and that when impregnation has taken place ovulation is suspended. It is, however,
certain that none of these are insuperable obstacles to a
second impregnation. The first was originally based on the
older and erroneous view which considered the decidua to be
an exudation lining the entire uterine cavity, and sealing up
the mouths of the Fallopian tubes and the aperture of the
internal os uteri. The decidua reflexa, however, does not
come into apposition with the decidua vera until about the
eighth week of pregnancy, and, therefore, until that time
there is a free space between the two membranes through
which the spermatozoa might pass to the open mouths of the
Fallopian tube, and in which a newly impregnated ovule
might graft itself. A reference to the accompanying figure
of a pregnancy in the third month, copied from Coste's work,
will readily show that, as far as the decidua is concerned,
there is no mechanical obstacle to the descent and lodgment
of another impregnated ovule (fig. 79). Then, as regards
the plug of mucus, it is pretty certain that this is in no way
different from the mucus filling the cervix in the non-
pregnant state, which offers no obstacle at all to the passage
of the spermatozoa. Lastly, respecting the cessation of
ovulation during pregnancy, this, no doubt, is the rule, and
probably satisfactorily explains the rarity of superfœtation.
There are, however, a sufficient number of authenticated
cases of menstruation during pregnancy to prove that ovulation is not always absolutely in abeyance; and, as long as it
occurs, there is unquestionably no positive mechanical obstruction, at least in the early months of pregnancy, in the

Objections to the admission of super-fœtation.

None of them seem an insur-mountable obstacle to its oc-currence.

The possibility of superfœtation must therefore be admitted.

way of the impregnation and lodgment of the ovules that are thrown off. The reasonable conclusion, therefore, seems to be that, although a large majority of the supposed cases are

Fig. 79.

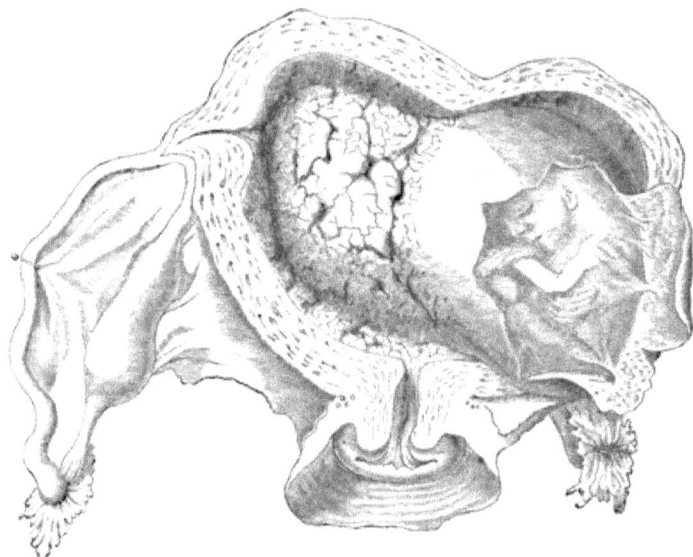

ILLUSTRATING THE CAVITY BETWEEN THE DECIDUA VERA AND THE DECIDUA REFLEXA DURING THE EARLY MONTHS OF PREGNANCY. (After Coste.)

explicable in other ways, it cannot be admitted that super-fœtation is either physiologically or mechanically impossible.

Extra-uterine pregnancy.

The most important of the abnormal varieties of pregnancy, if we consider the serious and very generally fatal results attending it, is the so-called *extra-uterine fœtation*, which consists in the arrest and development of the ovum outside the cavity of the uterus. Of late years this subject has received much well-merited attention, which, it is to be hoped, may lead to the establishment of some definite rules for the management of this most anxious and dangerous class of cases.

Site of extra-uterine pregnancy.

The ovum may be arrested and developed in various situations on its way to the uterus, most commonly in some part of the Fallopian tube, or it may be in the cavity of the abdomen, or even quite beyond it, as in a few rare cases in which the ovum has found its way into a hernial sac.

Extra-uterine gestation may be subdivided into the fol- Classifica-
tion.
lowing classes: 1st, and most common of all, *tubal* gestation, and as varieties of this, although by some made into
distinct classes, (*a*) *interstitial* and (*b*) *tubo-ovarian* gestation. In the former of these subdivisions the ovum is arrested in the part of the Fallopian tube that is situated in
the substance of the uterine parietes; in the latter, at or
near the fimbriated extremity of the tube—so that part of
its cyst is formed by the tube and part by the ovary. 2nd.
Abdominal gestation, in which an ovum, instead of finding
its way into the tube, falls into the peritoneal cavity, and
there becomes attached and developed; or the so-called
secondary abdominal gestation, in which an extra-uterine
pregnancy, originally tubal, becomes ventral, through rupture of its cysts and escape of its contents into the abdominal
cavity. 3rd. *Ovarian* gestation, the existence of which is Doubts as
to the ex-
istence of
denied by many writers of eminence, such as Velpeau and istence of
Arthur Farre, while it is maintained by others of equal cele- ovarian
brity, such as Kiwisch, Coste, and Hecker.   It must be ad- preg-
nancy.
mitted that it is extremely difficult to understand how an
ovarian pregnancy, in the strict sense of the word, can occur,
for it implies that the ovule has become impregnated before
the laceration of the Graafian follicle, through the coats of
which the spermatozoa must have passed. Coste, indeed, believes that this frequently happens; but, while spermatozoa
have been detected on the surface of the ovary, their penetration into the Graafian follicle has never been demonstrated.
Farre has also clearly shown that in many cases of supposed
ovarian pregnancy the surrounding structures were so altered
that it was impossible to trace their exact origin, and to say
to a certainty that the fœtus was really within the substance
of the ovary. Kiwisch gives a reasonable explanation of Possible
explana-
these cases by supposing that sometimes the Graafian fol- tion of
licle may rupture, but that the ovule may remain within it some
without being discharged. Through the rent in the walls cases.
of the follicle the spermatozoa may reach and impreg-
nate the ovule, which may develop in the situation in
which it has been detained. The subject has been recently
ably considered by Puech,[1] who admits two varieties of
ovarian pregnancy, according as the fœtus has developed in a

---

[1] *Annal. de Gynéc.* July 1878.

vesicle which has remained open or in one which has closed immediately after fecundation. He considers that most cases of so-called ovarian pregnancy are either dermoid cysts, ovario-tubal pregnancies, or abdominal pregnancies in which the placenta is attached to the ovary, and that even in the rare cases of true ovarian pregnancies the progress and results do not differ from that of abdominal pregnancy. While, therefore, it is impossible to deny the existence of ovarian pregnancy, it must be considered to be a very rare and exceptional variety, which, as far as treatment and results are concerned, does not differ from tubular or abdominal gestation. 4th. There are two rare varieties in which an ovum is developed either in the supplementary horn of a *bi-lobed uterus* or in a *hernial* sac.

For the sake of clearness, we may place these varieties of extra-uterine gestation in the following tabular form :—

  1st. *Tubal*—
    (*a*) Interstitial, (*b*) Tubo-ovarian.
  2nd. *Abdominal*—
    (*a*) Primary, (*b*) secondary.
  3rd. *Ovarian.*
  4th. In *bi-lobed uterus, hernial*, &c.

Causes. — The etiology of extra-uterine foetation in any individual case must necessarily be almost always obscure. Broadly speaking, it may be said that extra-uterine foetation may be produced by any condition which prevents or renders difficult the passage of the ovule to the uterus, while it does not prevent the access of the spermatozoa to the ovule. Thus inflammatory thickening of the coats of the Fallopian tubes by lessening their calibre, but not sufficiently so to prevent the passage of the spermatozoa, may interfere with the movements of the tube which propel the ovum forward, and so cause its arrest. A similar effect may be produced by various morbid conditions, such as inflammatory adhesions, from old-standing peritonitis, pressing on the tube ; obstruction of its calibre by inspissated mucus or small polypoid growths ; the pressure of uterine or other tumours, and the like. The fact that extra-uterine pregnancies occur most frequently in multi-paræ, and comparatively rarely in women under thirty years of age, tends to show that these conditions, which are clearly more likely to be met with in such women than in

Most common in multi-paræ.

young primiparæ, have considerable influence in its causation.
A curiously large proportion of cases occur in women who
have either been previously altogether sterile, or in whom a
long interval of time has elapsed since their last pregnancy.
The disturbing effects of fright, either during coition or a
few days afterwards, have been insisted on by many authors
as a possible cause. Numerous cases of this kind are recorded; and, although the influence of emotion in the production of this condition is not susceptible of proof, it is not
difficult to imagine that spasms of the Fallopian tubes might
be produced in this way, which would either interfere with
the passage of the ovum, or direct it into the abdominal
cavity. The occurrence of abdominal pregnancy is probably
less difficult to account for if we admit, with Coste, that the
ovule becomes impregnated on the surface of the ovary
itself, for there must be very many conditions which prevent
the proper adaptation of the fimbriated extremity of the tube
to the surface of the ovary, and failing this the ovum must
of necessity drop into the abdominal cavity. Kiwisch has
pointed out that this is particularly apt to occur when the
Graafian follicle develops on the posterior surface of the
ovary; and, indeed, it is probable that it may be of common
occurrence, and that the comparative rarity of abdominal
pregnancy is due to the difficulty with which the impregnated ovule engrafts itself on the surrounding viscera. Impregnation may actually occur in the abdominal cavity
itself, of which Keller [1] relates a remarkable instance. In
this case Koeberlé had removed the body of the uterus and
part of the cervix, leaving the ovaries. In the portion of
the cervix that remained there was a fistulous aperture
opening into the abdominal cavity, through which semen
passed and produced an abdominal gestation. Several
curious cases are also recorded, which have given rise to a
good deal of discussion, in which a tubal pregnancy existed
while the corpus luteum was on the opposite side (fig. 80).
The most probable explanation, however, is that the fimbriated extremity of the tube in which the ovum was found
had twisted across the abdominal cavity and grasped the
opposite ovary, in this way, perhaps, producing a flexion
which impeded the progress of the ovum it had received

Marginal notes: Causes producing abdominal pregnancy.

Cases in which the corpus luteum is in the ovary opposite to a tubal pregnancy.

---

[1] *Des Grossesses Extra-utérines*, Paris, 1872.

into its canal.   Tyler Smith suggested that such cases might
be explained by supposing that the ovum, after reaching the
uterus, failed to graft itself in the mucous membrane, but

Fig. 80.

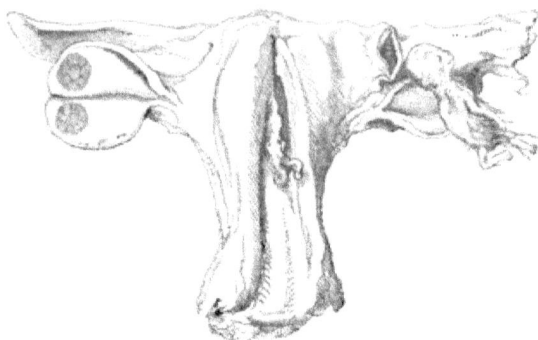

TUBAL PREGNANCY, WITH THE CORPUS LUTEUM IN THE OVARY OF THE OPPOSITE SIDE.

The decidua is represented in the process of detachment from the uterine cavity.

found its way into the opposite Fallopian tube.   Kussmaul[1]
thinks that such a passage of the ovum across the uterine
cavity may be caused by muscular contraction of the uterus,
occurring shortly after conception, squeezing the yet free
ovum upwards towards the opening of the opposite tube, and
possibly into the tube itself.

The history and progress of cases of extra-uterine pregnancy are materially different according to their site, and,
for practical purposes, we may consider them as forming
two great classes, the tubal (with its varieties), and the
abdominal.

Tubal pregnancies.   When the ovum is arrested in any part of the Fallopian
tube the chorion soon commences to develop villi, just as
in ordinary pregnancy, which engraft themselves into the
mucous lining of the tube, and fix the ovum in its new

Changes in the Fallopian tube.   position.   The mucous membrane becomes hypertrophied,
much in the same way as that of the uterus under similar
circumstances, so that it becomes developed into a sort of
pseudo-decidua.   Inasmuch, however, as the mucous coat of
the tubes is not furnished with tubular glands, a true decidua
can scarcely be said to exist ; nor is there any growth of
membrane around the ovum analogous to the decidua reflexa.

[1] *Mon. f. Geburt*, Oct. 1862.

The ovum is, therefore, comparatively speaking, loosely attached to its abnormal situation, and hence hæmorrhage from laceration of the chorion villi can very readily take place.

It is seldom that any development of the chorion villi into distinct placental structure is observed; this is probably owing to the fact that laceration and death generally occur before the period at which the placenta is normally formed. The muscular coat of the tube soon becomes hypertrophied, and as the size of the ovum increases the fibres are separated from each other, so that the ovum protrudes at certain points through them, and at these it is only covered by the stretched and attenuated mucous and peritoneal coats of the tube. At this time the tubal pregnancy forms a smooth oval tumour, which, as a rule, has not formed any adhesions to

Fig. 81.

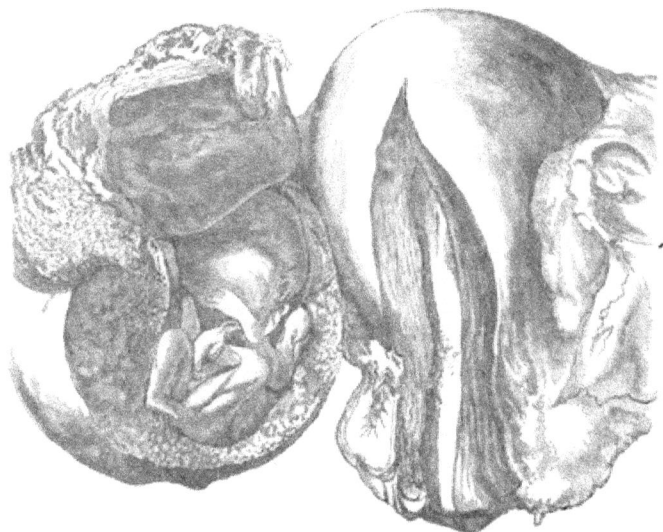

TUBAL PREGNANCY. (From a specimen in the Museum of King's College.)

the surrounding structures (fig. 81). The part of the tube unoccupied by the ovum may be found unaltered, and permeable in both directions; or, more frequently, it becomes so stretched and altered that its canal cannot be detected. Most frequently it is that part of the tube nearest the uterus which cannot be made out. The condition of the uterus in this, as in other forms of extra-uterine pregnancy, has been

Condition of the uterus.

the subject of considerable discussion. It is now universally admitted that the uterus undergoes a certain amount of sympathetic engorgement, the cervix becomes softened, as in natural pregnancy, and the mucous membrane develops into a true decidua. In many cases the decidua is found on post-mortem examination, in others it is not; and hence the doubts that some have expressed as to its existence. The most reasonable explanation of its absence is that given by Duguet,[1] who has shown that it is far from uncommon for the uterine decidua to be thrown off *en masse* during the hæmorrhagic discharges which so frequently precede the fatal issue of extra-uterine gestation.

Interstitial and false-ovarian pregnancy.

When the ovum is arrested in that portion of the tube passing through the uterus, in so-called interstitial pregnancy, the muscular fibres of the uterus become stretched and distended, and form the outer covering of the ovum. When, on the other hand, the site of arrest is in the fimbriated extremity of the tube, the containing cyst is formed partly of the fimbriæ of the tube, partly of ovarian tissue ; hence it is much more distensible, and the pregnancy may continue without laceration to a more advanced period, or

Progress and termination.

even to term, so that when the ovum is placed in this situation, the case much more nearly resembles one of abdominal pregnancy.

Period at which rupture occurs.

The termination of tubal pregnancy, in the immense majority of cases, is death, produced by laceration giving rise either to internal hæmorrhage or to subsequent intense peritonitis. Rupture usually occurs at an early period of pregnancy, most generally from the fourth to the twelfth week, rarely later. However, a few instances are recorded in which it did not take place until the fourth or fifth month, and Saxtorph and Spiegelberg have recorded apparently authentic cases in which the pregnancy advanced to term without laceration. It is generally effected by distension of the tube, which at last yields at the point which is most stretched ; and sometimes it seems to be hastened or determined by accidental circumstances, such as a blow or fall, or the excitement of sexual intercourse.

Symptoms of rupture.

The symptoms accompanying rupture are those of intense collapse, often associated with severe abdominal pain,

[1] *Annales de Gynécologie*. May 1874.

produced by the laceration of the cyst. The patient will be
found deadly pale, with a small, thready, and almost imper-
ceptible pulse, perhaps vomiting, but with mental faculties
clear. If the hæmorrhage be considerable, she may die ⟨Collapse
without any attempt at reaction. Sometimes, however— from
and this generally occurs in cases in which the tube tears, hæmor-
the ovum remaining intact—the hæmorrhage may cease on rhage.
account of the ovum protruding through the aperture and

Fig. 82.

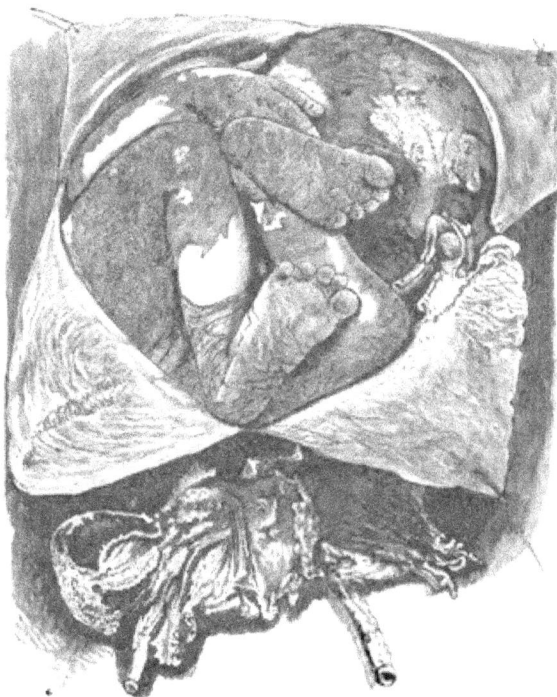

EXTRA-UTERINE PREGNANCY AT TERM OF THE TUBO-OVARIAN VARIETY. (After a Case of
Dr. A. Sibley Campbell's.)

acting as a plug. The patient may then imperfectly rally,
to be again prostrated by a second escape of blood, which
proves fatal. If the loss of blood is not of itself sufficient to
cause death from shock and anæmia, the fatal issue is generally only postponed, for the effused blood soon sets up a ⟨Secondary
violent general peritonitis, which rapidly carries off the peritoni-
patient. If she should survive the second danger, the case tis.
is transformed into one of abdominal pregnancy, the fœtus

becoming surrounded by a capsule produced by inflammatory exudation (fig. 82). The case is then subjected to the rules of treatment presently to be discussed when considering that variety of extra-uterine gestation.

Diagnosis.   The possibility of diagnosing tubal gestation before rupture occurs is a question of great and increasing interest, from the fact that, could its existence be ascertained, we might very fairly hope to avert the almost certainly fatal issue which is awaiting the patient. Unfortunately, the symptoms of tubal pregnancy are always obscure, and too often death occurs without the slightest suspicion as to the nature of the case having arisen. In the first place, it is to

Sympathetic disturbances of pregnancy are present.   be observed that all the usual sympathetic disturbances of pregnancy exist ; the breasts enlarge, the areolæ darken, and morning sickness is present. There is also an arrest of menstruation ; but, after the absence of one or more periods, there is often an irregular hæmorrhagic discharge. This is an important symptom, the value of which in indicating the existence of tubal pregnancy has of late years been much dwelt upon by various authors, both in this country and abroad. Barnes attributes it to partial detachment of the chorion villi, produced by the ovum growing out of proportion to the tube in which it is contained. Whether this is

Irregular metrorrhagia.   the correct explanation or not, it is a fact that irregular hæmorrhage very generally precedes the laceration for several days or more. Accompanying this hæmorrhage there is almost always more or less abdominal pain, produced by the stretching of the tissues in which the ovum is placed, and this is sometimes described as being of very intense and crampy character. If, then, we meet with a case in which the symptoms of early pregnancy exist, in which there are irregular losses of blood, possibly discharge of membranous

Abdominal pains.   shreds, and abdominal pain, a careful examination should be insisted on, and then the true nature of the case may possibly be ascertained. Should extra-uterine fœtation exist, we should expect to find the uterus somewhat enlarged, and the cervix softened, as in early pregnancy, but both these

Results of physical examination.   changes are doubtless generally less marked than in normal pregnancy. This fact, of itself, however, is of little diagnostic value, for slight differences of this kind must always be too indefinite to justify a positive opinion.

The existence of a peri-uterine tumour, rounded or oval *Presence of a peri-uterine tumour.* in outline, and producing more or less displacement of the uterus, in the direction opposite to that in which the tumour is situated, may point to the existence of tubular fœtation. By bi-manual examination, one hand depressing the abdominal wall, while the examining finger of the other acts in concert with it either through the vagina or rectum, the size and relations of the growth may be made out. There are various conditions, which give rise to very similar physical signs, such as small ovarian or fibroid growths, or the effusion of blood around the uterus ; and the differential diagnosis must always be very difficult, and often impossible. A curious example of the difficulty of diagnosis is recorded *Extreme uncertainty of diagnosis.* by Joulin, in which Huguier and six or seven of the most skilled obstetricians of Paris agreed on the existence of extra-uterine pregnancy, and had, in consultation, sanctioned an operation, when the case terminated by abortion, and proved to be a natural pregnancy. The use of the uterine sound, which might aid in clearing up the case, is necessarily contra-indicated unless uterine gestation is certainly disproved. Hence it must be admitted that positive diagnosis must always be very difficult. So that the most we can say is, that when the general signs of early pregnancy are present, associated with the other symptoms and signs alluded to, the suspicion of tubal pregnancy may be sufficiently strong to justify us in taking such action as may possibly spare the patient the necessarily fatal consequence of rupture.

If the diagnosis were quite certain, the removal of the *Treatment.* entire Fallopian tube and its contents by abdominal section would be quite justifiable, and probably would neither be more difficult, nor more dangerous, than ovariotomy ; for, at this stage of extra-uterine fœtation, there are no adhesions to complicate the operation. As yet, however, the uncertainty of the diagnosis has prevented the adoption of the practice.

Dr. Thomas, of New York,[1] has recently recorded a most instructive case, in which he saved the life of the patient by a bold and judicious operation. The nature of the case was rendered pretty evident by the signs above described, and

[1] *New York Med. Jour.* June 1875.

Opening of the sac by the galvano-caustic knife.

Thomas opened the cyst from the vagina by a platinum knife, rendered incandescent by a galvano-caustic battery, by which means he hoped to prevent hæmorrhage. Through the opening thus made he removed the fœtus. In subsequently attempting to remove the placenta very violent hæmorrhage took place, which was only arrested by injecting the cyst with a solution of persulphate of iron. The remains of the placenta subsequently came away piecemeal, after an attack of septicæmia, which was kept within bounds by freely washing out the cyst with antiseptic lotion, the patient eventually recovering. If I might venture to make a criticism on a case followed by so brilliant a success, it would be that, in another instance of this kind, it would be safer to follow the rule so strictly laid down with regard to gastrotomy in abdominal pregnancies, and leave the placenta untouched, trusting to the injection of antiseptics, and the thorough drainage of the cyst, to prevent mischief.

Means of destroying the vitality of the fœtus.

Another mode of managing these cases is to destroy the fœtus, so as to check its further growth, in the hope that it may remain inert and passive within its sac. Various operations have been suggested and practised for this purpose. Thus needles have been introduced into the tumour, through which currents of electricity have been passed, either the continuous current, or, as has been suggested by Duchenne, a spark of Franklinic electricity. Hicks, Allen, and others have endeavoured to destroy the fœtus by passing an electro-magnetic current through it by means of a needle. Lusk[1] relates several successful cases following the use of the Faradaic current, one pole being passed through the rectum to the site of the ovum, the other being placed on a point in the abdominal wall two or three inches above Poupart's ligament. The current should be passed daily for five or ten minutes, and continued for a week or two until the shrinking of the tumour gives satisfactory evidence of the death of the fœtus. In a case reported by Dr. Bachetti, in which the continuous current was used, the growth of the ovum was arrested, and the patient recovered. The same result, however, would probably have followed the simple puncture of the cyst. This has been successfully practised on several occasions, either with a small trocar and canula, or with a

[1] Science and Art of Midwifery, p. 321.

simple needle. A very interesting case, in which the development of a two months' tubal gestation was arrested in this way, is recorded by Greenhalgh,[1] and another by Martin of Berlin.[2] Joulin suggested that not only should the cyst be punctured, but that a solution of morphia should be injected into it, which, by its toxic influence, would insure the destruction of the fœtus; and this is probably one of the best means at our disposal for destroying the fœtus. Other means proposed for effecting the same object, such as pressure, or the administration of toxic remedies by the mouth, are far too uncertain to be relied on. The simplest and most effectual plan would be to introduce the needle of an aspirator, by which the liquor amnii would be drawn off, and the further growth of the fœtus effectually prevented. Parry,[3] indeed, is opposed to this practice, and has collected several cases in which the puncture of the cyst was followed by fatal results, either from hæmorrhage or septicæmia. In these, however, an ordinary trocar and canula were probably employed, which would necessarily admit air into the sac. It is difficult to imagine that a fine hair-like aspirating needle, rendered perfectly aseptic by carbolic acid, could have any injurious results; and it could do no harm, even if an error of diagnosis had been made, and the suspected extra-uterine fœtation turned out to be some other sort of growth. If the aspirator proves that an extra-uterine fœtation exists, then, if the cyst be of any considerable size, and the pregnancy advanced beyond the second month, we might, if deemed advisable, resort to a more radical operation, such as that so successfully practised by Thomas.

When the chance of arresting the growth of a tubular fœtation has never arisen, and we first recognise its existence after laceration has occurred, and the patient is collapsed from hæmorrhage, what course are we to pursue? Hitherto all that ever has been done is to attempt to rally the patient by stimulants, and, in the unlikely event of her surviving the immediate effects of laceration, endeavouring to control the subsequent peritonitis, in the hope that the effused blood may become absorbed, as in pelvic hæmatocele, This is, indeed, a frail reed to rest upon, and when laceration of a

Treatment when rupture has occurred.

[1] Lancet. 1867.   [2] Monat. f. Geburt. 1868.
[3] Parry on Extra-uterine Pregnancy, p. 204.

tubal gestation, advanced beyond a month, has occurred, death has been the most certain result. It is supposed by Bernutz, and his opinion is shared by Barnes, that rupture which does not prove fatal is probably not very rare in the first few days of extra-uterine gestation, and that it is not an uncommon cause of certain forms of pelvic hæmatocele. It has more than once been suggested that it would be perfectly justifiable when laceration has occurred to perform gastrotomy, to sponge away the effused blood, and to place a ligature around the lacerated tube and remove it, with its contents.

Question of gastro-tomy. This would no doubt be a bold and heroic procedure, but no one who is acquainted with the triumphs of modern abdominal surgery can say that it would be either impossible or hopeless. The sponging out of effused blood from the abdominal cavity is an every-day procedure in ovariotomy, nor is there any apparent difficulty in ligaturing and removing the sac of the extra-uterine pregnancy, for, as a rule, there are no adhesions formed to the surrounding parts. The history of these cases shows that death does not generally follow rupture for some hours, so that there would be usually time for the operation, and the extreme prostration might be, perhaps, temporarily counteracted by transfusion. Pressure on the abdominal aorta, resorted to when the patient is first seen, might possibly be employed with advantage to check further hæmorrhage, until the question of operation is decided. We must remember that the alternative is death, and hence any operation which would afford the slightest hope of success would be perfectly justifiable. I cannot, therefore, agree with those who hold that because the chances of success are so small, the operation should not be tried; and I do not doubt that it will yet fall to the lot of some one by this means, to snatch a patient from the jaws of death, and still further to extend the successes of abdominal surgery.

Abdominal pregnancy. In the second of the two classes into which, for practical convenience, we have divided extra-uterine gestation, the ovum is developed in the abdominal cavity. It is as yet an open question whether in some cases the pregnancy is primarily abdominal or not. Barnes believes that it probably never is so, on account of the difficulty of admitting that so minute a body as the ovum should be able to fix itself on the smooth peritoneal surface. He therefore thinks that all abdominal pregnancies are primarily either tubal or ovarian, By some believed not to exist as a primary condition. the sac in which they were contained having given way, and the ovum having retained its vitality through partial attachment to the original sac. This theory is opposed to that of the majority of writers, and, although it may perhaps render the facts less difficult to understand, it is purely hypothetical. There is no evidence to show that in most cases there is an early laceration of a tubal or ovarian sac. That the chorion villi do graft themselves upon the surrounding peritoneum is certain, and is observed in all cases of abdominal gestation. It is not more difficult to imagine them doing this from their very first development than a little later; for it must be allowed that if such laceration does occur, in most cases it can only be when pregnancy is very slightly advanced. On the whole, therefore, it seems not unreasonable to admit the usual explanation of these cases, that the ovule, already impregnated, escaped the grasp of the Fallopian tube, and fell into the abdominal cavity, where it rooted itself and developed. Some have, indeed, supposed that abdominal pregnancy may occasionally arise in consequence of spermatozoa finding their way into the peritoneal cavity, and there meeting and impregnating an ovule discharged from the Graafian follicle. Such an event one would suppose to be almost impossible, but Koeberlé's case, already quoted, proves that it has actually occurred. The probability is that it is by no means rare for impregnated ovules to drop into the peritoneal cavity, and that the majority of those that do so perish without doing any harm. When they do survive, however, the chorion Attachment of the ovum to surrounding structures. villi sprout, attach themselves to the surrounding structures, and eventually develop into a placenta. The mode in which the chorion villi are attached, and the arrangement of the maternal bloodvessels, have never yet been worked out, and would form a very interesting subject for investigation. The precise seat of attachment varies, and the placenta has been found fixed to most of the abdominal viscera, either those contained in the pelvis proper, or it may be the intestines, or to the iliac fossa; most frequently, apparently, the ovum finds its way into the retro-uterine cul-de-sac. .

The subsequent changes vary much. In the large Formation of a cyst round the ovum. majority of cases the ovum produces considerable irritation, resulting in the exudation of plastic material, which is thrown round it, so as to form a secondary cyst or capsule, in which maternal vessels are largely developed, and which stretches, *pari passu*, with the growth of the ovum (fig. 83).

Fig. 83.

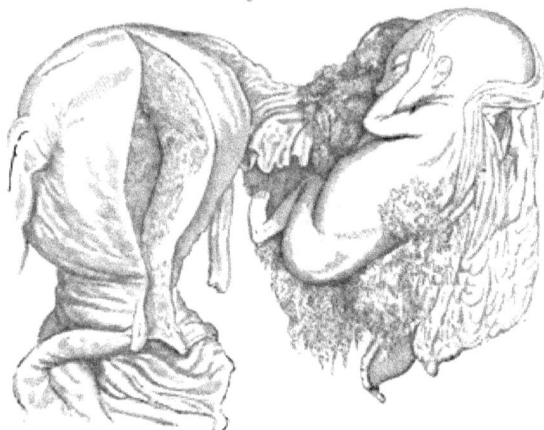

UTERUS AND FŒTUS IN A CASE OF ABDOMINAL PREGNANCY.

The density and strength of this cyst are found to be very different in different cases; sometimes it forms a complete and strong covering to the ovum, at others it is very thin and only partially developed, but it is rarely entirely absent. As there is ample space for the development of the ovum, and as the secondary cyst generally stretches and grows along with it, most cases of abdominal pregnancy progress without any very remarkable symptoms, beyond occasional severe attacks of pain, until the full term of pregnancy has been reached. Sometimes, however, the cyst lacerates, and there is an escape of blood into the abdominal cavity, accompanied by more or less prostration and collapse, which may prove fatal, but from which the patient more generally rallies. The fœtus, now dead, will remain in the abdomen, and will undergo changes and produce results similar to those which we shall presently describe as occurring in cases progressing to the full period.

Occasional rupture of the cyst.

Pseudo-labour sometimes comes on.

In most cases, at the natural termination of pregnancy a strange series of phenomena occur; pseudo-labour comes on, there are more or less frequent and strong uterine contractions, possibly an escape of blood from the vagina, the discharge of the broken-down uterine decidua, and even the establishment of lactation. Sometimes the contractions of the abdominal muscles produced by this ineffective labour have been so strong as to cause the laceration of the adventitious cyst surrounding the fœtus, and the escape of blood and liquor amnii into the abdominal cavity, with a rapidly fatal result. More frequently laceration does not occur, and the spurious labour pains continue at intervals, until the Death of fœtus dies, possibly from pressure, but more often from the fœtus. effusion of blood into the tissue of the placenta, and consequent asphyxia. Occasionally the fœtus has apparently lived a considerable time, in some cases even for several months, after the natural limit of pregnancy has been reached.

It is after the death of the fœtus that the dangers of Changes abdominal pregnancy generally commence, and they are after the death of numerous and various. The subsequent changes that occur the fœtus. are well worthy of study. Occasionally the fœtus has been retained for a length of time, even until the end of a long life, without producing any serious discomfort, and in many cases of this kind several normal pregnancies and deliveries have subsequently taken place. Even when the extra-uterine Patient is gestation appears to be tolerated, and has remained for long always subjected without producing any bad effects, serious symptoms may be to risks as suddenly developed; so that no woman, under such circum-long as the fœtus is stances, can be considered safe. The condition of these re-retained. tained fœtuses varies much. Most commonly the liquor amnii Changes undergone is absorbed, the fœtus shrinks and dies, all its soft structures by the are changed into adipocere, and the bones only remain un-retained fœtus. altered. Sometimes this change occurs with great rapidity. I have elsewhere [1] recorded a case of extra-uterine fœtation in which at the full term of pregnancy the fœtus was alive, and the woman died in less than a year afterwards. On postmortem the fœtus was found entirely transformed into a greasy mass of adipocere, studded with fœtal bones, in which not a trace of any of the soft parts could be detected. On the other hand, the fœtus may remain unchanged; in the Museum of the College of Surgeons there is one which was retained in the abdomen for fifty-two years, and which was found to be as fresh and unaltered as a new-born child. In other cases the sac and its contents atrophy and shrink, and

[1] *Obst. Trans.* vol. vii.

calcareous matter is deposited in them, so that the whole becomes converted into a solid mass known as *lithopædion* (fig. 84). The cases, however, in which the retention of the fœtus gives rise to no mischief are quite exceptional. Generally the fœtus putrefies, and this may either immediately cause fatal peritonitis or septicæmia, or, as more commonly happens, secondary inflammation and suppuration of the sac. Under the influence of the latter the sac opens externally, either directly at some point of the abdominal walls, or indirectly through the vagina, the bowels, or even the bladder.

In most cases the fœtus is discharged piecemeal. Through the aperture or apertures thus formed (for there are often several fistulous openings), pus, and the bones and other parts of the broken-down fœtus are discharged; and this may go on for months, and even years, until at last, if the patient's strength does not give way, the whole contents of the cyst are expelled, and recovery takes place. From various statistical observations it appears that the chances of recovery are best when the cyst opens through the abdominal walls, next through the vagina or bladder, and that the fœtus is discharged with most difficulty and danger when the aperture is formed into the bowel. At the best, however, the process is long, tedious, and full of danger; and the patient too often sinks, during the attempt at expulsion, through the irritation and exhaustion produced by the abundant and long-continued discharge.

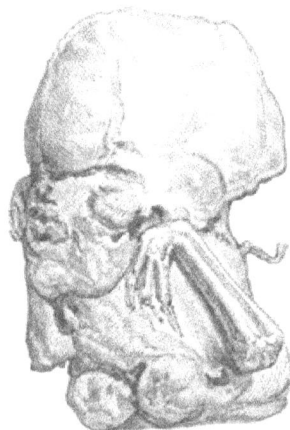

Fig. 84.

LITHOPÆDION.

(From a preparation in the Museum of the College of Surgeons.

Diagnosis. The diagnosis of abdominal gestation is by no means so easy as might be thought, and the most experienced practitioners have been mistaken with regard to it.

The most characteristic symptom, although this is not so common as in tubal gestation, is metrorrhagia combined with the general signs of pregnancy. Very severe and frequently

repeated attacks of abdominal pain are rarely absent, and
should at once cause suspicion, especially if associated with
hæmorrhage, and the discharge of a decidual membrane
from the uterus. They are supposed by some to depend on
intercurrent attacks of peritonitis, by which the fœtal cyst is
formed. Parry doubts this explanation, and attributes them
partly to the distension of the cyst by the growing fœtus,
and partly to pressure on the surrounding structures. On
palpation the form of the abdomen will be observed to
differ from that of normal pregnancy, being generally more
developed in the transverse direction, and the rounded outline of the gravid uterus cannot be detected. When development has advanced nearly to term, the extreme distinctness
with which the fœtal limbs can be felt will arouse suspicion.
Per vaginam the os and cervix will be felt softened, as in
ordinary pregnancy, but often displaced by the pressure of
the cyst, and sometimes fixed by peri-metritic adhesions;
either of these signs is of great diagnostic value.

By bi-manual examination it may be possible to make
out that the uterus is not greatly enlarged, and that it is
distinctly separate from the bulk of the tumour; these facts,
if recognised, would of themselves disprove the existence of
uterine gestation. The diagnosis, if the fœtal limbs or heart-
sounds could be detected, would be cleared up in any case
by the uterine sound, which would show that the uterus was
empty and only slightly elongated. But we must be careful
not to resort to this test unless the existence of uterine gestation is positively disproved by other means. As, however,
it places the diagnosis beyond a doubt, it should always be
employed whenever operative procedure is in contemplation.
Quite recently I have seen a remarkable case which illus-
trates the importance of this rule. The case had been diagnosed as abdominal pregnancy by no less than six experienced
practitioners, and was actually on the operating table for the
performance of laparatomy. As a precaution, having some
doubts of the diagnosis, I suggested the passage of the
sound, which entered into a gravid uterus, the case proving to be one of small ovarian tumour jammed down into
Douglas's space, and displacing the cervix forwards. Had it
not been for this precaution its true nature would certainly
not have been detected.

Treatment.
Abdominal pregnancy should not be interfered with until the fœtus is fully developed.

The treatment of abdominal gestation will always be a subject of anxious consideration, and there is much difference of opinion as to the proper course to pursue. It is pretty generally admitted that it is not advisable to adopt any active measures until the full term of development is reached. Puncturing the cyst, with the view of destroying the fœtus and arresting its further growth, has been practised, but there are good grounds for rejecting it, for there is not the same imminent risk of death from rupture of the cyst as in tubal fœtation; and, even if the destruction of the fœtus could be brought about, there would still be formidable dangers from subsequent attempts at elimination, or from internal hæmorrhage.

Question as to the performance of primary gastrotomy.

When the full period has arrived, the child being still alive, as proved by auscultation, we have to consider whether it may not be advisable to perform gastrotomy before the fœtus perishes, and so at least save the life of the child. There are few questions of greater importance and more difficult to settle. The tendency of medical opinion is rather in favour of immediate operation, which is recommended by Velpeau, Kiwisch, Koeberlé, Schroeder, and many other writers, whose opinion necessarily carries great weight. The

Arguments in favour of the operation.

arguments used in favour of immediate operation are that while it affords a probability of saving the child, the risks to the mother, great though they undoubtedly are, are not greater than those which may be anticipated by delay. If we put off interference the cyst may rupture during the ineffectual efforts at labour, and death at once ensue; or, if this does not take place, other risks, which can never be foreseen, are always in store for the patient. She may sink from peritonitis, or from exhaustion, consequent on the efforts at elimination, which in the majority of cases are sooner or later set up, so that, as Barnes properly says, 'the patient's life may be said to be at the mercy of accidents of

Advantages of delay.

which we have no sufficient warning.' On the other hand, if we delay, while we sacrifice all hope of saving the child, we at least give the mother the chance of the fœtation remaining quiescent for a length of time, as certainly not unfrequently occurs. Thus, Campbell collected 62 cases of ultimate recovery after abdominal gestation, in 21 of which the fœtus was retained without injury for a number of years.

Then there is the question of secondary gastrotomy, which consists in operating after the death of the fœtus when urgent symptoms have arisen, a course which is advocated by Mr. Hutchinson. In favour of this procedure it is urged that by delay the inflammation taking place about the cyst will have greatly increased the chance of adhesions having formed between it and the abdominal parietes, so as to shut off its contents from the cavity of the peritoneum. The more effectually this has been accomplished, the greater are the chances of recovery. When the fœtus has been dead for some time, the vascularity of the cyst will also be lessened, and the placental circulation will have ceased, so that the danger of hæmorrhage will be much diminished.

Arguments in favour of secondary gastrotomy.

It will be seen, therefore, that there are arguments in favour of each of these views. The results of the primary operation are far less favourable than we should have, à priori, supposed. Since the first edition of this work appeared the subject has been carefully studied by Dr. Parry in his exhaustive treatise on Extra-uterine Fœtation. He has there shown that when the case is left until nature has shown the channel through which elimination is to be effected, the mortality is 17·35 per cent. less than in the cases in which the primary operation was performed. His conclusion is that 'the primary operation cannot be too forcibly condemned. It is not too much to say that this operation adds only another danger to a life already trembling in the balance, which the delusive hope of saving the uncertain life of a child does not warrant us in assuming.' It is only just to remember, as is forcibly pointed out by Keller, that in these days of advanced abdominal surgery a better result might be anticipated than when gastrotomy was performed in the haphazard way which was usual before we had gained experience from ovariotomy. No doubt, minute care in the performance of the operation, a due attention to its details, studiously avoiding, as much as possible, the passage of blood and the contents of the cyst into the peritoneal cavity, and a free use of antiseptics, would materially lessen its peril. This conclusion is well illustrated in a recent interesting paper by Thomas, who relates three successful cases of laparotomy in abdominal pregnancy.[1]

[1] *Am. Journ. of Med. Sci.* Jan. 1879.

The operation, then, should be performed with all the precautions with which we surround ovariotomy. The incision, best made in the linea alba, should not be greater than is necessary to extract the fœtus, and may be lengthened as occasion requires. It has been suggested that should the head be felt presenting above the vagina, the intervening structures should be divided, and the fœtus withdrawn by the forceps. This procedure was actually adopted with success in 1816, by Dr. John King, of Edisto Island, South Carolina. If there are no adhesions the walls of the cyst should be stitched to the margin of the incision, so as to shut it off as completely as possible from the peritoneal cavity. This has been specially insisted on by Braxton Hicks, and should never be omitted. The special risk is not so much the wounding of the peritoneum as the subsequent entrance of septic matter from the cyst into its cavity.

Import-
ance of not
interfer-
ing with
the pla-
centa.
Another cardinal rule, both in primary and secondary gastrotomy, is to make no attempt to remove the placenta. Its attachments are generally so deep-seated and diffused that any endeavour to separate it is likely to be attended with profuse and uncontrollable hæmorrhage, or with serious injury to the structure to which it is attached. Many of the failures after operating can be traced to a neglect of this rule. The best subsequent course to pursue, after removing the fœtus and arresting all hæmorrhage, either by ligature or the actual cautery, is to sponge out the cyst as gently as possible, sprinkle the cavity with iodoform, or with equal parts of tannin and salicylic acid, as recommended by Freund,[1] and then to bring the upper part of the wound into apposition with sutures, leaving the lower open, with the cord protruding so as to insure an outlet for the escape of the placenta as it slips down. The subsequent treatment must be specially directed to favour the escape of the discharge, and to prevent the risk of septicæmia. These objects may be much aided by injections of antiseptic fluids, such as solution of carbolic acid, or diluted Condy's fluid; and it would probably be advisable to place a drainage-tube in the lower angle of the wound. It may be well to point out that there is no operation in which a scrupulous following of the antiseptic method, on Sir Joseph Lister's principles, is so likely to be useful.

[1] *Edin. Med. Journ.* Dec. 1883.

As long as the placenta is retained the danger is necessarily great, and it may be many days, or even weeks, before
it is discharged. When once this is effected the sac may be
expected to contract, and eventually to close entirely.

When the fœtus is dead, or when we have determined Treat-
not to attempt primary gastrotomy, it is advisable to wait, ment
very carefully watching the patient, until either the gravity when the
of her general symptoms, or some positive indication of the fœtus is
channel through which nature is about to attempt to elimi- dead.
nate the fœtus, shows us that the time for action has arrived.
If there be distinct bulging of the cyst in the vagina, or in
the retro-vaginal cul-de-sac, especially if an opening has
formed there, we may properly content ourselves with aiding
the passage of the fœtus through the channel thus indicated,
and removing the parts that present piecemeal as they come
within reach, cautiously enlarging the aperture if necessary.
If the sac have opened into the intestines, the expulsion of
the fœtus through this channel is so tedious and difficult,
the exhaustion attending it so likely to prove fatal, and the
danger from decomposition of the fœtus through passage of
intestinal gas so great, that it would probably be best to
attempt to remove it by gastrotomy, especially if it is only
recently dead, and the greater portion is still retained.

If an opening forms at the abdominal parietes, or if the Mode of
symptoms determine us to resort to secondary gastrotomy perform-
before this occurs, the operation must be performed in the condary
same way, and with the same precautions, as primary gas- gastro-
trotomy. Here, as before, the safety of the operation must tomy.
greatly depend on the amount and firmness of the adhesions:
for if the cyst be not completely shut off from the peritoneal
cavity, the risks of the operation will be little less than those
of primary gastrotomy. It would obviously materially influence our decision and prognosis if we could determine
this point before operating. Unfortunately, it is impossible, Detection
as the experience of ovariotomists proves, to ascertain the of adhe-
existence of adhesions with any certainty. If, however, we sions.
find that the abdominal parietes do not move freely over
the cyst, and if the umbilicus be depressed and immovable,
the presumption is that considerable adhesions exist. If
they are found not to be present, the cyst walls should be

stitched to the margin of the incision, in the manner already indicated, before the contents are removed.

If the fœtus has been long dead, and its tissues greatly altered, its removal may be a matter of difficulty. In the case under my own care, already alluded to, the fœtal structures formed a sticky mass of such a nature that I believe it would have been impossible to empty the cyst had an operation been attempted. This would be, to some extent, a further argument in favour of the primary operation.

Opening of cyst by caustics.

The importance of adhesions has led some practitioners to recommend the opening of the cyst by potassa fusa or some other caustic, in the hope that it would set up adhesive inflammation around the aperture thus formed. Several successful operations by this method are recorded, and it would be worth trying, should the extreme mobility of the cyst lead us to suspect that no adhesions existed. If we have to deal with a case in which fistulous openings leading to the cyst have already formed, it may, perhaps, be advisable to dilate the apertures already existing, rather than make a fresh incision ; but, in determining this point, the surgeon will naturally be guided by the nature of the case, and the character and direction of the fistulous openings.

General treatment.

It is almost needless to say anything of general treatment in these trying cases ; but the administration of opiates to allay the sufferings of the patient, and the endeavour to support the severely taxed vital energies by appropriate food and medication will form a most important part of the management. Freund specially insists on the importance of careful regulation of the bowels, and on making milk the staple article of diet, as important points in the management of cases prior to operation.

Gestation in a bi-lobed uterus.

A few words may be said as to gestation in the rudimentary horn of a bi-lobed uterus, to which considerable attention has of late years been directed by the writings of Kussmaul and others. It appears certain that many cases of supposed tubal gestation are really to be referred to this category. Although such cases are of interest pathologically, they scarcely require much discussion from a practical point of view, inasmuch as their history is pretty nearly identical with that of tubal pregnancy. The rudimentary horn is distended by the enlarging ovum, and after a time, when further distension is impossible, laceration takes place. As a matter of fact, all the thirteen cases collected by Kussmaul terminated in this way; and even on post-mortem examination it is often extremely difficult to distinguish them from tubal pregnancies. The best way of doing so is probably by observing the relations of the round ligaments to the tumour, for, if the gestation be tubal, they will be found attached to the uterus on the inner or uterine side of the cyst; whereas, if the pregnancy be in a rudimentary horn of the uterus, they will be pushed outwards, and be external to the sac. In the latter case, moreover, the sac will be probably found to contain a true decidua, which is not the case in tubal pregnancy. The only point in which they differ is that in cornual pregnancy rupture may be delayed to a somewhat later period than in tubal, on account of the greater distensibility of the supplementary horn.

The term ' *missed labour* ' is applied to an exceedingly rare class of cases in which, at the full period of pregnancy, labour has either not come on at all, or, having commenced, the pains have subsequently passed off, and the foetus is retained in utero for a very considerable length of time. Under such circumstances it has usually happened that the membranes have ruptured at or about the proper term, and the access of air to the foetus in utero has been followed by decomposition. A putrid and offensive discharge has then commenced, and eventually portions of the disintegrating foetus have been expelled per vaginam. This discharge may go on until the entire foetus is gradually thrown off; or, more frequently, the patient dies from septicæmia, or other secondary result of the presence of the decomposing mass in utero. Thus McClintock relates one case,[1] in which symptoms of labour came on in a woman, 45 years of age, at the expected period of delivery, but passed off without the expulsion of the foetus. For a period of sixty-seven weeks a highly offensive discharge came away, with some few bones, and she eventually died with symptoms of pyæmia. He also cites another case in which the patient died in the same way, after the foetus had been retained for eleven years.

Sometimes, when the foetus has been retained for a

Missed labour.

Its symptoms.

[1] *Dublin Quart. Journ.* Feb. and May 1864.

Ulceration of the uterine walls sometimes occurs.

length of time, a further source of danger has been added by ulceration or destruction of the uterine walls, probably in consequence of an ineffectual attempt at its elimination. This occurred in Dr. Oldham's case (fig. 85), in which the contained mass is said to have nearly worn through the anterior wall of the uterus; and also in one reported by Sir James Simpson,[1] in which a patient died three months after

Fig. 85.

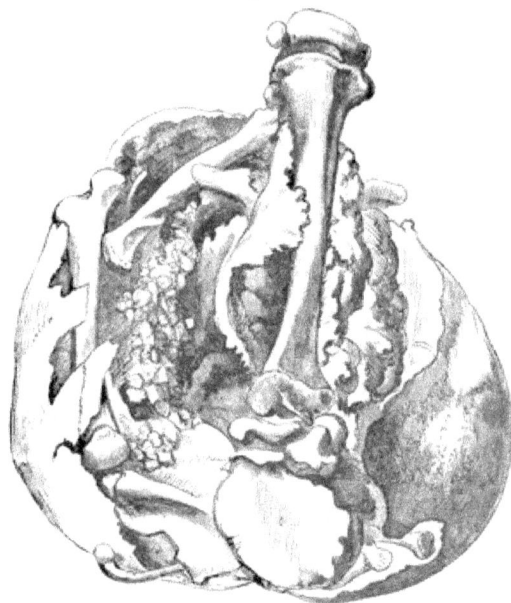

CONTENTS OF THE CYST IN DR. OLDHAM'S CASE OF MISSED LABOUR.

term, the fœtus having undergone fatty metamorphosis, an opening the size of half-a-crown having formed between the transverse colon and the uterine cavity. It is also stated that 'the uterine walls were as thin as parchment.'

In some few cases, however, probably when the entrance of air has been prevented, the fœtus has been retained for a length of time without decomposing, and without giving rise to any troublesome symptoms. Such a case is reported by Dr. Cheston,[2] in which the fœtus remained in utero for fifty-two years.

The causes of this strange occurrence are altogether un-

---

[1] *Edin. Med. Journ.* 1865.   [2] *Med.-Chir. Trans.* 1814.

known. Generally the fœtus seems to have died some time before the proper term for labour, and this may have influenced the character of the pains. It is probably also most apt to occur in women of feeble and inert habit of body, possibly where there was some obstacle to the dilatation of the cervix, which the pains were unable to overcome. Barnes suggests [1] that some presumed examples of missed labour 'were really cases of interstitial gestation, or gestation in one horn of a two-horned uterus.' In several of the cases, however, the details of the post-mortem examination are too minute to admit of the possibility of mistake having been made.

Its causes are not properly understood.

Müller, of Nancy, has recently attempted to prove, by a critical examination of published cases, that most examples of so-called 'missed labour' were in reality cases of extra-uterine fœtation, in which an ineffectual attempt at parturition took place, the fœtus being subsequently retained.

Sometimes confounded with extra-uterine fœtation.

From what has been said, it will be seen that the dangers arising from this state are very considerable, and when once the full term has passed beyond doubt, especially if the presence of an offensive discharge shows that decomposition of the fœtus has commenced, it would be proper practice to empty the uterus as soon as possible. The necessary precaution, however, is not to decide too quickly that the term has really passed; and, therefore, we must either allow sufficient time to elapse to make it quite certain that the case really falls under this category, or have unequivocal signs of the death of the fœtus, and injury to the mother's health. If we had to deal with the case before any extensive decomposition of the fœtus had occurred, we probably should find little difficulty in its management, for the proper course then would be to dilate the cervix with fluid dilators, and remove the fœtus by turning; or, before doing so, we might endeavour to excite uterine action by pressure and ergot. If the case did not come under observation until disintegration of the fœtus had begun, it would be more difficult to deal with. If the fœtus had become so much broken up that it was being discharged in pieces, Dr. McClintock says that 'in regard to treatment, our measures should consist mainly of palliatives, viz., rest and hip-baths, to subdue uterine irrita-

Its dangers are serious.

Treatment.

[1] *Diseases of Women,* p. 445.

tion; vaginal injections, to secure cleanliness and prevent excoriation; occasional digital examination, so as to detect any fragments of bone that might be presenting at the os, and to assist in removing them. These are plain rational measures, and beyond them we shall scarcely, perhaps, be justified in venturing. Nevertheless, under certain circumstances, I would not hesitate to dilate the cervical canal so as to permit of examining the interior of the womb, and of extracting any fragments of bone that may be easily accessible; but unless they could thus be easily reached and removed, the safer course would be to defer, for the present, interfering with them.[1]

It may be doubted, I think, whether, considering the serious results which are known to have followed so many cases, it would not, on the whole, be safer to make at least one decided effort, under chloroform, to remove as much as possible of the putrefying uterine contents, after the os has been fully dilated. Such a procedure would be less irritating than frequently repeated endeavours to pick away detached portions of the fœtus, as they present at the os uteri. When once the os is dilated, antiseptic intra-uterine injections, as of diluted Condy's fluid, might safely and advantageously be used. Unquestionably, it would be better practice to interfere and empty the uterus as soon as we are quite satisfied of the nature of the case, rather than to delay until the fœtus has been disintegrated.

[1] *Dublin Quart. Journ.* vol. xxxvii. p. 314.

CHAPTER VII.

DISEASES OF PREGNANCY.

THE diseases of pregnancy form a subject so extensive that they might well of themselves furnish ample material for a separate treatise. The pregnant woman is, of course, liable to the same diseases as the non-pregnant; but it is only necessary to allude to those whose course and effects are essentially modified by the existence of pregnancy, or which have some peculiar effect on the patient in consequence of her condition. There are, moreover, many disorders which can be distinctly traced to the existence of pregnancy. Some of them are the direct results of the sympathetic irritations which are then so commonly observed; and, of these, several are only exaggerations of irritations which may be said to be normal accompaniments of gestation. These functional derangements may be classed under the head of neuroses, and they are sometimes so slight as merely to cause temporary inconvenience, at others so grave as seriously to imperil the life of the patient. Another class of disorders is to be traced to local causes in connection with the gravid uterus, and are either the mechanical results of pressure, or of some displacement or morbid state of the uterus; while the origin of others may be said to be complex, being partly due to sympathetic irritation, partly to pressure, and partly to obscure nutritive changes produced by the pregnant state.

Among the sympathetic derangements there are none which are more common, and none which more frequently produce distress, and even danger, than those which affect the digestive system. Under the heading of 'The Signs of Pregnancy,' the frequent occurrence of nausea and vomiting has already been discussed, and its most probable causes considered (p. 152). A certain amount of nausea is, indeed, so

Diseases of pregnancy.

Many are only sympathetic derangements.

Others are mechanical or complex in their origin.

Derangements of the digestive system.

Excessive nausea or vomiting.

common an accompaniment of pregnancy, that its consideration as one of the normal symptoms of that state is fully
justified. We need here only discuss those cases in which
the nausea is excessive and long-continued, and leads to
serious results, from inanition and from the constant distress
it occasions. Fortunately a pregnant woman may bear a surprising amount of nausea and sickness without constitutional
injury, so that apparently almost all aliments may be rejected,
without the nutrition of the body very materially suffering.
At times the vomiting is limited to the early part of the
day, when all food is rejected, and when there is a frequent
retching of glairy transparent fluid, in several cases mixed
with bile, while at the latter part of the day the stomach
may be able to retain a sufficient quantity of food, and the
nausea disappears. In other cases the nausea and vomiting
are almost incessant. The patient feels constantly sick, and
the mere taste or sight of food may bring on excessive and
painful vomiting. The duration of this distressing accompaniment of pregnancy is also variable. Generally it commences between the second and third months, and disappears
after the woman has quickened. Sometimes, however, it
begins with conception, and continues unabated until the
pregnancy is over.

Symptoms of the graver cases.   In the worst class of cases, when all nourishment is rejected, and when the retching is continuous and painful,
symptoms of very great gravity, which may even prove fatal,
develop themselves. The countenance becomes haggard
from suffering, the tongue dry and coated, the epigastrium
tender on pressure, and a state of extreme nervous irritability,
attended with restlessness and loss of sleep, becomes established. In a still more aggravated degree, there is general
feverishness, with a rapid, small, and thready pulse. Extreme emaciation supervenes, the result of wasting from
lack of nourishment. The breath is intensely fetid, and the
tongue dry and black. The vomited matters are sometimes
mixed with blood. The patient becomes profoundly exhausted, a low form of delirium ensues, and death may follow
if relief is not obtained.

Prognosis.   Symptoms of such gravity are fortunately of extreme
rarity, but they do from time to time arise, and cause much
anxiety. Gueniot collected 118 cases of this form of the

disease, out of which 46 died; and out of the 72 that recovered, in 42 the symptoms only ceased when abortion, either spontaneous or artificially produced, had occurred. When pregnancy is over the symptoms occasionally cease with marvellous rapidity. The power of retaining and assimilating food is rapidly regained, and all the threatening symptoms disappear.

In the milder forms of obstinate vomiting, one of the first indications will be to remedy any morbid state of the primæ viæ. The bowels will not unfrequently be found to be obstinately constipated, the tongue loaded, and the breath offensive; and when attention has been paid to the general state of the digestive organs by general aperient medicines and antacid remedies, such as bismuth and soda and liquor pepticus after meals, the tendency to vomiting may abate without further treatment.

Treatment.

The careful regulation of the diet is very important. Great benefit is often derived from recommending the patient not to rise from the recumbent position in the morning until she has taken something. Half a cup of milk and lime-water, or a cup of strong coffee, or a little rum and milk, or cocoa and milk, a glass of sparkling koumiss, or even a morsel of biscuit, taken on waking, often has a remarkable effect in diminishing the nausea. When any attempt at swallowing solid food brings on vomiting, it is better to give up all pretence at keeping to regular meals, and to order such light and easily assimilated food, at short intervals, as can be retained. Iced milk, with lime or soda-water, given frequently, and not more than a mouthful at a time, will frequently be retained when nothing else will. Cold beef-jelly, a spoonful at a time, will also be often kept down. Sparkling koumiss has been strongly recommended as very useful in such cases, and is worthy of trial. It is well, however, to bear in mind, in regulating the diet, that the stomach is fanciful and capricious, and that the patient may be able to retain strange and apparently unlikely articles of food; and that, if she express a desire for such, the experiment of letting her have them should certainly be tried.

Regulation of diet.

The medicines that have been recommended are innumerable, and the practitioner will often have to try one after the other unsuccessfully; or may find, in an individual case, that

Medicinal treatment.

a remedy will prove valuable which, in another, may be altogether powerless. Amongst those most generally useful are effervescing draughts, containing from three to five minims of dilute hydrocyanic acid; the creasote mixture of the Pharmacopœia; tincture of nux vomica, in doses of five or ten minims; single minim doses of vinum ipecacuanhæ, every hour in severe cases, three or four times daily in those which are less urgent; salicine, in doses of three to five grains three times a day, recommended by Tyler Smith; oxalate of cerium, in the form of pill, of which three to five grains may be given three times a day—a remedy strongly advocated by Sir James Simpson, and which occasionally is of undoubted service, but more often fails; the compound pyroxylic spirit of the London Pharmacopœia, in doses of five minims every four hours, with a little compound tincture of cardamoms, a drug which is comparatively little known, but which occasionally has a very marked and beneficial effect in checking vomiting; opiates in various forms—which sometimes prove useful, more often not—may be administered either by the mouth, in pills containing from half a grain to a grain of opium, or in small doses of the solution of the bi-meconate of morphia or of Battley's sedative solution, or subcutaneously, a mode of administration which is much more often successful. If there is much tenderness about the epigastrium, one or two leeches may be advantageously applied, or one third of a grain of morphia may be sprinkled on the surface of a small blister, or cloths saturated in laudanum may be kept over the pit of the stomach. The administration per rectum of twenty grains of chloral, combined with the same amount of bromide of potassium, in a small enema, is said to be very useful. In many cases I have found that the application of a spinal ice-bag to the cervical vertebræ, in the manner recommended by Dr. Chapman, has checked the vomiting when all drugs have failed. The ice may be placed in one of Chapman's spinal ice-bags, and applied for half an hour or an hour, twice or three times a day. It invariably produces a comforting sensation of warmth, which is always agreeable to the patient. Ice may be given to suck *ad libitum*, and is very useful; while, if there be much exhaustion, small quantities of iced champagne may also be given from time to time.

Inasmuch as the vomiting unquestionably has its origin Local
treatment. in the uterus, it is only natural that practitioners should endeavour to check it by remedies calculated to relieve the irritability of that organ. Thus morphia in the form of pessaries per vaginam, or belladonna applied to the cervix, have been recommended, and—the former especially—are often of undoubted service. A pessary containing one third to half a grain of morphia may be introduced night and morning, without interfering with other methods of treatment. Dr. Henry Bennet directs especial attention to the cervix, which, he says, is almost always congested and inflamed, and covered with granular erosions. This condition he recommends to be treated by the application of nitrate of silver through the speculum. Dr. Clay, of Manchester, corroborates this view, and strongly advocates, especially when vomiting continues in the latter months, that one or two leeches should be applied to the cervix. Exception may fairly be taken to both these methods of treatment as being somewhat hazardous, unless other means have been tried and failed. I have little doubt, however, that in many cases a state of uterine congestion is an important factor in keeping up the unduly irritable condition of the uterine fibres, and an endeavour should always be made to lessen it by insisting on absolute rest in the recumbent posture. Of the importance of this precaution in obstinate cases there can be no question. Dr. Chapman, of Norwich, strongly recommended dilatation of the cervix by the finger, and stated that he found it very serviceable in checking nausea. It is obvious that this treatment must be adopted with great caution, as, roughly performed, it might lead to the production of abortion. Dr. Hewitt's views as to the dependence of sickness on flexions of the uterus have already been adverted to, and reasons have been given for doubting the general correctness of his theory. It is quite likely, however, that well-marked displacements of the uterus, either forwards or backwards, may serve to intensify the irritability of the organ. Cazeaux mentions an obstinate case immediately cured by replacing a retroverted uterus. A careful vaginal examination should, therefore, be instituted in all intractable cases, and if distinct displacement be detected, an endeavour should be made to support the uterus in its normal axis. If retroverted, a

Hodge's pessary may be safely employed ; if anteverted, a small air-ball pessary, as recommended by Hewitt, should be inserted. I believe, however, that such displacements are the exception, rather than the rule, in cases of severe sickness.

Import-ance of promoting the nutri-tion of the patient.
The importance of promoting nutrition by every means in our power should always be borne in mind. The effer-vescing koumiss, which can now be readily obtained, I have found of great value, as it can often be retained when all other aliment is rejected. The exhaustion produced by want of food soon increases the irritable state of the nervous system, and, if the stomach will not retain anything, we can only combat it by occasional nutrient enemata of strong beef-tea, yolk of egg, and the like.

The pro-duction of artificial abortion.
Finally, in the worst class of cases, when all treatment has failed, and when the patient has fallen into the condition of extreme prostration already described, we may be driven to consider the necessity of producing abortion. Fortunately cases justifying this extreme resource are of great rarity, but nevertheless there is abundant evidence that every now and then women do die from uncontrollable vomiting whose lives might have been saved had the pregnancy been brought to an end. The value of artificial abortion has been abun-dantly proved. Indeed, it is remarkable how rapidly the serious symptoms disappear when the uterus is emptied, and the tension of the uterine fibres lessened. It has fortunately but rarely fallen to my lot to have to perform this operation for intractable vomiting. In one such case the patient was reduced to a state of the utmost prostration, having kept hardly any food on her stomach for many weeks, and when I first saw her she was lying in a state of low muttering delirium. Within a few hours after abortion was induced all the threatening symptoms had disappeared, the vomiting had entirely ceased, and she was next day able to retain and absorb all that was given to her. The value of the operation, therefore, I believe to be undoubted. Where it has failed, it seems to have been on account of undue delay. Owing to the natural repugnance which all must feel towards this plan, it has generally been postponed until the patient has been too exhausted to rally. If, therefore, it is done at all, it should be before prostration has advanced so far as to render the operation useless. In these cases the obvious indication is to

lessen the tension of the uterus at once, and therefore the Method of operating. membranes should be punctured by the uterine sound, so as to let the liquor amnii drain away, and this may of itself be sufficient to accomplish the desired effect. It is almost need-less to add, that no one would be justified in resorting to this expedient without having his opinion fortified by con-sultation with a fellow-practitioner.

Other disorders of the digestive system may give rise to Other disorders of the digestive system. considerable discomfort, but not to the serious peril attend-ing obstinate vomiting. Amongst them are loss of appetite, acidity and heartburn, flatulent distension, and sometimes a capricious appetite, which assumes the form of longing for strange and even disgusting articles of diet. Associated with these conditions there is generally derangement of the whole intestinal tract, indicated by furred tongue and slug-gish bowels, and they are best treated by remedies calculated to restore a healthy condition of the digestive organs, such as a light easily digested diet, mineral acids, vegetable bitters, occasional aperients, bismuth and soda, and pepsine. The indications for treatment are not different from those which accompany the same symptoms in the non-pregnant state.

Diarrhœa is an occasional accompaniment of pregnancy, Diarrhœa. often depending on errors of diet. When excessive and continuous it has a decided tendency to induce uterine con-tractions, and I have frequently observed premature labour to follow a sharp attack of diarrhœa. It should, therefore, not be neglected; and, if at all excessive, should be checked by the usual means, such as chalk mixture with aromatic confection, and small doses of laudanum or chlorodyne. The possibility of apparent diarrhœa being associated with actual constipation, the fluid matter finding its way past the solid materials blocking up the intestines, should be borne in mind.

Constipation is much more common, and is indeed a very Constipa-tion. general accompaniment of pregnancy, even in women who do not suffer from it at other times. It partly depends on the mechanical interference of the gravid uterus with the proper movements of the intestines, and partly on defective innervation of the bowels resulting from the altered state of the blood. The first indication will be to remedy this defect by appropriate diet, such as fresh fruits, brown bread, oatmeal porridge, etc. Some medicinal treatment will also be necessary, and, in selecting the drugs to be used, care should be taken to choose such as are mild and unirritating in their action, and tend to improve the tone of the muscular coat of the intestine. A small quantity of aperient mineral water in the early morning, such as the Hunyadi, Friedrichshalle, or Pullna water, often answers very well; or an occasional dose of the confection of sulphur; or a pill containing three or four grains of the extract of colocynth, with a quarter of a grain of the extract of nux vomica, and a grain of extract of hyoscyamus at bedtime; or a teaspoonful of the compound liquorice powder in milk at bedtime. Constipation is also sometimes effectually combated by administering, twice daily, a pill containing a couple of grains of the inspissated ox-gall, with a quarter of a grain of extract of belladonna. Enemata of soap and water are often very useful, and have the advantage of not disturbing the digestion. In the latter months of pregnancy, especially in the few weeks preceding delivery, the irritation produced by the collection of hardened fæces in the bowel is a not infrequent cause of the annoying false pains which then so commonly trouble the patient. In order to relieve them, it will be necessary to empty the bowels thoroughly by an aperient, such as a good dose of castor-oil, in which fifteen or twenty minims of laudanum may be advantageously added. Should the rectum become loaded with scybalous masses, it may be necessary to break down and remove them by mechanical means, provided we are unable to effect this by copious enemata.

Hæmor- The loaded state of the rectum so common in pregnancy,
rhoids. combined with the mechanical effect of the pressure of the gravid uterus on the hæmorrhoidal veins, often produces very troublesome symptoms from piles. In such cases a regular and gentle evacuation of the bowels should be secured daily, so as to lessen as much as possible the congestion of the veins. Any of the aperients already mentioned, especially the sulphur electuary, may be used. Dr. Fordyce Barker[1] insists that, contrary to the usual impression, one of the best remedies for this purpose is a pill containing a grain or a grain and a half of powdered aloes, with a quarter of a grain of extract of nux vomica, and that castor-oil is distinctly prejudicial,

[1] *The Puerperal Diseases*, p. 33.

and apt to increase the symptoms.   I have certainly found it
answer well in several cases.   When the piles are tender and
swollen, they should be freely covered with an ointment consisting of four grains of muriate of morphia to an ounce of
simple ointment, or with the Ung. Gallæ c. opio of the
Pharmacopœia ; and, if protruded, an attempt should be made
to push them gently above the sphincter, by which they are
often unduly constricted.   Relief may also be obtained by
frequent hot fomentations, and sometimes, when the piles are
much swollen, it will be found useful to puncture them, so
as to lessen the congestion, before any attempt at reduction
is made.

A profuse discharge from the salivary glands is an occa-   Ptyalism.
sional distressing accompaniment of pregnancy.   It is generally confined to the early months, but it occasionally continues
during the whole period of gestation, and resists all treatment, only ceasing when delivery is over.   Under such circumstances the discharge of saliva is sometimes enormous,
amounting to several quarts a day, and the distress and
annoyance to the patient are very great.   In one case under
my care the saliva poured from the mouth all day long, and
for several months the patient sat with a basin constantly
by her side, incessantly emptying her mouth, until she was
reduced to a condition giving rise to really serious anxiety.
This profuse salivation is, no doubt, a purely nervous disorder,
and not readily controlled by remedies.   Astringent gargles,
containing tannin and chlorate of potass, frequent sucking
of ice, or of tannin lozenges, inhalation of turpentine and
creasote, counter-irritation over the salivary glands, by blis-
ters or iodine, the bromides, opium internally, small doses of
belladonna or atropine, may all be tried in turn, but none
of them can be depended on with any degree of confidence.

Severe dental neuralgia is also a frequent accompaniment   Toothache
of pregnancy, especially in the early months.   When purely   and caries
neuralgic, quinine in tolerably large doses is the best remedy   teeth.
at our disposal ; but not unfrequently it depends on actual
caries of the teeth, and attention should always be paid to
the condition of the teeth when facial neuralgia exists.
There is no doubt that pregnancy predisposes to caries, and
the observation of this fact has given rise to the old
proverb, ' For every child a tooth.'   Mr. Oakley Coles, in an

interesting paper [1] on the condition of the mouth and teeth
during pregnancy, refers the prevalence of caries to the co-
existence of acid dyspepsia, causing acidity of the oral secretions. There is much unreasonable dread amongst practitioners as to interfering with the teeth during pregnancy,
and some recommend that all operations, even stopping,
should be postponed until after delivery. It seems to me
certain that the suffering of severe toothache is likely to give
rise to far more severe irritation than the operation required
for its relief, and I have frequently seen badly decayed teeth
extracted during pregnancy, and with only a beneficial result.

Affections of the respiratory organs. Cough.

Amongst the derangements of the respiratory organs, one
of the most common is spasmodic cough, which is often excessively troublesome. Like many other of the sympathetic
derangements accompanying gestation, it is purely nervous
in character, and is unaccompanied by elevated temperature,
quickened pulse, or any distinct auscultatory phenomena.
In character it is not unlike whooping-cough. The treatment must obviously be guided by the character of the
cough. Expectorants are not likely to be of service, while
benefit may be derived from some of the anti-spasmodic
class of drugs, such as belladonna, hydrocyanic acid, opiates,
or bromide of potassium. Such remedies may be tried in
succession, but will often be found to be of little value in
arresting the cough. Dyspnœa may also be nervous in character, and sometimes symptoms not unlike those of spasmodic asthma are produced. Like the other sympathetic
disorders, it, as well as nervous cough, is most frequently
observed during the early months. There is another form of
dyspnœa, not uncommonly met with, which is the mechanical result of the interference with the action of the diaphragm and lungs by the pressure of the enlarged uterus.
Hence this is most generally troublesome in the latter
months, and continues unrelieved until delivery, or until
the sinking of the uterine tumour which immediately precedes it. Beyond taking care that the pressure is not increased by tight lacing or injudicious arrangement of the
clothes, there is little that can be done to relieve this form
of breathlessness.

Dyspnœa.

[1] Trans. of the Odontological Society.

Palpitation, like dyspnœa, may be due either to sympa-Palpita-
thetic disturbance, or to mechanical interference with the tion.
proper action of the heart.   When occurring in weakly
women it may be referred to the functional derangements
which accompany the chlorotic condition of the blood often
associated with pregnancy, and is then best remedied by a
general tonic regimen, and the administration of ferruginous
preparations.   At other times anti-spasmodic remedies may
be indicated, and it is seldom sufficiently serious to call for
much special treatment.

Attacks of fainting are not rare, especially in delicate Syncope.
women of highly developed nervous temperament, and are,
perhaps, most common at or about the period of quickening.
In most cases these attacks cannot be classed as cardiac, but
are more probably nervous in character, and they are rarely
associated with complete abolition of consciousness.   They
rather, therefore, resemble the condition described by the
older authors as *lypothemia*.   The patient lies in a semi-
unconscious condition with a feeble pulse and widely dilated
pupils, and this state lasts for varying periods, from a few
minutes to half an hour or more.   In one very troublesome
case under my care they often recurred as frequently as three
or four times a day.   I have observed that they rarely occur
when the more common sympathetic phenomena of pregnancy, especially vomiting, are present.   Sometimes they
terminate with the ordinary symptoms of hysteria, such as
sobbing.   The treatment should consist during the attack in
the administration of diffusable stimulants, such as ether,
sal-volatile, and valerian, the patient being placed in the
recumbent position, with the head low.   If frequently repeated it is unadvisable to attempt to rally the patient by
the too free administration of stimulants.   In the intervals
a generally tonic regimen, and the administration of ferruginous remedies, are indicated.   If they recur with great
frequency, the daily application of the spinal ice-bag has
proved of much service.

In connection with disorders of the circulatory system Extreme
may be noticed those which depend on the state of the anæmia
and
blood.   The altered condition of the blood, which has already chlorosis.
been described as a physiological accompaniment of pregnancy (p. 143), is sometimes carried to an extent which may

fairly be called morbid'; and either on account of the deficiency of blood-corpuscles, or from the increase in its watery
constituents, a state of extreme anæmia and chlorosis may
be developed. This may be sometimes carried to a very
serious extent. Thus Gusserow [1] records five cases, in which
nothing but excessive anæmia could be detected, all of which
ended fatally. Generally when such symptoms have been
carried to an extreme extent, the patient has been in a state

Treat-
ment.

of chlorosis before pregnancy. The treatment must, of course,
be calculated to improve the general nutrition, and enrich
the impoverished blood ; a light and easily assimilated diet,
milk, eggs, beef-tea, and animal food—if it can be taken—
attention to the proper action of the bowels, a due amount
of stimulants, and abundance of fresh air, will be the chief
indications in the general management of the case. Medi-
cinally, ferruginous preparations will be required. Some
practitioners object, apparently without sufficient reason, to
the administration of iron during pregnancy, as liable to
promote abortion. This unfounded prejudice may probably
be traced to the supposed emmenagogue properties of the
preparations of iron ; but, if the general condition of the
patient indicate such medication, they may be administered
without any fear. Preparations of phosphorus, such as the
phosphide of zinc, or free phosphorus, also promise favourably, and are well worthy of trial.

Œdema
associated
with
hydræmia.

Some of the more aggravated cases are associated with a
considerable amount of serous effusion into the cellular tissue,
generally limited to the lower extremities, but occasionally
extending to the arms, face, and neck, and even producing
ascites and pleuritic effusion. Under the latter circumstances
this complication is, of course, of great gravity, and it is said
that after delivery the disappearance of the serous effusion
may be accompanied by metastasis of a fatal character to
the lungs or the nervous centres. This form of œdema must
be distinguished from the slight œdematous swelling of the
feet and legs so commonly observed as a mechanical result
of the pressure of the gravid uterus, and also from those
cases of œdema associated with albuminuria. The treatment must be directed to the cause, while the disappearance
of the effusion may be promoted by the administration of

[1] *Arch. f. Gyn.* ii. 2, 1871.

diuretic drinks, the occasional use of saline aperients, and rest in the horizontal position.

The existence of albumen in the urine of pregnant women has for many years attracted the attention of obstetricians, and it is now well known to be associated, in ways still imperfectly understood, with many important puerperal diseases. Its presence in most cases of puerperal eclampsia was long ago pointed out by Lever in this country and Rayer in France, and its association with this disease gave rise to the theory of the dependence of the convulsion on uræmia, which is generally still entertained. It has been shown of late years, especially by Braxton Hicks, that this association is by no means so universal as was supposed; or rather, that in some cases the albuminuria follows and does not precede the convulsions, of which it might therefore be supposed to be the consequence rather than the cause; so that further investigations as to these particular points are still required. Modern researches have shown that there is an intimate connection between many other affections and albuminuria; as, for example, certain forms of paralysis, either of special nerves, as puerperal amaurosis, or of the spinal system; cephalalgia and dizziness; puerperal mania; and possibly hæmorrhage. It cannot, therefore, be doubted that albuminuria in the pregnant woman is liable, at any rate, to be associated with grave disease, although the present state of our knowledge does not enable us to define very distinctly its precise mode of action. *[margin: Albuminuria.]*

The presence of albumen in the urine of pregnant women is far from a rare phenomenon. Blot and Litzman met with albuminuria in 20 per cent. of pregnant women, which is, however, far above the estimate of other authors; Fordyce Barker [1] thinks it occurs in about one out of 25 cases, or 4 per cent., while Hofmier [2] found it in 137 out of 5,000 deliveries in the Berlin Gynœcological Institution, or 2·74 per cent. As in the large majority of these cases it rapidly disappears after delivery, it is obvious that its presence must, in a large proportion of cases, depend on temporary causes, and has not always the same serious importance as in the non-pregnant state. This is further proved by the undoubted *[margin: Causes of puerperal albuminuria.]*

[1] *American Journal of Obstetrics*, July 1878.
[2] *Berlin Klin. Woch.* Sept. 1878.

fact that albumen, rapidly disappearing after delivery, is often found in urine of pregnant women who go to term, and pass through labour without any unfavourable symptoms.

Pressure by the gravid uterus.

The obvious facts that in pregnancy the vessels supplying the kidneys are subjected to mechanical pressure from the gravid uterus, and that congestion of the venous circulation of those viscera must necessarily exist to a greater or less degree, suggest that here we may find an explanation of the frequent occurrence of albuminuria. This view is further strengthened by the fact that the albumen rarely appears until after the fifth month, and therefore, not until the uterus has attained a considerable size; and also that it is comparatively more frequently met with in primiparæ, in whom the resistance of the abdominal parietes, and consequent pressure, must be greater than in women who have already borne children. It is, indeed, probable that pressure and consequent venous congestion of the kidneys have an important influence in its production; but there must be, as a rule, some other factors in operation, since an equal or even greater amount of pressure is often exerted by ovarian and fibroid tumours, without any such consequences. They are probably complex. One important condition is doubtless the increased amount of work the kidneys have to do in excreting the waste products of the fœtus, as well as those of the mother. The increased arterial tension throughout the body associated with hypertrophy of the heart, known to exist in pregnancy, also operates in the same direction. But in the large majority of cases, although these conditions are present, no albuminuria exists, and they must, therefore, be looked upon as predisposing causes, to which some other is

Other causes probably also in operation.

added before the albumen escapes from the vessels. What this is generally escapes our observation, but probably any condition producing sudden hyperæmia of the kidneys, and giving rise to a state analogous to the first stage of Bright's disease—such, for example, as sudden exposure to cold and impeded cutaneous action—may be sufficient to set a light to the match already prepared by the existence of pregnancy. It has more recently been pointed out that a transient albuminuria, disappearing in a few days, is very common after delivery, and probably depends on a catarrhal condition of the urinary tract. Ingersten observed this in

50 out of 153 deliveries, and in 15 only had any albumen existed before the confinement.[1] In addition to these temporary causes it must not be forgotten that pregnancy may supervene in a patient already suffering from Bright's disease, when of course the albumen will exist in the urine from the commencement of gestation.

The various diseases associated with the presence of albumen in the urine will require separate consideration. Some of these, especially puerperal eclampsia, are amongst the most dangerous complications of pregnancy. Others, such as paralysis, cephalalgia, dizziness, may also be of considerable gravity. The precise mode of their production, and whether they can be traced, as is generally believed, to the retention of urinary elements in the blood, either urea or free carbonate of ammonia produced by its decomposition, or whether the two are only common results of some undetermined cause, will be considered when we come to discuss puerperal convulsions. Whatever view may ultimately be taken on these points, it is sufficiently obvious that albuminuria in a pregnant woman must constantly be a source of much anxiety, and must induce us to look forward with considerable apprehension to the termination of the case.

The effects of puerperal albuminuria.

We are scarcely in possession of a sufficiently large number of observations to justify any very accurate conclusions as to the risk attending albuminuria during pregnancy, but it is certainly by no means slight. Hofmier believes that albuminuria is a most severe complication both for woman and child, even when uncomplicated with eclampsia. The prognosis, he thinks, depends on whether it is acute in its onset, that is, coming on within a few days of labour, or is extended over several weeks. The former is more likely to pass entirely away after delivery, while in the latter there is more risk of the morbid state of the kidneys becoming permanent, and leading to the establishment of Bright's disease after the pregnancy is over. Goubeyre estimated that 49 per cent. of primiparæ who have albuminuria, and who escape eclampsia, die from morbid conditions traceable to the albuminuria. This conclusion is probably much exaggerated, but, if it even approximate to the truth, the danger must be very great.

Prognosis.

[1] *Zeitschrift f. Geburt. Band. v. Heft. 2.*

Tendency to produce abortion.

Besides the ultimate risk to the mother, albuminuria strongly predisposes to abortion, no doubt on account of the imperfect nutrition of the fœtus by blood impoverished by the drain of albuminous materials through the kidneys. This fact has been observed by many writers. A good illustration of it is given by Tanner,[1] who states that four out of seven women he attended suffering from Bright's disease during pregnancy, aborted, one of them three times in succession.

Symptoms. Anasarca.

The symptoms accompanying albuminuria in pregnancy are by no means uniform or constantly present. That which most frequently causes suspicion is the anasarca—not only the œdematous swelling of the lower limbs which is so common a consequence of the pressure of the gravid uterus, but also of the face and upper extremities. Any puffiness or infiltration about the face, or any œdema about the hands or arms, should always give rise to suspicion, and lead to a careful examination of the urine. Sometimes this is carried to an exaggerated degree, so that there is anasarca of the whole body.

Nervous phenomena.

Anomalous nervous symptoms—such as headache, transient dizziness, dimness of vision, spots before the eyes, inability to see objects distinctly, sickness in women not at other times suffering from nausea, sleeplessness, irritability of temper—are also often met with, sometimes to a slight degree, at others very strongly developed, and should always arouse suspicion. Indeed, knowing as we do that many morbid states may be associated with albuminuria, we should make a point of carefully examining the urine of all patients in whom any unusually morbid phenomena show themselves during pregnancy.

Character of the urine.

The condition of the urine varies considerably, but it is generally scanty and highly coloured, and, in addition to the albumen, especially in cases in which the albuminuria has existed for some time, we may find epithelium cells, tube casts, and occasionally blood-corpuscles.

Treatment.

The treatment must be based on what has been said as to the causes of the albuminuria. Of course it is out of our power to remove the pressure of the gravid uterus, except by inducing labour; but its effects may at least be lessened

[1] *Signs and Diseases of Pregnancy.* p. 428.

by remedies tending to promote an increased secretion of
urine, and thus diminishing the congestion of the renal
vessels. The administration of saline diuretics, such as the
acetate of potash, or bitartrate of potash, the latter being
given in the form of the well-known imperial drink, will
best answer this indication. The action of the bowels may
be solicited by purgatives producing watery motions, such as
occasional doses of the compound jalap powder. Dry cupping
over the loins, frequently repeated, has a beneficial effect in
lessening the renal hyperæmia. The action of the skin
should also be promoted by the use of the vapour bath, and
with this view the Turkish bath may be employed with great
benefit and perfect safety. Jaborandi and pilocarpin have
been given for this purpose, but have been found by Fordyce
Barker to produce a dangerous degree of depression. The
next indication is to improve the condition of the blood by
appropriate diet and medication. A very light and easily
assimilated diet should be ordered, of which milk should
form the staple. Tarnier[1] has recorded several cases in which
a purely milk diet was very successful in removing albuminuria. With the milk, which should be skimmed, we may
allow white of egg, or a little white fish. The tincture of
the perchloride of iron is the best medicine we can give,
and it may be advantageously combined with small doses of
tincture of digitalis, which acts as an excellent diuretic.

Finally, in obstinate cases we shall have to consider the
advisability of inducing premature labour. The propriety
of this procedure in the albuminuria of pregnancy has of
late years been much discussed. Spiegelberg[2] is opposed to
it, while Barker[3] thinks it should only be resorted to ' when
treatment has been thoroughly and perseveringly tried without success for the removal of symptoms of so grave a
character that their continuance would result in the death
of the patient.' Hofmier,[4] on the other hand, is in favour of
the operation, which he does not think increases the risk
of eclampsia, and may avert it altogether. I believe that,
having in view the undoubted risks which attend this complication, the operation is unquestionably indicated, and is
perfectly justifiable, in all cases attended with symptoms of

Question of inducing labour.

---

[1] *Annal. de Gynéc.* Jan. 1876.   [2] *Lehrbuch des Geburt.*
[3] *Amer. Jour. of Obstet.* July 1878.   [4] *Op. cit.*

serious gravity. It is not easy to lay down any definite rules to guide our decision ; but I should not hesitate to adopt this resource in all cases in which the quantity of albumen is considerable and progressively increasing, and in which treatment has failed to lessen the amount ; and, above all, in every case attended with threatening symptoms, such as severe headache, dizziness, or loss of sight. The risks of the operation are infinitesimal compared to those which the patient would run in the event of puerperal convulsions supervening, or chronic Bright's disease becoming established. As the operation is seldom likely to be indicated until the child has reached a viable age, and as the albuminuria places the child's life in danger, we are quite justified in considering the mother's safety alone in determining on its performance.

Diabetes. The occurrence of pregnancy in a woman suffering from diabetes may lead to serious consequences, and has recently been specially investigated by Dr. Matthew Duncan.[1] This must be carefully distinguished from the physiological glycosuria commonly present at the end of pregnancy, and during lactation. It is probable that diabetic patients are inapt to conceive, but when pregnancy does occur under such conditions, the case cannot be considered devoid of anxiety. From the cases collected by Dr. Duncan it would appear that pregnancy is very liable to be interrupted in its course, generally by the death of the fœtus, which has very often occurred. In some instances no bad results have been observed, while in others the patient has collapsed after delivery. Diabetic coma does not seem to have been observed. Out of twenty-two pregnancies in diabetic women four ended fatally, so that the mortality is obviously very large. Too little is known on this subject to justify positive rules of treatment ; but if the symptoms are serious and increasing, it would probably be justifiable to induce labour prematurely, so as to lessen the strain to which the patient's constitution is subjected.

[1] *Obst. Trans.* vol. 24.

CHAPTER VIII.

DISEASES OF PREGNANCY (CONTINUED).

THERE are many disorders of the nervous system met with during the course of pregnancy. Among the most common are morbid irritability of temper, or a state of mental despondency and dread of the results of the labour, sometimes almost amounting to insanity, or even progressing to actual mania. These are but exaggerations of the highly susceptible state of the nervous system generally associated with gestation. Want of sleep is not uncommon, and, if carried to any great extent, may cause serious trouble from the irritability and exhaustion it produces. In such cases we should endeavour to lessen the excitable state of the nerves, by insisting on the avoidance of late hours, overmuch society, exciting amusements, and the like; while it may be essential to promote sleep by the administration of sedatives, none answering so well as the chloral hydrate, in combination with large doses of the bromide of potassium or sodium, which greatly intensify its hypnotic effects.

Disorders of the nervous system.

Insomnia.

Severe headaches and various intense neuralgiæ are common. Amongst the latter the most frequently met with are pain in the breasts, due to the intimate sympathetic connection of the mammæ with the gravid uterus; and intense intercostal neuralgia, which a careless observer might mistake for pleuritic or inflammatory pain. The thermometer, by showing that there is no elevation of temperature, would prevent such a mistake. Neuralgia of the uterus itself, or severe pains in the groins or thighs—the latter being probably the mechanical results of dragging on the attachments of the abdominal muscles—are also far from uncommon. In the treatment of such neuralgic affections attention to the state of the general health, and large doses of quinine

Headaches and neuralgiæ.

and ferruginous preparations whenever there is much debility, will be indicated. Locally sedative applications, such as belladonna and chloroform liniments; friction with aconite liniment when the pain is limited to a small space; and, in the worst cases, the subcutaneous injection of morphia, will be called for. Those pains which apparently depend on mechanical causes may often be best relieved by lessening the traction on the muscles, by wearing a well-made elastic belt to support the uterus.

Paralysis depending on pregnancy. Among the most interesting of the nervous diseases are various paralytic affections. Almost all varieties of paralysis have been observed, such as paraplegia, hemiplegia (complete or incomplete), facial paralysis, and paralysis of the nerves of special sense, giving rise to amaurosis, deafness, and loss of taste. Churchill records 22 cases of paralysis during pregnancy, collected by him from various sources. A large number have also been brought together by Imbert Goubeyre, in an interesting memoir on the subject, and others are recorded by Fordyce Barker, Joulin, and other authors; so that there can be no doubt of the fact that paralytic *Generally associated with albuminuria.* affections are common during gestation. In a large proportion of the cases recorded the paralyses have been associated with albuminuria, and are doubtless uræmic in origin. Thus in 19 cases, related by Goubeyre, albuminuria was present in all; Darcy,[1] however, found no albuminuria in 5 out of 14 cases. The dependency of the paralysis on a transient cause explains the fact that in the large majority of these cases the paralysis was not permanent, but disappeared shortly after labour. In every case of paralysis, whatever be its nature, special attention should be directed to the state of the urine, and, should it be found to be albuminous, labour *In such cases labour should be at once induced.* should be at once induced. This is clearly the proper course to pursue, and we should certainly not be justified in running the risk that must attend the progress of a case in which so formidable a symptom has already developed itself. When the cause has been removed, the effect will also generally rapidly disappear, and the prognosis is therefore, on the whole, favourable. Should the paralysis continue after delivery, the treatment must be such as we would adopt in the non-pregnant state; and small doses of strychnia, along with faradi-

[1] *Thèse de Paris*, 1877.

sation of the affected limbs, would be the best remedies at our disposal.

There are, however, unquestionably some cases of puer- Paralyses which are not uræmic in their origin. peral paralysis which are not uræmic in their origin, and the nature of which is somewhat obscure. Hemiplegia may doubtless be occasioned by cerebral hæmorrhage, as in the non-pregnant state. Other organic causes of paralysis, such as cerebral congestion, or embolism, may, now and again, be met with during pregnancy, but cases of this kind must be of comparative rarity. Other cases are functional in their origin. Tarnier relates a case of hemiplegia which he could only refer to extreme anæmia. Some, again, may be hysterical. Paraplegia is apparently more frequently unconnected with albuminuria than the other forms of paralysis ; and it may either depend on pressure of the gravid uterus on the nerves as they pass through the pelvis, or on reflex action, as is sometimes observed in connection with uterine disease. When, in such cases, the absence of albuminuria is ascertained by frequent examination of the urine, there is obviously not the same risk to the patient as in cases depending on uræmia, and therefore it may be justifiable to allow pregnancy to go on to term, trusting to subsequent general treatment to remove the paralytic symptoms. As the loss of power here depends on a transient cause, a favourable prognosis is quite justifiable. Partial paralysis of one lower extremity, generally the left, sometimes occurs, from pressure of the fœtal occiput, and may continue for days or weeks, with a gradual improvement, after parturition.

Chorea is not unfrequently observed, and forms a serious Chorea. complication. It is generally met with in young women of delicate health, and in the first pregnancy. In a large proportion of the cases the patient has already suffered from the disease before marriage. On the occurrence of pregnancy, the disposition to the disease again becomes evoked, and choreic movements are re-established. This fact may be explained partly by the susceptible state of the nervous system, partly by the impoverished condition of the blood.

That chorea is a dangerous complication of pregnancy is Prognosis. apparent by the fact that out of 56 cases collected by Dr. Barnes[1] no less than 17, or 1 to 3, proved fatal. Nor is it

[1] *Obst. Trans.* vol. x.

danger to life alone that is to be feared, for it appears
certain that chorea is more apt to leave permanent mental
disturbance when it occurs during pregnancy than at other
times. It has also an unquestionable tendency to bring on
abortion or premature labour, and in most cases the life of
the child is sacrificed.

Treat-
ment.

The treatment of chorea during pregnancy does not differ
from that of the disease under more ordinary circumstances ;
and our chief reliance will be placed on such drugs as the
liquor arsenicalis, bromide of potassium, and iron. In the
severe form of the disease, the incessant movements, and the
weariness and loss of sleep, may very seriously imperil the
life of the patient, and more prompt and radical measures
will be indicated. If, in spite of our remedies, the paroxysms
go on increasing in severity, and the patient's strength appears to be exhausted, our only resource is to remove the
most evident cause by inducing labour. Generally the symptoms lessen and disappear soon after this is done. There
can be no question that the operation is perfectly justifiable,
and may even be essential under such circumstances. It
should be borne in mind that the chorea often recurs in a
subsequent pregnancy, and extra care should then always be
taken to prevent its development.

Disorders
of the
urinary
organs.
Reten-
tion of
urine.

Disorders of the urinary organs are of frequent occurrence. Retention of urine may be met with, and this is
often the result of a retroverted uterus. The treatment,
therefore, must then be directed to the removal of the cause.
This subject will be more particularly considered when we
come to discuss that form of displacement (p. 242); but we
may here point out that retention of urine, if long continued,
may not only lead to much distress, but to actual disease of
the coats of the bladder. Several cases have been recorded
in which cystitis, resulting from urinary retention in pregnancy, eventually caused the exfoliation of the entire mucous
membrane of the bladder,[1] which was cast off, sometimes
entire, sometimes in shreds, and occasionally with portions
of the muscular coat attached to it. The possibility of this
formidable accident should teach us to be careful not to allow
any undue retention of urine, but, by a timely use of the

---

[1] *Obst. Trans.* vol. xi.

catheter, to relieve the symptoms, while we, at the same time,
endeavour to remove the cause.

Irritability of the bladder is of frequent occurrence. In the early months it seems to be the consequence of sympathetic irritation of the neck of the bladder, combined with pressure, while in the later months it is, probably, solely produced by mechanical causes. When severe it leads to much distress, the patient's rest being broken and disturbed by incessant calls to micturate, and the suffering induced may produce serious constitutional disturbances. I have elsewhere pointed out[1] that irritability of the bladder in the later months of pregnancy is frequently associated with an abnormal position of the fœtus, which is placed transversely or obliquely. The result is either that undue pressure is applied to the bladder, or that it is drawn out of its proper position. The abnormal position of the fœtus can readily be detected by palpation, and is readily altered by external manipulation. In some of the cases I have recorded, altering the position of the fœtus was immediately followed by relief; the symptoms recurring after a time, when the fœtus had again assumed an oblique position. Should the fœtus frequently become displaced, an endeavour may be made to retain it in the longitudinal axis of the uterus by a proper adaptation of bandages or pads. In cases not referable to this cause we should attempt to relieve the bladder symptoms by appropriate medication, such as small doses of liquor potassæ, if the urine be very acid; tincture of belladonna; the decoction of triticum repens, an old but very serviceable remedy; and vaginal sedative pessaries containing morphia or atropine.

Irritability of the bladder.

Women who have borne many children are often troubled with incontinence of urine during pregnancy, the water dribbling away on the slightest movement. Through this much irritation of the skin surrounding the genitals is produced, attended with troublesome excoriations and eruptions. Relief may be partially obtained by lessening the pressure on the bladder by an abdominal belt, while the skin is protected by applications of simple ointment or glycerine.

Incontinence of urine.

Dr. Tyler Smith has directed attention to a phosphatic condition of the urine occurring in delicate women, whose

Phosphatic deposit.

[1] *Obst. Trans.* vol. xiii.

constitutions are severely tried by gestation. This condition can easily be altered by rest, nutritious diet, and a course of restorative medicines, such as steel, mineral acids, and the like.

Leucor-rhœa.

A profuse whitish leucorrhœal discharge is very common during pregnancy, especially in its latter half. The discharge frequently alarms the patient, but, unless it is attended with disagreeable symptoms, it does not call for special treatment. When at all excessive, it may lead to much irritation of the vagina and external generative organs. The labia may become excoriated and covered with small aphthous patches, and the whole vulva may be hot, swollen, and tender. Warty growths, similar in appearance to syphilitic condylomata, are occasionally developed in pregnant women, unconnected with any specific taint, and associated with the presence of an irritating leucorrhœal discharge. According to Thibièrge,[1]

Treat-ment.

these resist local applications, such as sulphate of copper or nitrate of silver, but spontaneously disappear after delivery. Inasmuch as the leucorrhœal discharge is dependent on the congested condition of the generative organs accompanying pregnancy, we can hope to do little more than alleviate it. In the severer forms, as has been pointed out by Henry Bennet, the cervix will be found to be abraded or covered with granular erosion, and it may be, from time to time, cautiously touched with the nitrate of silver or a solution of carbolic acid. Generally speaking, we must content ourselves with recommending the patient to wash the vagina out gently with diluted Condy's fluid; or with a solution of the sulpho-carbolate of zinc, of the strength of four grains to the ounce of water; or with plain tepid water. For obvious reasons frequent and strong vaginal douches are to be avoided, but a daily gentle injection, for the purpose of ablution, can do no harm.

Pruritus.

A very distressing pruritus of the vulva is frequently met with along with leucorrhœa, especially when the discharge is of an acrid character, which in some cases leads to intense and protracted suffering, forcing the patient to resort to incessant friction of the parts. Pruritus, however, may exist

It may exist inde-pendently of leucor-rhœa.

without leucorrhœa, being apparently sometimes of a neur-algic character, at others associated with aphthous patches

[1] *Arch. Gén. de Méd.* 1856.

on the mucous membrane, ascarides in the rectum, or pedi-
culi in the hairs of the mons veneris and labia. Cases are
even recorded in which the pruritic irritation extended over
the whole body. The treatment is difficult and unsatisfactory. Treat-
Various sedative applications may be tried, such as weak ment.
solutions of Goulard's lotion ; or a lotion composed of an
ounce of the solution of the muriate of morphia, with a
drachm and a half of hydrocyanic acid, in six ounces of
water ; or one formed by mixing one part of chloroform with
six of almond oil. A very useful form of medication consists
in the insertion into the vagina of a pledget of cotton-wool,
soaked in equal parts of the glycerine of borax and sulphu-
rous acid ; this may be inserted at bedtime, and withdrawn
in the morning by means of a string attached to it. Smear-
ing the parts with an ointment consisting of boracic acid
and vaseline often answers admirably. In the more obstinate
cases, the solid nitrate of silver may be lightly brushed over
the vulva ; or, as recommended by Tarnier, a solution of
bichloride of mercury, of about the strength of two grs. to
the ounce, may be applied night and morning. The state
of the digestive organs should always be attended to, and
aperient mineral water may be usefully administered. When
the pruritus extends beyond the vulva, or even in severe
local cases, large doses of bromide of potassium may perhaps
be useful in lessening the general hyperæsthetic state of the
nerves.

Some of the disorders of pregnancy are the direct results   Œdema of
of the mechanical pressure of the gravid uterus. The most   the lower
                                                             limbs.
common of these are œdema and a varicose state of the veins
of the lower extremities, or even of the vulva. The former   Effects of
is of little consequence, provided we have assured ourselves pressure.
that it is really the result of pressure, and not of albuminuria, and it can generally be relieved by rest in the
horizontal position. A varicose state of the veins of the   Varicose
lower limbs is very common, especially in multiparæ, in whom veins.
it is apt to continue after delivery. Occasionally the veins
of the vulva, and even of the vagina, are also enlarged and
varicose, producing considerable swelling of the external
genitals. Rest in the recumbent position, and the use of
an abdominal belt, so as to take the pressure off the veins as
much as possible, are all that can be done to relieve this

troublesome complication. If the veins of the legs are much swollen some benefit may be derived from an elastic stocking or a carefully applied bandage.

Occasional serious results from laceration of the veins.

Serious and even fatal consequences have followed the accidental laceration of the swollen veins. When laceration occurs during or immediately after delivery—a not uncommon result of the pressure of the head—it gives rise to the formation of a vaginal thrombus. It has occasionally happened from an accidental injury during pregnancy, as in the cases recorded by Simpson, in which death followed a kick on the pudenda, producing laceration of a varicose vein, or in one mentioned by Tarnier, where the patient fell on the edge of a chair. Severe hæmorrhage has followed the accidental rupture of a vein in the leg.

Treatment.

The only satisfactory treatment is pressure, applied directly to the bleeding parts by means of the finger, or by compresses saturated in a solution of the perchloride of iron. The treatment of vaginal thrombus following labour must be considered elsewhere. Occasionally the varicose veins inflame, become very tender and painful, and coagula form in their canals. In such cases absolute rest should be insisted on, while sedative lotions, such as the chloroform and belladonna liniments, should be applied to relieve the pain.

Displacements of the gravid uterus.

Certain displacements of the gravid uterus are met with, which may give rise to symptoms of great gravity.

Prolapse of the gravid uterus.

Prolapse, which is rare, is almost always the result of pregnancy occurring in a uterus which had been previously more or less procident. Under such circumstances the increasing weight of the uterus will at first necessarily augment the previously existing tendency to prolapse of the womb, which may come to protrude partially or entirely beyond the vulva. In the great majority of cases, as pregnancy advances, the prolapse cures itself, for at about the fourth or fifth month the uterus will rise above the pelvic brim. It has been said that in some cases of complete procidentia pregnancy has gone even to term, with the uterus lying entirely outside the vulva. Most probably these cases were imperfectly observed, the greater part of the uterus being in reality above the pelvic brim, a portion only of its lower segment protruding externally; or, as has sometimes been the case, the protruding portion has been an old-standing hypertrophic elongation of the cervix, the internal os uteri and fundus being normally situated. Should a prolapsed uterus not rise into the abdominal cavity as pregnancy advances, serious symptoms will be apt to develop themselves; for, unless the pelvis be unusually capacious, the enlarging uterus will get jammed within its bony walls, the rectum and urethra will be pressed upon, defæcation and micturition will be consequently impeded, and severe pain and much irritation will result. In all probability such a state of things would lead to abortion. The possibility of these consequences should, therefore, teach us to be careful in the management of every case of prolapse, however slight, in which pregnancy occurs. Absolute rest, in the horizontal position, should be insisted on; while the uterus should be supported in the pelvis by a full-sized Hodge's pessary, which should be worn until at least the sixth month, when the uterus would be fully within the abdominal cavity. After delivery, prolonged rest should be recommended, in the hope that the process of involution may be accompanied by a cure of the prolapse. There can be no doubt that pregnancy carried to term affords an opportunity of curing even old-standing displacements, which should not be neglected. *Its treatment.*

Anteversion of the gravid uterus seldom produces symptoms of consequence. In all probability it is common enough when pregnancy occurs in a uterus which is more than usually anteverted, or is anteflexed. Under such circumstances, there is not the same risk of incarceration in the pelvic cavity as in cases in which pregnancy exists in a retroflexed uterus, for, as the uterus increases in size, it rises without difficulty into the abdominal cavity. In the early months the pressure of the fundus on the bladder may account for the irritability of that viscus then so commonly observed. It will be remembered that Graily Hewitt attributes great importance to this condition as explaining the sickness of pregnancy—a theory, however, which has not met with general acceptation. *Anteversion is of comparatively little consequence.*

Extreme anteversion of the uterus, at an advanced period of pregnancy, is sometimes observed in multiparæ with very lax abdominal walls, occasionally to such an extent that the uterus falls completely forwards and downwards, so that the fundus is almost on a level with the patient's knees. This *Anteversion of the gravid uterus in advanced pregnancy.*

form of pendulous belly may be associated with a separation
of the recti muscles, between which the womb forms a ventral hernia, covered only by the cutaneous textures. When
labour comes on this variety of displacement may give rise
to trouble by destroying the proper relation of the uterine
and pelvic axes. The treatment is purely mechanical, keeping the patient lying on her back as much as possible, and
supporting the pendulous abdomen by a properly adjusted
bandage. A similar forward displacement is observed in
cases of pelvic deformity, and in the worst forms, in rachitic
and dwarfed women, it exists to a very exaggerated degree.

Retrover-
sion.

    The most important of the displacements, in consequence
of its occasional very serious results, is retroversion of the
gravid uterus. It was formerly generally believed that this
was most commonly produced by some accident, such as a
fall, which dislocated a uterus previously in a normal position. Undue distension of the bladder was also considered
to have an important influence in its production, by pressing
the uterus backwards and downwards.

Its causes.

    It is now almost universally admitted that, although the
above-named causes may possibly sometimes produce it, in
the very large proportion of cases it depends on pregnancy
having occurred in a uterus previously retroverted or retroflexed. The merit of pointing out this fact unquestionably
belongs to the late Dr. Tyler Smith, and further observations
have fully corroborated the correctness of his views.

    In the large majority of cases in which pregnancy occurs
in a uterus so displaced, as the womb enlarges it straightens
itself, and rises into the abdominal cavity, without giving
any particular trouble ; or, as not unfrequently happens, the
abnormal position of the organ interferes so much with its
enlargement as to produce abortion. Sometimes, however,
the uterus increases without leaving the pelvis until the
third or fourth month, when it can no longer be retained in
the pelvic cavity without inconvenience. It then presses on
the urethra and rectum, and eventually becomes completely
incarcerated within the rigid walls of the bony pelvis, giving
rise to characteristic symptoms.

Symp-
toms.

    The first sign which attracts attention is generally some
trouble connected with micturition, in consequence of pressure on the urethra. On examination, the bladder will often

be found to be enormously distended, forming a large, fluctuating abdominal tumour, which the patient has lost all
power of emptying. Frequently small quantities of urine
dribble away, leading the woman to believe that she has
passed water, and thus the distension is often overlooked.
Sometimes the obstruction to the discharge of urine is so
great as to lead to dropsical effusion into the cellular tissue
of the arms and legs. This was very well marked in one of
my cases, and disappeared rapidly after the bladder had been
emptied. Difficulty in defæcation, tenesmus, obstinate constipation, and inability to empty the bowels, becomes established about the same time. These symptoms increase,
accompanied by some pelvic pain and a sense of weight and
bearing down, until at last the patient applies for advice, and
the true nature of the case is detected. When the retroversion occurs suddenly, all these symptoms develop with
great rapidity, and are sometimes very serious from the first.

The further progress is various. Sometimes, after the *Progress and termination.* uterus has been incarcerated in the pelvis for more or less
time, it may spontaneously rise into the abdominal cavity,
when all threatening symptoms will disappear. So happy a
termination is quite exceptional, and if the practitioner
should not interfere and effect reposition of the organ, serious
and even fatal consequences may ensue, unless abortion
occurs.

The extreme distension of the bladder, and the impossi- *Termination if reduction is not effected.* bility of relieving it, may lead to lacerations of its coats
and fatal peritonitis; or the retention of urine may produce
cystitis, with exfoliation of the coats of the bladder; or, as
more commonly happens, retention of urinary elements may
take place, and death occur with all the symptoms of
uræmic poisoning. At other times the impacted uterus
becomes congested and inflamed, and eventually sloughs, its
contents, if the patient survive, being discharged by fistulous
communications into the rectum and vagina. It need hardly
be said that such terminations are only posssible in cases
which have been grossly mismanaged, or the nature of which
has not been detected till a late period.

The diagnosis is not difficult. On making a vaginal *Diagnosis.* examination, the finger impinges on a smooth round elastic
swelling, filling up the lower part of the pelvis, stretching

and depressing the posterior vaginal wall, which occasionally protrudes beyond the vulva. On passing the finger forwards and upwards we shall generally be able to reach the cervix, high up behind the pubes, and pressing on the urethral canal. In very complete retroversion it may be difficult or impossible to reach the cervix at all. On abdominal examination the fundus uteri cannot be felt above the pelvic brim; this, as the retroversion does not give rise to serious symptoms until between the third and fourth months, should, under natural circumstances, always be possible. By bi-manual examination we can make out, with due care, the alternate relaxation and contraction of the uterine parietes characteristic of the gravid uterus, and so differentiate the swelling from any other in the same situation. The accompanying phenomena of pregnancy will also prevent any mistake of this kind.

Retroversion going on to term; its explanation. In some few cases retroversion has been supposed to go on to term. Strictly speaking, this is impossible; but in the supposed examples, such as the well-known case recorded by Oldham, part of a retroflexed uterus remained in the pelvic cavity, while the greater part developed in the abdominal cavity. The uterus is, therefore, divided, as it were, into two portions; one, which is the flexed fundus, remaining in the pelvis, the other, containing the greater part of the foetus, rising above it. Under these circumstances, a tumour in the vagina would exist in combination with an abdominal tumour, and pregnancy might go on to term. Considerable difficulty may even arise in labour, but the malposition generally rectifies itself before it gives rise to any serious results.

Treatment. The treatment of retroversion of the gravid uterus should be taken in hand as soon as possible, for every day's delay involves an increase in the size of the uterus, and, therefore, greater difficulty in reposition. Our object is to restore the natural direction of the uterus, by lifting the fundus above the promontory of the sacrum. The first thing to be done is **The bladder should first be emptied.** to relieve the patient by emptying the bladder, the retention of urine having probably originally called attention to the case. For this purpose it is essential to use a long elastic male catheter of small size, as the urethra is too elongated and compressed to admit of the passage of the ordinary silver instrument. Even then it may be extremely difficult to introduce the catheter, and sometimes it has been found to be quite impossible. Under such circumstances, provided reposition cannot be effected without it, the bladder may be punctured an inch or two above the pubes by means of the fine needle of an aspirator, and the urine drawn off. Dieulafoy's work on aspiration proves conclusively that this may be done without risk, and the operation has been successfully performed by Schatz and others. It very rarely happens, however, and in long-neglected cases only, that the withdrawal of the urine is found to be impossible.

The bladder being emptied, and the bowels being also opened, if possible, by copious enemata, we proceed to attempt reduction. For this purpose various procedures are adopted. If the case is not of very long standing, I am inclined to think that the gentlest and safest plan is the continuous pressure of a caoutchouc bag, filled with water, placed in the vagina. The good effects of steady and long-continued pressure of this kind was proved by Tyler Smith, who effected in this way the reduction of an inverted uterus of long standing, and it is not difficult to understand that it may succeed when a more sudden and violent effort fails. I have tried this plan successfully in two cases, a pyriform india-rubber bag being inserted into the vagina, and distended as far as the patient could bear by means of a syringe. The water must be let out occasionally to allow the patient to empty the bladder, and the bag immediately refilled. In both my cases reposition occurred within twenty-four hours. Barnes has failed with this method; but it succeeded so well in my cases, and is so obviously less likely to prove hurtful than forcible reposition with the hand, that I am inclined to consider it the preferable procedure, and one that should be tried first. Failing with the fluid pressure, we should endeavour to replace the uterus in the following way. The patient should be placed at the edge of the bed, in the ordinary obstetric position, and thoroughly anæsthetised. This is of importance, as it relaxes all the parts, and admits of much freer manipulation than is otherwise possible. One or more fingers of the left hand are then inserted into the rectum; if the patient be deeply chloroformed, it is quite possible, with due care, even to pass the whole hand, and an attempt is then made to lift or push the fundus

Mode of effecting reduction.

above the promontory of the sacrum. At the same time reposition is aided by drawing down the cervix with the ·fingers of the right hand per vaginam. It has been insisted that the pressure should be made in the direction of one or other sacro-iliac synchondrosis rather than directly upwards, so that the uterus may not be jammed against the projection of the promontory of the sacrum. Failing reposition through the rectum, an attempt may be made per vaginam, and for this some have advised the upward pressure of the closed fist passed into the canal. Others recommend the hand and knee position as facilitating reposition, but this prevents the administration of chloroform, which is of more assistance than any change of position can possibly be. Various complex instruments have been invented to facilitate the operation, but they are all more or less dangerous, and are unlikely to succeed when manual pressure has failed.

As soon as the reduction is accomplished, subsequent descent of the uterus should be prevented by a large-sized Hodge's pessary, and the patient should be kept at rest for some days, the state of the bladder and bowels being particularly attended to. When reposition has been fairly effected, a relapse is unlikely to occur.

Treatment when reduction is found impossible.

In cases in which reduction is found to be impossible, our only resource is the artificial induction of abortion. Under such circumstances this is imperatively called for. It is best effected by puncturing the membranes, the discharge of the liquor amnii of itself lessening the size of the uterus, and thus diminishing the pressure to which the neighbouring parts are subjected. After this reposition may be possible, or we may wait until the fœtus is spontaneously expelled. It is not always easy to reach the os uteri, although we can generally do so with a curved uterine sound. If we cannot puncture the membranes, the liquor amnii may be drawn off through the uterine walls by means of the aspirator, inserted through either the rectum or vagina. The injury to the uterine walls thus inflicted is not likely to be hurtful, and the risk is certainly far less than leaving the case alone. Naturally so extreme a measure would not be adopted until all the simpler means indicated have been tried and failed.

The pregnant woman is, of course, liable to contract the same diseases as in the non-pregnant state, and pregnancy

may occur in women already the subject of some constitu- Diseases co-exist-ing with preg-nancy.
tional disease. There is no doubt yet much to be learned as
to the influence of co-existing disease on pregnancy. It is
certain that some diseases are but little modified by pregnancy, and that others are so to a considerable extent; and
that the influence of the disease on the fœtus varies much.
The subject is too extensive to be entered into at any length,
but a few words may be said as to some of the more important affections that are likely to be met with.

The eruptive fevers have often very serious consequences, Eruptive fevers. Small-pox.
proportionate to the intensity of the attack. Of these variola
has the most disastrous results, which are related in the
writings of the older authors, but which are, fortunately,
rarely seen in these days of vaccination. The severe and
confluent forms of the disease are almost certainly fatal to
both the mother and child. In the discrete form, and in
modified small-pox after vaccination, the patient generally
has the disease favourably, and although abortion frequently
results, it does not necessarily do so.

If scarlet fever of an intense character attacks a pregnant Scarlet fever.
woman, abortion is likely to occur, and the risks to the
mother are very great. The milder cases run their course
without the production of any untoward symptoms. Should
abortion occur, the well-known dangerous effect of this
zymotic disease after delivery will gravely influence the
prognosis. Cazeaux was of opinion that pregnant women
are not apt to contract the disease; while Montgomery
thought that the poison when absorbed during pregnancy
might remain latent until delivery, when its characteristic
effects were produced.

Measles, unless very severe, often runs its course without Measles.
seriously affecting the mother or child. I have myself seen
several examples of this. De Tourcoing, however, states
that out of fifteen cases the mother aborted in seven, these
being all very severe attacks. Some cases are recorded in
which the child was born with the rubeolous eruption
upon it.

The pregnant woman may be attacked with any of the Continued fevers.
continued fevers, and, if they are at all severe, they are apt
to produce abortion. Out of 22 cases of typhoid, 16 aborted,
and the remaining six, who had slight attacks, went on to

term; out of 63 cases of relapsing fever, abortion or premature labour occurred in 23. According to Schweden the
main cause of danger to the fœtus in continued fevers is
the hyperpyrexia, especially when the maternal temperature
reaches 104° or upwards. The fevers do not appear to be
aggravated as regards the mother, and the same observation
has been made by Cazeaux with regard to this class of disease
occurring after delivery.

**Pneumo
nia.**   Pneumonia seems to be specially dangerous, for of 15
cases collected by Grisolle [1] 11 died—a mortality immensely
greater than that of the disease in general. The larger
proportion also aborted, the children being generally dead,
and the fatal result is probably due, as in the severe continued fevers, to hyperpyrexia. The cause of the maternal
mortality does not seem quite apparent, since the same
danger does not appear to exist in severe bronchitis, or
other inflammatory affections.

Phthisis.   Contrary to the usually received opinion it appears certain
that pregnancy has no retarding influence on co-existing
phthisis, nor does the disease necessarily advance with greater
rapidity after delivery. Out of 27 cases of phthisis, collected
by Grisolle,[2] 24 showed the first symptoms of the disease
after pregnancy had commenced. Phthisical women are not
apt to conceive; a fact which may probably be explained by
the frequent co-existence, in such cases, of uterine disease,
especially severe leucorrhœa. The entire duration of the
phthisis seems to be shortened, as it averaged only nine and
a half months in the 27 cases collected—a fact which proves,
at least, that pregnancy has no material influence in arresting its progress. If we consider the tax on the vital powers
which pregnancy naturally involves, we must admit that this
view is more physiologically probable than the one generally
received, and apparently adopted without any due grounds.

**Heart-
disease.**   The evil effects of pregnancy and parturition on chronic
heart-disease have of late received much attention from
Spiegelberg, Fritsch, Peter, and other writers. The subject
has been ably discussed [3] in a series of elaborate papers by
Dr. Angus MacDonald, which are well worthy of study. Out
of 28 cases collected by him, 17, or 60 per cent., proved

[1] *Arch. Gén de Méd.* vol. xiii. p. 291.
[2] *Ibid.* vol. xxii.   [3] *Obst. Journ.* 1877.

fatal. This, no doubt, is not altogether a reliable estimate of the probable risk of the complication; but, at any rate, it shows the serious anxiety which the occurrence of pregnancy in a patient suffering from chronic heart-disease must cause. Dr. MacDonald refers the evils resulting from pregnancy in connection with cardiac lesions to two causes: first, destruction of that equilibrium of the circulation which has been established by compensatory arrangements; secondly, the occurrence of fresh inflammatory lesions upon the valves of the heart already diseased.

The dangerous symptoms do not usually appear until after the first half of the pregnancy has passed, and the pregnancy seldom advances to term. The pathological phenomena generally met with in fatal cases are pulmonary congestion, especially of the bronchial mucous membrane, and pulmonary œdema, with occasional pneumonia and pleurisy. Mitral stenosis seems to be the form of cardiac lesion most likely to prove serious, and, next to this, aortic incompetency. The obvious deduction from these facts is, that heart-disease, especially when associated with serious symptoms, such as dyspnœa, palpitation, and the like, should be considered a strong contra-indication of marriage. When pregnancy has actually occurred, all that can be done is to enjoin the careful regulation of the life of the patient, so as to avoid exposure to cold, and all forms of severe exertion.

The important influence of syphilis on the ovum is fully considered elsewhere. As regards the mother, its effects are not different from those at other times. It need only, therefore, be said that, whenever indications of syphilis in a pregnant woman exist, the appropriate treatment should be at once instituted and carried on during her gestation, not only with the view of checking the progress of the disease, but in the hope of preventing or lessening the risk of abortion, or of the birth of an infected infant. So far from pregnancy contra-indicating mercurial treatment, there rather is a reason for insisting on it more strongly. As to the precise medication, it is advisable to choose a form that can be exhibited continuously for a length of time without producing serious constitutional results. Small doses of the bichloride of mercury, such as one-sixteenth of a grain, thrice daily, or of the iodide of mercury, or of the hydrargyrum cum creta, in

Syphilis.

combination with reduced iron, answer this purpose well : or in the early stages of pregnancy, the mercurial vapour bath, or cutaneous inunction, may be employed.

Dr. Weber, of St. Petersburg,[1] has made some observations showing the superiority of the latter methods, which he found did not interfere with the course of pregnancy ; the contrary was the case when the mercury was administered by the mouth, probably, as he supposes, from disturbance of the digestive system. It must be borne in mind that in married women it may sometimes be expedient to prescribe an anti-syphilitic course without their knowledge of its nature, so that inunction is not always feasible.

Epilepsy. The influence of pregnancy on epilepsy does not appear to be as uniform as might perhaps be expected. In some cases the number and intensity of the fits have been lessened, in others the disease becomes aggravated. Some cases are even recorded in which epilepsy appeared for the first time during gestation. On account of the resemblance between epilepsy and eclampsia there is a natural apprehension that a pregnant epileptic may suffer from convulsions during delivery. Fortunately, this is by no means necessarily the case, and labour often goes on satisfactorily without any attack.

Icterus: acute yellow atrophy of liver. Jaundice, the result of acute yellow atrophy of the liver, is occasionally observed, and is said to have been sometimes epidemic. Independently of the grave risks to the mother, it is most likely to produce abortion or the death of the fœtus. According to Davidson,[2] it originates in catarrhal icterus, the excretion of the bile-products being impeded in consequence of pregnancy, and their retention giving rise to the fatal blood-poisoning which accompanies the severer forms of the disease. Slight and transient attacks of jaundice may occur, without being accompanied by any bad consequences. Their production is probably favoured by the mechanical pressure of the gravid uterus on the intestines and the bile-ducts.

Carcinoma. The occurrence of pregnancy in a woman suffering from malignant disease of the uterus is by no means so rare as might be supposed, and must naturally give rise to much

[1] Allgem. Med. Cent. Zeit. Feb. 1875.
[2] Monat. f. Geburt. 1867.

anxiety as to the result.   The obstetrical treatment of these cases will be discussed elsewhere.   Should we be aware of the existence of the disease during gestation, the question will arise whether we should not attempt to lessen the risks of delivery by bringing on abortion or premature labour.   The question is one which is by no means easy to settle.   We have to deal with a disease which is certain to prove fatal to the mother before long, and the progress of which is probably accelerated after labour, while the manipulations necessary to induce delivery may very unfavourably influence the diseased structures.   Again, by such a measure we necessarily sacrifice the child, while we are by no means certain that we materially lessen the danger to the mother.   The question cannot be settled except on a consideration of each particular case.   If we see the patient early in pregnancy, by inducing abortion we may save her the dangers of labour at term— possibly of the Cæsarean section—if the obstruction be great. Under such circumstances, the operation would be justifiable. If the pregnancy has advanced beyond the sixth or seventh month, unless the amount of malignant deposit be very small indeed, it is probable that the risks of labour would be as great to the mother as at term, and it would then be advisable to give her the advantage of the few months' delay.

Cases are occasionally met with in which pregnancy occurs in women who are suffering from ovarian tumour, and their proper management has given rise to considerable discussion. There can be no doubt that such cases are attended with very dangerous and often fatal consequences, for the abdomen cannot well accommodate the gravid uterus and the ovarian tumour, both increasing simultaneously.   The result is that the tumour is subject to much contusion and pressure, which have sometimes led to the rupture of the cyst, and the escape of its contents into the peritoneal cavity ; at others to a low form of inflammation, attended with much exhaustion, the death of the patient supervening either before or shortly after delivery.   The danger during delivery from the same cause, in the cases which go on to term, is also very great.   Of 13 cases of delivery by the natural powers, which I collected in a paper on 'Labour Complicated with Ovarian Tumour,'[1] far more than one-half proved fatal.   Another

Ovarian tumour.

----

[1] *Obst. Trans.* vol. ix.

Explana-
tion of the
dangers. source of danger is twisting of the pedicle, and consequent strangulation of the cyst, of which several instances are recorded. It is obvious, then, that the risks are so manifold that in every case it is advisable to consider whether they can be lessened by surgical treatment.

Methods
of treat-
ment.
The means at our disposal are either to induce labour prematurely, to treat the tumour by tapping, or to perform ovariotomy. The question has been particularly discussed by Spencer Wells in his works on 'Ovariotomy,' and by Barnes in his 'Obstetric Operations.' The former holds that the proper course to pursue is to tap the tumour when there is any chance of its being materially lessened in size by that procedure, but that when it is multilocular, or when its contents are solid, ovariotomy should be performed at as early a period of pregnancy as possible. Barnes, en the other hand, maintains that the safer course is to imitate the means by which nature often meets this complication, and bring on premature labour without interfering with the tumour. He thinks that ovariotomy is out of the question, and that tapping may be insufficient and leave enough of the tumour to interfere seriously with labour. So far as recorded cases go, they unquestionably seem to show that tapping is not more dangerous than at other times, and that ovariotomy may be practised during pregnancy with a fair amount of success. Wells records 10 cases which were surgically interfered with. In 1 tapping was performed, and in 9 ovariotomy; and of these 8 recovered, the pregnancy going on to term in 5. On the other hand, 5 cases were left alone, and either went to term, or spontaneous premature labour supervened; and of these 3 died. The cases are not sufficiently numerous to settle the question, but they certainly favour the view taken by Wells rather than that by Barnes. It is to be observed that, unless we give up all hope of saving the child, and induce abortion, the risk of induced premature labour, when the pregnancy is sufficiently advanced to hope for a viable child, would almost be as great as that of labour at term; for the question of interference will only have to be considered with regard to large tumours, which would be nearly as much affected by the pressure of a gravid uterus at seven or eight months, as by one at term. Small tumours generally escape attention, and are more apt to be impacted

before the presenting part in delivery. The success of ovariotomy during pregnancy has certainly been great, and we have to bear in mind that the woman must necessarily be subjected to the risk of the operation sooner or later, so that we cannot judge of the case as one in which abortion termi-nates the risk. Even if the operation should put an end to the pregnancy—and there is at least a fair chance that it will not do so—there is no certainty that that would increase the risk of the operation to the mother, while as regards the child we should only have the same result as if we in-tentionally produced abortion. On the whole, then, it seems that the best chance to the mother, and certainly the best to the child, is to resort to the apparently heroic practice recom-mended by Wells. The determination must, however, be to some extent influenced by the skill and experience of the operator. If the medical attendant has not gained that ex-perience which is so essential for a successful ovariotomist, the interests of the mother would be best consulted by the induction of abortion at as early a period as possible. One or other procedure is essential; for, in spite of a few cases in which several successive pregnancies have occurred in women who have had ovarian tumours, the risks are such as not to justify an expectant practice. Should rupture of the cyst occur, there can be no doubt that ovariotomy should at once be resorted to, with the view of removing the lacerated cyst and its extravasated contents.

Pregnancy may occur in a uterus in which there are one or more fibroid tumours. If these are situated low down and in a position likely to obstruct the passage of the fœtus, they may very seriously complicate delivery. When they are situated in the fundus or body of the uterus they may give rise to risk from hæmorrhage, or from inflammation of their own structure. Inasmuch as they are structurally similar to the uterine walls, they partake of the growth of the uterus during pregnancy, and frequently increase remarkably in size. Cazeaux says—'I have known them in several instances to acquire a size in three or four months which they would not have done in several years in the non-pregnant condition.' Conversely, they share in the involution of the uterus after delivery, and often lessen greatly in size, or even entirely disappear. Of this fact I have elsewhere recorded several

Fibroid tumours.

curious examples;[1] and many other instances of the complete disappearance of even large tumours have been described by authors whose accuracy of observation cannot be questioned.

Treatment.

The treatment will vary with the position of the tumour. If it is such as to be certain to obstruct the passage of the child, abortion should be induced as soon as possible. If the tumour is well out of the way, this is not so urgently called for. The principal danger then is that the tumour will impede the post-partum contraction of the uterus, and favour hæmorrhage. Even if this should happen, the flooding could be controlled by the usual means, especially by the injection of the perchloride of iron. I have seen several cases in which delivery has taken place under such circumstances without any untoward accident. The danger from inflammation and subsequent extrusion of the fibroid masses would probably be as great after abortion or premature labour as after delivery at term. It seems, therefore, to be the proper rule to interfere when the tumours are likely to impede delivery, and in other cases to allow the pregnancy to go on, and be prepared to cope with any complications as they arise. The risks of pregnancy should be avoided in every case in which uterine fibroids of any size exist, the patients being advised to lead a celibate life.

[1] *Obst. Trans.* vols. v., xiii., and xix.

CHAPTER IX.

PATHOLOGY OF THE DECIDUA AND OVUM.

COMPARATIVELY little is, unfortunately, known of the patho- Pathology
logical changes which occur in the mucous membrane of the of the
decidua.
uterus during pregnancy.   It is probable that they are of
much more consequence than is generally believed to be the
case ; and it is certain that they are a frequent cause of
abortion.

One of the most generally observed probably depends on Endometritis.
endometritis antecedent to conception.   When the impregnated ovule reached the uterus, it engrafted itself on the
inflamed mucous membrane, which was in an unfit condition
for its reception and growth.   A not uncommon result, under
such circumstances, is the laceration of some of the decidual
vessels, extravasation of the blood between the decidua and
the uterine walls, and consequent abortion at an early stage
of pregnancy.   As this morbid state of the uterine mucous
membrane is likely to continue after abortion is completed,
the same history repeats itself on each impregnation, and
thus we may have constant early miscarriages produced.   It
does not necessarily follow, however, that the pregnancy is
immediately terminated when this state of things is present.
Sometimes a condition of hyperplasia of the decidua is produced, the membrane becomes much thickened and hypertrophied in consequence of proliferation of its interstitial
connective tissue, and the decidual cells are greatly increased
in size (fig. 86).   In other instances the internal surface of
the decidua becomes studded with rough polypoid growths,[1]
depending on proliferation of its interstitial tissue.   Duncan
has found that the hypertrophied decidua is always in a
state of fatty degeneration, more advanced in some places
than in others.[2]   The result of these alterations is frequently

---

[1] Virchow's *Archiv. für Path.* 1861, 1st edition.

[2] *Researches in Obstetrics*, p. 293.

to produce dwindling or death of the ovum, which, however, retains its connection with the decidua, until, after a lapse of time, the decidua is expelled in the form of a thick triangular

Fig. 86.

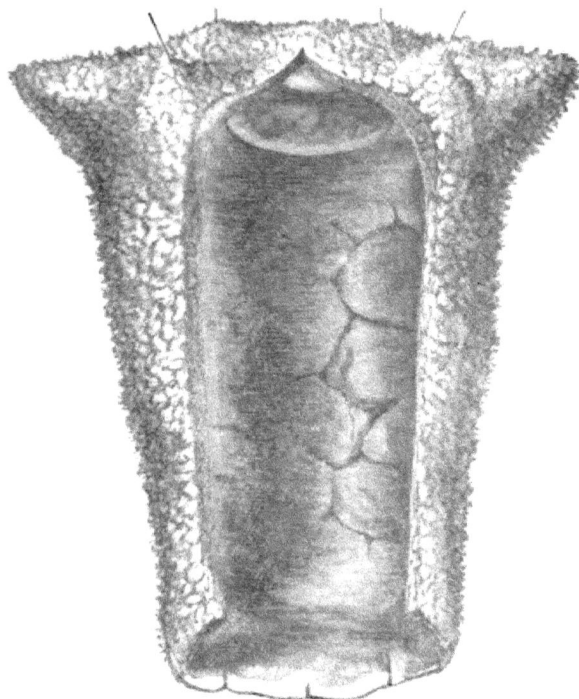

HYPERTROPHIED DECIDUA LAID OPEN, WITH THE OVUM ATTACHED TO ITS FUNDAL PORTION. (After Duncan.)

fleshy substance, with the atrophied ovum attached to some part of its inner surface. In other cases, in which the hyperplasia has advanced to a less extent, the nutrition of the fœtus is not interfered with, and pregnancy may continue to term, the changes in the decidua being recognisable after delivery. Other diseases besides endometritis may give rise to similar alterations in the decidua, one of these being, as Virchow maintains, syphilis. The converse condition, an imperfect development of the decidua, especially of the decidua reflexa, has also been noted as a cause of abortion. The ovum will then hang loosely in the uterine cavity, without the support which the growth of the decidua reflexa

Syphilis.

around it ought to afford, and its premature expulsion readily follows (fig. 87).

The peculiar condition known as *hydrorrhœa gravidarum* most probably depends on some obscure morbid state Hydrorrhœa gravidarum.

Fig. 87.

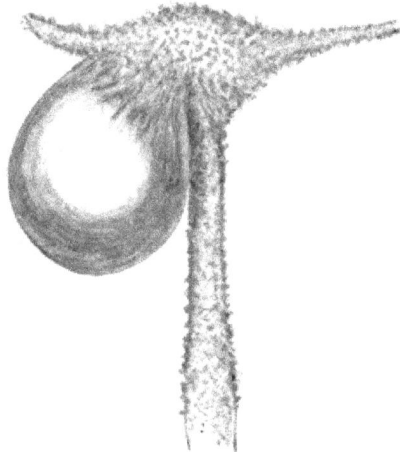

IMPERFECTLY DEVELOPED DECIDUA VERA, WITH THE OVUM. (After Duncan.)

of the uterine mucous membrane. By it is meant a discharge of clear watery fluid at intervals during pregnancy. It may happen at any period of gestation, but is most commonly met with in the latter months. It may commence with a mere dribbling, or there may be a sudden and copious discharge of fluid. Afterwards the watery fluid, which is generally of a pale yellowish colour and transparent like the liquor amnii, may continue to escape at intervals for many weeks, and sometimes in very great abundance, so as to saturate the patient's clothes. Very frequently it is expelled in gushes, and at night, when the patient is lying quietly in bed ; its escape is then probably due to uterine contraction.

Many theories have been held as to its cause. By some it is attributed to the rupture of a cyst placed between the ovum and the uterine walls; Baudelocque referred it to a transudation of the liquor amnii through the membranes ; while Burgess and Dubois believed it to depend on a laceration of the membranes at a distance from the os uteri. Mattei more recently has attributed it to the existence of a

sac between the chorion and the amnion. It may be that in
some instances a single discharge of fluid may come from one
of the two last-mentioned causes. But if it be continuous
or repeated, another source must be sought for. Hegar[1]
maintains that it is the result of abundant secretion from the
glands of the mucous membrane, which accumulates between
the decidua and chorion, and escapes through the os uteri.
If this occur the decidua is probably in an hypertrophied
and otherwise morbid state. Hydrorrhœa is chiefly of interest from the error of diagnosis it is likely to give rise to ;
for, on being summoned to a case in which watery discharge
has occurred for the first time, we are naturally apt to suppose that the membranes have ruptured, and that labour is
imminent. Nor is there any very certain means of deciding
if this be so. In hydrorrhœa, we find that pains are absent,
the os uteri unopened, and ballottement may be made out.
Even if the membranes be ruptured, there will be no indication for interference unless labour has actually commenced ;
and the repetition of the discharge, and the continuance of
the pregnancy, will soon clear up the diagnosis. Hydrorrhœa,
although apt to alarm the patient, need not give rise to any
anxiety. The pregnancy generally progresses favourably to
the full period ; although, in exceptional cases, premature
labour may supervene. No treatment is necessary, nor is
there any that could have the least effect in controlling the
discharge.

Pathology
of the
chorion.
The only important disease of the chorion with which
we are acquainted is the well-known condition which is
variously described as *uterine hydatids, cystic disease of the
ovum, hydatiform degeneration of the chorion,* or *vesicular
mole.* The name of uterine hydatids was long given to it
on the supposition that the grape-like vesicles which characterise the disease were true hydatids, similar to those
which develop in the liver and other structures. This idea
has long been exploded, and it is now known as a certainty
that the disease originates in the villi of the chorion. The
precise mode and the causes of its production are, however,
not yet satisfactorily settled. The disease is characterised
by the existence in the cavity of the uterus of a large

[1] *Monat. f. Geburt,* 1863.

number of translucent vesicles, containing a clear limpid fluid, which has been found on analysis to bear close resemblance to the liquor amnii. These small bladder-like bodies, which vary in size from that of a millet-seed to an acorn, are often described as resembling a bunch of grapes or currants. On more minute examination, they are found not to be each attached to independent pedicles, as is the case in a bunch of grapes, but some of them grow from other vesicles, while others have distinct pedicles attached to the chorion, the pedicles themselves sometimes being distended by fluid (fig. 88). This peculiar arrangement of the vesicles is explained by their mode of growth.

Fig. 88.

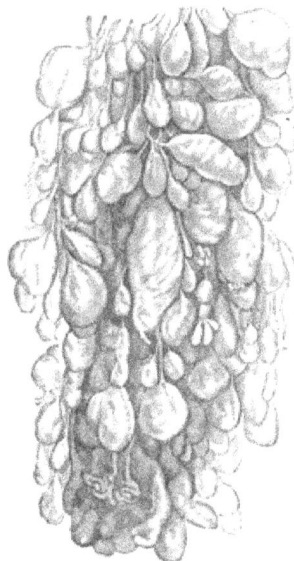

HYDATIFORM DEGENERATION OF THE CHORION.

There has been considerable discussion as to the etiology of this disease. By some it is supposed always to follow death of the fœtus; and the whole developmental energy being expended on the chorion, which retains its attachment to the decidua, the result is its abnormal growth and cystic degeneration. This is the view maintained by Gierse and Graily Hewitt, and it is favoured by the undoubted fact that in almost all cases the fœtus has entirely disappeared; and by the occasional occurrence of cases of twin conceptions in which one chorion has degenerated, the other remaining healthy until term. On the other hand, it is maintained that the starting-point is connected with the maternal organism. Virchow thinks it originates in a morbid state of the decidua; while others have attributed it to some blood dyscrasia on the part of the mother, such as syphilis. There are many reasons for believing that causes of this nature may originate the affection. Thus it is often found to occur more than once in the same person; and alterations of a similar kind, although limited in extent, are not unfre-

[margin: Causes of cystic degeneration.]

quently found in connection with the placenta and membranes of living children.   On this theory the death of the
fœtus is secondary, the consequence of impaired nutrition
from the morbid state of the chorion.   The probability is
that both views may be right, the disease sometimes following the death of the embryo, and at others being the result
of obscure maternal causes.

Its
pathology.
    The degeneration of the chorion villi generally commences
at an early period of pregnancy, before the placenta has
commenced to form.   In that case the entire superficies of
the chorion becomes affected.   The disease, however, may
not begin until after the greater part of the chorion villi
have atrophied, and then it is limited to the placenta.   The
epithelium of the villi appears to be the part first affected,
and the whole interior of the diseased villus becomes filled
with cells.   The connective tissue of the villus undergoes a
remarkable proliferation, and collects in masses at individual
spots, the remainder of the villus being unaffected.   By the
growth of these elements the villus becomes distended, and
many of the cells liquefy, the intercellular fluid, thus produced, widely separating the connective tissue, so as to form
a network in the interior of the villus.[1]   Thus are formed
the peculiar grape-like bodies which characterise the disease.
When once the degeneration has commenced, the diseased
tissue has a remarkable power of increase, so that it sometimes forms a mass as large as a child's head, and several
pounds in weight.

    The nutrition of the altered chorion is maintained by
its connections with the decidua, which is also generally
diseased and hypertrophied.   Sometimes the adhesion of the
mass to the uterine walls is very firm, and may interfere with
its expulsion; while, in a few rare cases, it has been found
that the villi have forced their way into the substance of the
uterus, chiefly through the uterine sinuses, and thus caused
atrophy and thinning of its muscular structure.   Cases of
this kind are related by Volkmann, Waldeyer,[2] and Barnes,
and it is obvious that the intimate adhesion thus affected
must seriously add to the gravity of the prognosis.

    Taking this view of the etiology of this disease, it is

[1] Braxton Hicks, *Guy's Hospital Reports*, vol. ii. Third Series, p. 380.
[2] Virchow's *Archiv*, vol. xliv. p. 86.

obvious that it is essentially connected with pregnancy, and that there is no valid ground for maintaining, as has sometimes been done, that it may occur independently of conception. It is just possible, however, that true entozoa may form in the substance of the uterus, which, being expelled per vaginam, might be taken for the results of cystic disease, and thus give rise to groundless suspicions as to the patient's chastity. Hewitt has related one case in which true hydatids, originally formed in the liver, had extended to the peritoneum, and were about to burst through the vagina at the time of death. This occurred in an unmarried woman. One or two other examples of true hydatids forming in the substance of the uterus are also recorded. A very interesting case is also related by Hewitt,[1] in which undoubted acephalocysts were expelled from the uterus of a patient who ultimately recovered. A careful examination of the cyst and its contents would show their true nature, as the echinococci heads, with their characteristic hooklets, would be discoverable by the microscope.

It is also possible that unfounded suspicions might arise from the fact of a patient expelling a mass of hydatids long after impregnation. In the case of a widow, or woman living apart from her husband, serious mistakes might thus be made. This has been specially pointed out by McClintock,[2] who says: 'Hydatids may be retained in utero for many months or years, or a portion only may be expelled, and the residue may throw out a fresh crop of vesicles, to be discharged on a future occasion.'

The symptoms of cystic disease of the ovum are by no means well marked. At first there is nothing to point to the existence of any morbid condition, but as pregnancy advances its ordinary course is interfered with. There is more general disturbance of the health than there ought to be, and the reflex irritations, such as vomiting, may be unusually developed. The first physical sign remarked is rapid increase of the uterine tumour, which soon does not correspond in size to the supposed period of pregnancy. Thus, at the third month, the uterus may be found to reach up to, or beyond, the umbilicus. About this time there generally are more or less profuse watery and sanguineous discharges, which have

---

[1] *Obs. Trans.* vol. xii.   [2] McClintock's *Diseases of Women*, p. 398.

been described as resembling currant juice. They no doubt depend on the breaking down and expulsion of the cysts, caused by painless uterine contractions. They are sometimes excessive in amount, recur with great frequency, and often reduce the patient extremely. Portions of cysts may now generally be found mingled with the discharge, and sometimes large masses of them are expelled from time to time. Indeed, the discovery of portions of cysts is the only certain diagnostic sign. Vaginal examination, before the os has dilated, will give no information except the absence of ballottement. An unusual hardness or density of the uterus—described by Leishman, who attributes much importance to it, as 'a peculiar doughy, boggy feeling'—has been pointed out by several writers. The contour of the uterine tumour, moreover, is often irregular. In addition, we, of course, fail to discover the usual auscultatory signs of pregnancy. All this may aid in diagnosis, but nothing, except the presence of cysts in the watery bloody discharge, will enable us to pronounce with certainty as to the nature of the disease.

Treatment.

As soon as the diagnosis is established, the indications for treatment are obvious. The sooner the uterus is cleared of its contents the better. Ergot may be given with advantage to favour uterine contraction, and the expulsion of the diseased ovum. Should this fail, more especially if the hæmorrhage be great, the fingers, or the whole hand, must be introduced into the uterus, and as much as possible of the mass removed. As the os is likely to be closed, its preliminary dilatation by sponge or laminaria tents, or by a Barnes's bag, if it be already opened to some extent, will in most cases be required. If chloroform be then administered, the remaining steps of the operation will be easy. On account of the occasional firm adhesion of the cystic mass to the uterus, too energetic attempts at complete separation should be avoided. Any severe hæmorrhage after the operation can be controlled by swabbing out the uterine cavity with the perchloride of iron solution.

Myxoma fibrosum.

Under the name of *Myxoma fibrosum*, a more rare degeneration of the chorion has been described by Virchow and Hildebrandt,[1] characterised, not by vesicular, but fibroid

[1] *Monat. f. Geburt*, May 1865.

degeneration of the connective tissue of the chorion. This is, however, too little understood to require further observation.

The pathology of the placenta has of late years attracted much attention, and it has an important practical bearing, in consequence of its effect on the child.

Pathology of the placenta.

Placentæ vary considerably in shape. They may be crescentic, or spread over a considerable surface, in consequence of the chorion villi entering into communication with a larger portion of the decidua than usual (*Placenta membranacea*). Such forms, however, are merely of scientific interest. The only anomaly of shape of any practical importance is the formation of what have been called *placentæ succenturiæ*. These consist of one or more separate masses of placental tissue, produced by the development of isolated patches of chorion villi. Hohl believes that they always form exactly at the junction of the anterior and posterior walls of the uterus, which in early pregnancy is a mere line. As the uterus expands, the portions of placenta on each side of this become separated from each other. They are only of consequence from the possibility of their remaining unnoticed in the uterus after delivery, and giving rise to secondary post-partum hæmorrhage. The rare form of double placenta with a single cord, figured in the accompanying woodcut (fig. 89), was probably formed in this way, and the supplementary portion, in such a case, might readily escape notice.

The placenta may also vary in dimensions. Sometimes it is of excessive size, generally when the child is unusually big; but not unfrequently in connection with hydramnios, the child being dead and shrivelled. In other cases it is remarkably small, or at least appears to be so. If the child be healthy, this is probably of no pathological importance, as its smallness may be more apparent than real, depending on its vessels not being distended with blood. When true atrophy of the placenta exists, the vitality of the fœtus may be seriously interfered with. This condition may depend either on a diseased state of the chorion villi, or of the decidua in which they are implanted.[1] The latter is the more common of the two; and it generally consists in hyperplasia of the connective tissue of the decidua, which presses

---

[1] Whittaker, *Amer. Jour. of Obst.* vol. iii. p. 229.

on the villi and vessels, and gives rise to general or local atrophy. This change is similar in its nature to that observed in cirrhosis of the liver, and certain forms of Bright's

Fig. 89.

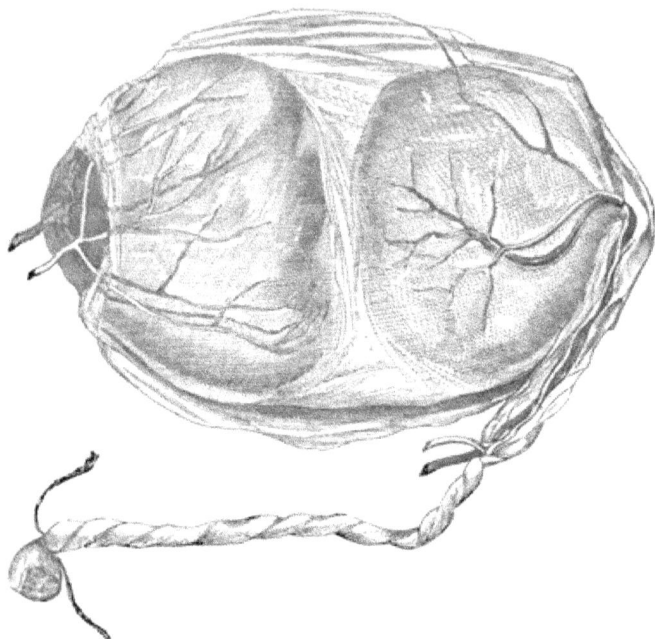

DOUBLE PLACENTA, WITH SINGLE CORD.

Placen-
titis.

disease. It has generally been ascribed to inflammatory changes, and, under the name of *placentitis*, has been described by many authors, and has been considered to be a common disease. To it are attributed many of the morbid alterations which are commonly observed in placentæ, such as hepatisations, circumscribed purulent deposits and adhesions to the uterine walls. Many modern pathologists have doubted whether these changes are in any proper sense inflammatory. Whittaker observes on this point : ' The disposition to reject placentitis altogether increases in modern times. Indeed, it is impossible to conceive of inflammation on the modern theory (Cohnheim) of that process, since there are no capillaries, in the maternal portion at least, through whose walls a " migration " might occur, and there are no nerves to regulate the contractility of the vessel-walls in the entire structure.' Robin thus explains the various pathological changes above alluded to: ' What has been taken for inflammation of the placenta is nothing else than a condition of transformation of blood-clots at various periods. What has been regarded as pus is only fibrine in the course of disorganisation, and in those cases where true pus has been found the pus did not come from the placenta, but from an inflammation of the tissue of the uterine vessels and an accidental disposition in the tissue of the placenta.' The extravasations of blood here alluded to are of very common occurrence, and they are found in all parts of the organ; in its substance, on its decidual surface, or immediately below the amnion, where they serve as points of origin for the cysts that are there often observed. The

Blood extravasations.

Fig. 90.

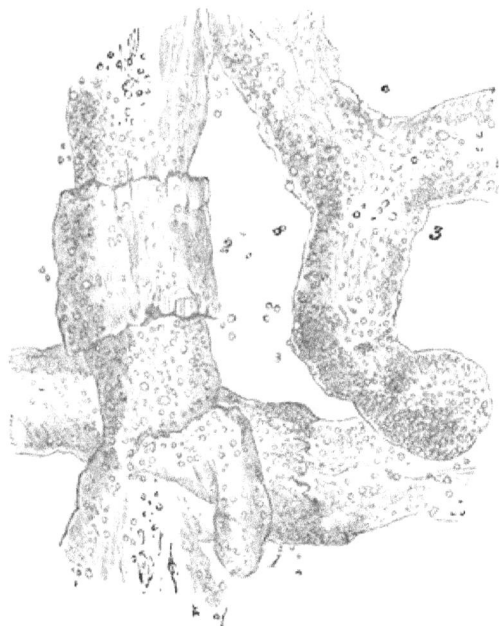

FATTY DEGENERATION OF THE PLACENTA.

fibrine thus deposited undergoes retrograde metamorphosis as in other parts of the body; it becomes decolourised, it undergoes fatty degeneration, or becomes changed into calcareous masses; and in this way, it is supposed, may be explained the various pathological changes which are so commonly observed.   The amount of retrograde metamorphosis, and the precise appearance presented, will, of course, depend on the time that has elapsed since the blood extravasations took place.

Fatty degeneration.
Fatty degeneration of the placenta, and its influence on the nutrition of the fœtus, have been specially studied in this country by Barnes and Druitt.   Yellowish masses of varying sizes are very commonly met with in placentæ, and these are found to consist, in great part, of molecular fat, mixed with a fine network of fibrous tissue.   The true fatty degeneration, however, specially affects the chorion villi (fig. 90).   On microscopic examination they are found to be altered and misshaped in their contour, and to be loaded with fine granular fat-globules. Similar changes are observed in the cells of the decidua.   The influence on the fœtus will, of course, depend on the extent to which the functions of the villi are interfered with.   The probable cause of this degeneration is, no doubt, some obscure alteration in the nutrition of the tissue, depending on the state of the mother's health.   Barnes believes that syphilis has much influence in its production.   Druitt has pointed out that some amount of fatty degeneration is always present in a mature placenta, and is probably connected with the physiological separation of the organ ; and Goodell has more recently suggested that an unusual amount of this change may be merely an anticipation of the natural termination of the life of the placenta.[1]

Other morbid states.
Other morbid states of the placenta, of greater rarity, are occasionally met with, as an œdematous infiltration of its tissue, always occurring, according to Lange, in cases of hydramnios, pigmentary and calcareous deposits, and tumours of various kinds ; but these require only a passing mention.

Pathology of the umbilical cord.
The umbilical cord may be of excessive length, varying from 18 to 20 inches, which is its average measurement, up to 50 or 60 inches, and a case is recorded in which it even reached the extraordinary length of nine feet.   If unusually long it may be twisted round the limbs or neck of the child, and the latter position may, in exceptional instances, prove injurious during labour.

[1] *American Journal of Obstetrics*, vol. ii. p. 535.

Some authors refer cases of spontaneous amputation of fœtal limbs in utero to constrictions by the umbilical cord, but this accident is more probably produced by filamentous adnexa of the amnion. Knots in the cord are not uncommon, and they result from the fœtus, in its movements, passing through a loop of the cord (fig. 91). If there is an average amount of Wharton's jelly in the cord the vessels are protected from pressure, and no bad effects follow. Géry, in a recent paper on this subject,[1] attempts to show that such knots are more important than is generally believed, and relates two cases in which he believes them to have caused the death of the fœtus.

Fig. 91.

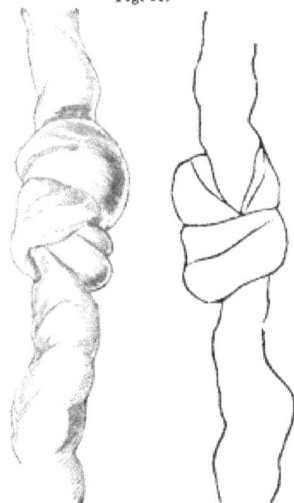

KNOTS OF THE UMBILICAL CORD.

Extreme torsion of the cord, an exaggeration of the spiral twists generally observed, may prove injurious, and even fatal, to the child by obstructing the circulation in the vessels. Spaeth mentions three cases in which this caused the death of the fœtus, the cord being twisted until it was reduced to the thickness of a thread.

Anomalies in the distribution of the vessels of the cord are of common occurrence. The cord may be attached to the edge, instead of to the centre, of the placenta (*battledore placenta*). It may break up into its component parts before reaching the placenta, the vessels running through the membranes; and if, in such a case, traction on the cord be made, the separate vessels may lacerate, and the cord become detached. There may be two veins and one artery, or only one vein and one artery, or there may be two separate cords to one placenta. These and other anomalies that might be mentioned are of little practical importance.

The principal pathological condition of the amnion with which we are acquainted is that which is associated with

Pathology of the amnion.

---

[1] *L'Union Médicale*, Oct. 1876.

excessive secretion of liquor amnii, and is generally known under the name of *hydramnios*, which term Kidd[1] limits to cases in which more than two quarts of amniotic fluid exist. Its precise cause is still a matter of doubt. By some it is referred to inflammation of the amnion itself; at other times it is apparently connected with some morbid state of the decidua, which may be found diseased and hypertrophied. The fœtus is very often dead and shrivelled, and the placenta enlarged and œdematous. It does not necessarily follow, however, that hydramnios causes the death of the child. Out of 33 cases McClintock found that nine children were born dead;[2] and of the 24 born alive, 10 died within a few hours; the remainder survived. There does not appear to be any marked relation between the state of the mother's health and the occurrence of this disease; and it is certainly not necessarily present when the mother is suffering from dropsical effusions in other parts of the body. The theory that the disease is of purely local origin is favoured by the fact that when hydramnios occurs in twin pregnancy one ovum only is generally affected. Its effects, as regards the mother, are chiefly mechanical. It rarely begins to show itself before the fifth or sixth month of pregnancy, but when once it has commenced it rapidly produces a feeling of discomfort and enlargement, altogether beyond that which should exist at the period of pregnancy which has been reached. In advanced stages the distress produced is often very great, the enlarged uterus pressing upon the diaphragm, and producing much embarrassment of respiration. Premature expulsion of the fœtus very often supervenes. Four out of McClintock's patients died after labour, showing that the maternal mortality is high, a result which he refers to the debilitated state of the women who were the subjects of the disease.

The diagnosis is not, as a rule, difficult. It has to be distinguished from ascitic distension of the abdomen, from enlargement of the uterus from twin pregnancy, and from ovarian tumour, or pregnancy complicated with ovarian tumour. The first will be recognised by the superficial

---

[1] 'On the Diagnosis of Dropsy of the Amnion.' *Proceedings of the Obstetrical Society of Dublin*, May 11, 1878.

[2] *Diseases of Women*, p. 383.

position of the fluid; the difficulty of feeling the contour
of the uterus, which is obscured by the surrounding fluid,
and the results of percussion which show that the fluid is
free in the peritoneal cavity; and by the co-existence of
dropsical effusions in other parts of the body. The second
may be difficult, and even impossible, to diagnose from it;
generally, however, in hydramnios the uterine tumour is
more distinctly tense or fluctuating; the fœtal limbs cannot
be felt on palpation; and the lower segment of the uterus,
as felt per vaginam, is unusually distended, the presenting
part not being appreciable. Ovarian tumours alone, or complicating pregnancy, may also be difficult to distinguish
from dropsy of the amnion. The general history of the
case, and the presence or absence of signs of pregnancy, may
enable us to arrive at a diagnosis; and Kidd points out that
the position of the uterus, whether gravid or not, is usually
low down in the pelvis in ovarian dropsy, while in dropsy of
the amnion it is drawn high up, and reached with difficulty
on vaginal examination.

During labour an excessive amount of liquor amnii is
often a cause of deficient uterine action and delay, the pains
being feeble and ineffective. This, of course, tells chiefly in
the first stage, which is often much prolonged, unless the
membranes are punctured early, and the superabundant
fluid allowed to escape. *Its effect on labour.*

No treatment is known to have any effect on the disease.
If the discomfort and distension are very great, it may be
absolutely necessary to puncture the membranes, and allow
the water to escape. This inevitably brings on labour. If
the pregnancy be not sufficiently advanced to give hope for
the birth of a living child, we would not, of course, resort to
this expedient unless the mother's health was seriously imperilled. It is possible that in such cases the patient might
be relieved by inserting the minute needle of an aspirator
through the os, and removing a certain quantity of the
liquor amnii by aspiration, without inducing the labour. I
have never had an opportunity of trying this expedient, but
it seems a possibility. *Treatment.*

A defective amount of liquor amnii is said to favour
certain malformations, by allowing the uterus to compress
the fœtus unduly. It certainly occasionally gives rise to *Deficiency of liquor amnii.*

adhesion between the fœtus and the membranes, and to the formation of amniotic bands which are capable of producing certain fœtal deformities (pp. 269 and 276).

Appearance of the liquor amnii.

The liquor amnii itself varies much in appearance. It is sometimes thick and treacly, instead of limpid, and it may be offensive in odour. The cause of these variations is not well understood.

Pathology of the fœtus.

There is abundant evidence that the fœtus in utero is subject to many diseases, some of which cause its death, and others leave distinct traces of their existence, although not proving fatal. The subject is of great importance, and is well worthy of study. There is still much to be done in this direction, which may lead to important practical results. I can, however, do little more than enumerate some of the principal affections which have been observed.

Blood diseases transmitted through the mother. Small-pox.

It is a well-established fact that the various eruptive fevers from which the mother may suffer may be communicated to the fœtus in utero. When the mother is attacked with confluent small-pox she almost always aborts, but not necessarily so when it is discrete or modified. In such cases it has often happened that the fœtus has been born with evident marks of small-pox. Cases are on record which prove that the fœtus was attacked subsequently to the mother. Thus a mother attacked with small-pox has miscarried, and has given birth to a living child showing no trace of the disease, which, however, showed itself in two or three days; proving that it had been contracted, and had run through its usual period of incubation, when the fœtus was still in utero. It does not follow, however, that the fœtus is affected, as Serres has collected 22 cases in which women suffering from small-pox gave birth to children who had not contracted the disease. It has been supposed that in such cases the child is protected from small-pox, though it has shown no symptom of having had the disease. Tarnier, however, cites two instances in which such children had small-pox two years after birth. Madge and Simpson record cases in which vaccination performed on the mother during pregnancy protected the fœtus, on whom all subsequent attempts at vaccination failed. There is evidence also to prove that the disease may be transmitted to the fœtus through a mother who is herself unsusceptible of contagion ;

the child having been covered with small-pox eruption, the mother being quite free from it. It is probable that the same facts which have been observed with regard to small-pox hold true with reference to other zymotic diseases, such as scarlet fever and measles, although there is not sufficient evidence to justify a positive assertion to that effect.

Amongst other maternal diseases, malaria and lead-poisoning are known to affect the fœtus in utero. Dr. Stokes relates cases in which the mother suffered from tertian ague, the child having also attacks, as evidenced by its convulsive movements, appreciable by the mother, which took place at the regular intervals, but at a different time from the mother's paroxysms. In other cases the febrile paroxysm comes on at the same time in the fœtus as in the mother; and the fact has been verified by the observation that the paroxysms continued to recur simultaneously after delivery. The fœtus has also been born with distinct malarious enlargement of the spleen. From the frequency with which largely hypertrophied spleens are seen in mere infants in malarious districts, I imagine that the intra-uterine disease must be common. I have frequently observed this fact in India, although, of course, without any possibility of ascertaining if the mothers had suffered from intermittent fever during pregnancy. Lead-poisoning is also known to have a most prejudicial effect on the fœtus, and frequently to lead to abortion. M. Paul has collected 81 cases[1] in which it caused the death of the fœtus, in some not until after birth; and occasionally it seems to have affected the fœtus even when the mother escaped.

Of all blood-dyscrasiæ transmitted to the fœtus, the most important is syphilis. Its influence in producing repeated abortion has been elsewhere described (p. 285). It may unquestionably be transmitted to the fœtus without producing abortion, and at term the mother may be either delivered of a living child, bearing evident traces of the disease; of a dead child similarly affected; or of an apparently healthy child in whom the disease develops itself after a lapse of a month or two. These varying effects probably depend on the intensity of the poison; and the longer the time that has elapsed since the origin of the disease in the affected parents, the better

Measles and scarlet fever.

Malaria and lead-poisoning.

Syphilis.

[1] *Arch. Gén. de Med.* 1860.

will be the chance for the child. The disease is, no doubt, generally transmitted through the mother, and if she be affected at the time of conception, the infection of the fœtus seems certain. If, however, she contracts the disease at an advanced period of pregnancy, the child may entirely escape. Ricord even believes that syphilis, contracted after the sixth month of pregnancy, never affects the child. The father alone may transmit the disease to the ovum; and Hutchinson has recorded cases to show that the mother may become secondarily affected through the diseased fœtus. The evidences of syphilitic taint in a living or dead child are sufficiently characteristic. The child is generally puny and ill-developed. An eruption of pemphigus is common, either fully developed bullæ, or their early stage, when they form circular copper-coloured patches. This eruption is always most marked on the hands and feet, and a child born with such an eruption may be certainly considered syphilitic. On post-mortem examination the most usual signs are small patches of suppuration in the thymus, similar localised suppurations in the tissues of the lungs, indurated yellowish patches in the liver, and peritonitis, the importance of which in causing the death of syphilitic children has been specially dwelt on by Simpson.[1]

Inflammatory diseases.

The most important of the inflammatory diseases affecting the fœtus is peritonitis. Simpson has shown that traces of it are very frequently met with, and that it is not always syphilitic. Sometimes it has been observed when the mother has been in bad health during pregnancy, and at others it seems to have resulted from some morbid condition of the fœtal viscera. Pleurisy with effusion is another inflammatory affection which has been noticed.

Dropsies.

The dropsical affections most generally met with are ascites and hydrocephalus, which may both have the effect of impeding delivery. Of these hydrocephalus is the more common, and may give rise to much difficulty in labour. Its causes are uncertain, but it probably depends on some altered state of the mother's health, as it is apt to recur in several successive pregnancies, and is not infrequently associated with an imperfectly developed vertebral column and spina bifida. The fluid collects in the ventricles, which it

[1] *Obst. Works*, vol. i. p. 117.

greatly distends, and these then produce expansion and thinning of the cranium, the bones of which are widely separated from each other at the sutures, which are prominent and fluctuating. In a few cases internal hydrocephalus may be complicated, and the diagnosis in labour consequently obscured by the co-existence of what has been called 'external hydrocephalus.' This consists of a collection of fluid between the skull and the scalp, which may be either formed there originally or may collect from a rupture of one of the sutures or fontanelles during labour, through which the intracranial fluid escapes.

Ascites is generally associated with hydramnios, and sometimes with hydro-thorax, or other dropsical effusions. It is a rare affection, and according to Depaul [1] extreme distension of the bladder is not infrequently mistaken for it.

Tumours of different kinds may be met with in various parts of the child's body, which sometimes grow to a great size and impede delivery. Tarnier records cases of meningocele larger than a child's head, and large cystic growths have been observed attached to the nates, pectoral region, or other parts of the body. Cancerous tumours of considerable size, either external or of the viscera, have also been met with. Other fœtal tumours may be produced by congenital deformities, such as projection of the liver or other abdominal viscera through a deficiency of the abdominal wall ; or spina bifida, from imperfectly developed vertebræ. The amount of dystocia produced by such causes will, of course, vary much in proportion to the size, consistency, and accessibility of the tumour. *Tumours.*

Accidents of serious gravity to the fœtus may happen from violence, to which the mother has been subjected, such as falls or blows, without necessarily interfering with gestation. Many curious examples of this kind are on record. Thus a child has been born presenting a severe lacerated wound extending the whole length of the spine, where both the skin and the muscles had been torn, and which seems to have resulted from the mother having fallen in the last month of pregnancy. Similar lacerations and contusions have been observed in other parts of the body, the wounds being in various stages of cicatrisation, corresponding to the lapse of *Wounds and injuries of the fœtus*

[1] Tarnier's *Cazeaux*, p. 855.

time since the accident had occurred. Intra-uterine fractures
are not rare, apparently arising from similar causes. In some
of these cases the broken ends of the bones had united, but,
from want of accurate apposition, at an acute angle, so as to
give rise to much subsequent deformity. Chaussier records
two cases in which there were many fractures in the same
child, in one 113, and in another 42, which were in different
stages of repair. He attributes this curious occurrence to
some congenital defect in the nutrition of the bones, possibly allied to mollities ossium.[1]

Intra-
uterine
amputa-
tions of
fœtal
limbs.

Intra-uterine amputations of fœtal limbs have not unfrequently been observed. Children are occasionally born
with one extremity more or less
completely absent, and cases are
known in which the whole four
extremities were wanting (fig. 92).
The mode in which these malformations are produced has given rise
to much discussion. At one time
it was supposed that the deficiency
of the limb was due to gangrene of
the extremity, and subsequent separation of the sphacelated parts.
Reuss, who has studied the whole
subject very minutely,[2] considers
gangrene in the unruptured ovum
to be an impossibility, for that
change cannot occur unless there is
access of oxygen, and when portions
of the separated extremity are found in utero, as is often the
case, they show evidences of maceration, but not of decomposition. The general belief is that these intra-uterine amputations depend on constriction of the limb by folds or bands
of the amnion—most often met with when the liquor amnii
is deficient in quantity—which obstruct the circulation and
thus give rise to atrophy of the part below the constriction.
It has been supposed that the umbilical cord might, by en-
circling the limb, produce a like result. It appears doubtful,
however, whether this cause is sufficient to produce complete
separation of the limb, as any great amount of constriction

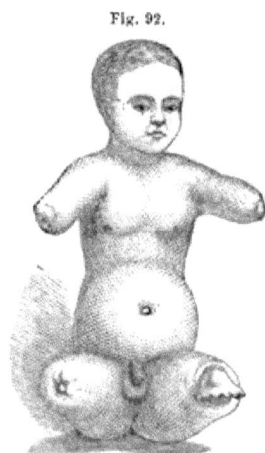

Fig. 92.

INTRA-UTERINE AMPUTATION OF
BOTH ARMS AND LEGS.

[1] *Gazette Hebdom.* 1860.   [2] Scanzoni's *Beiträge*, 1869.

would interfere with the circulation through the cord. Sometimes, when intra-uterine amputation occurs, the separated portion of the limb is found lying loose in the amniotic cavity, and is expelled after the child. Cases of this kind have been recorded by Martin, Chaussier, and Watkinson. More often no trace of the separated extremity can be found. The explanation probably depends upon the period of utero-gestation at which amputation took place. If it occurred at a very early period of pregnancy, before the third month, the detached portion would be minute and soft, and would easily disappear by solution. If at a later period, this could hardly happen, and the detached portion would remain in utero. In cases of the latter kind cicatrisation of the stump has often been observed to be incomplete. Simpson pointed out the occasional existence of rudimentary fingers or toes on the stump of an amputated limb, such as are seen on the thighs in fig. 92. These he attributed to an abortive reproduction of the separated extremity, analogous to what is observed in some of the lower animals. This explanation has been contested with much show of reason. Martin believes that the reproduction is only apparent, and that the rudimentary extremities are, in reality, instances of arrested development. The constricting agents interfered with the circulation sufficiently to arrest the growth of the limb below the site of constriction, but not sufficiently to effect complete separation. If constriction occurred at a very early stage of development, an appearance similar to that observed by Simpson would be produced. It does not follow, however, that all cases of absence of limbs depend on intra-uterine amputations. In some cases they would appear to be the result of a spontaneous arrest of development, or of congenital monstrosity. Mr. Scott[1] relates a case in which a distinct hereditary tendency was evident, and here the deformity certainly could not have resulted from the constriction of amniotic bands. In this family the grandfather had both forearms wanting, with rudimentary fingers attached; the next generation escaped; but the grandchild had a deformity precisely similar to the grandfather.

When from any cause the fœtus has died during pregnancy, it may be either soon expelled, or it may be retained

Death of fœtus.

[1] *Obst. Trans.* vol. xiii. p. 94.

in utero for a longer or shorter time, or even to the full period. The changes observed in such fœtuses vary considerably according to the age of the fœtus at the time of death, or the time that it has been retained in utero. If it die at an early period, when the tissues are very soft, it may entirely dissolve in the liquor amnii, and no trace of it may be found when the membranes are expelled. Or it may shrivel or mummify; and if this happen in a twin pregnancy, as sometimes occurs, the growing fœtus may compress and flatten the dead one against the uterine wall.

Appearance of a putrid fœtus.

At a later period of pregnancy a dead fœtus undergoes changes ascribed to putrefaction, but which produce appearances different from those of decomposition in animal textures exposed to the atmosphere. There is no offensive smell, as in ordinary decay. The tissues are all softened and flaccid. The more manifest changes are in the skin, the epidermis of which is separated from the cutis vera, which has a deep reddish colour. This is especially apparent on the abdomen, which is flaccid, and hollow in the centre. The internal organs are much altered. The brain is diffluent and pulpy, and the cranial bones loose within the scalp. The structures of the muscles and viscera are in various stages of transformation, many having undergone fatty changes, and contain crystals of margarin and cholesterin. The extent to which these changes occur depends, in a great measure, on the length of time the fœtus has been dead, but they do not admit of our estimating with any degree of accuracy what that time has been.

Symptoms and diagnosis of the death of the fœtus.

The symptoms and diagnosis of the death of the fœtus may here be considered. They are, unfortunately, not very reliable. The cessation of the fœtal movements cannot be depended on, as they are frequently unfelt for days or weeks, when the child is alive and well. Sometimes the death of the fœtus is preceded by its irregular and tumultuous movements, and, in women who have been delivered of several dead children in succession, this sensation may guide us in our diagnosis. This suspicion may be confirmed by auscultation. The mere fact that we are unable, at any given time, to hear the fœtal heart will not justify an opinion that the fœtus is dead. If, however, the fœtal heart has been distinctly heard, and after one or two careful examinations, repeated at separate times, it cannot again be made out, the probability of the child being dead may be assumed. Certain changes in the mother's health have been noted in connection with the death of the fœtus, such as depression and lowness of spirits, a feeling of coldness and weight about the lower parts of the abdomen, paleness of the face, a livid circle round the eyes, irregular shiverings and feverishness, shrinking of the breasts, and diminution in the size of the abdominal tumour. All these, however, are too indefinite to justify a positive diagnosis, and they are not infrequently altogether absent. At most they can do no more than cause a suspicion as to what has happened.

CHAPTER X.

ABORTION AND PREMATURE LABOUR.

Import-
ance and
frequency
of abor-
tion.
THE premature expulsion of the foetus is an event of great frequency. The number of foetal lives thus lost is enormous. There are few multiparæ who have not aborted at one time or other of their lives. Hegar estimates that about one abortion occurs to every 8 or 10 deliveries at term. White-head has calculated that at least 90 per cent. of married women, who lived to the change of life, had aborted. The influence of this incident on the future health of the mother is also of great importance. It rarely, indeed, proves directly fatal, but it often produces great debility from the profuse loss of blood accompanying it ; and it is one of the most prolific causes of uterine disease in after-life, possibly because women are apt to be more careless during convalescence than after delivery, and the proper involution of the uterus is thus more frequently interfered with.

Defini-
tion.
A not uncommon division of the subject is into *abortion*, *miscarriage*, and *premature labour*, the first name being applied to expulsion of the ovum before the end of the fourth month of utero-gestation ; miscarriage, to expulsion from the end of the fourth to the end of the sixth month ; and premature labour, to expulsion from the end of the sixth month to the term of pregnancy. This is, however, a needless and confusing subdivision, which leads to no practical result. It suffices to apply the term abortion or miscarriage indiscriminately to all cases in which pregnancy is terminated before the foetus has arrived at a viable age, and premature labour to those in which there is a possibility of its survival.

Age at
which the
foetus is
viable.
There is little or no hope of a foetus living before the 28th week or seventh lunar month, and this period is therefore generally fixed on as the limit between premature labour and

abortion. The rule is, however, not without an occasional, although very rare, exception. Dr. Keiller, of Edinburgh, has recorded an instance in which a fœtus was born alive at the fourth month, nine days after the mother had experienced the sensation of quickening. I myself recently attended a lady who miscarried in the fifth month of pregnancy, the child being born alive, and living for three hours. Several cases are on record in which after delivery in the sixth month the child survived and was reared. The possibility of the birth of a living child under such circumstances should be recognised, as it may give rise to legal questions of importance; but the exceptions to the ordinary rule are so rare that they need not interfere with the division of the subject usually made.

Multiparæ abort far more frequently than primiparæ. This is contrary to the statement in many obstetrical works. Thus, Tyler Smith says 'there seems to be a greater danger of this accident in the first pregnancy.' Schroeder,[1] however, states that 23 multiparæ abort to 3 primiparæ; and Dr. Whitehead, of Manchester, who has particularly studied the subject, believes that abortion is more apt to occur after the third and fourth pregnancies, especially when these take place towards the time for the cessation of menstruation. *Abortion is most common in multiparæ.*

There can be no doubt that women who have aborted more than once are peculiarly liable to a recurrence of the accident. This can generally be traced to the existence of some predisposing cause which persists through several pregnancies, as, for example, a syphilitic taint, a uterine flexion, or a morbid state of the lining membrane of the uterus. It is probable that in many women a recurrence of the accident induces a habit of abortion, or perhaps it might be more accurate to say, a peculiar irritable condition of the uterus, which renders the continuance of pregnancy a matter of difficulty, independently of any recognisable organic cause. *Liability to a recurrence of abortion.*

The frequency of abortion varies much at different periods of pregnancy; and it occurs much more often in the early months, because of the comparatively slight connection then existing between the chorion and the decidua. At a very early period of pregnancy the ovum is cast off with such facility, and is of such minute size, that the fact of abortion *Very early abortions are often unrecognised.*

[1] Schroeder, *Manual of Midwifery*, p. 149.

having occurred passes unrecognised. Very many cases, in which the patient goes one or two weeks over her time, and then has what is supposed to be merely a more than usually profuse period, are probably instances of such early miscarriages. Velpeau detected an ovum, of about fourteen days, which was not larger than an ordinary pea, and it is easy to understand how so small a body should pass unnoticed in the blood which escapes along with it.

Before the end of the third month the ovum is generally expelled entire.

Up to the end of the third month, when miscarriage occurs, the ovum is generally cast off *en masse*, the decidua subsequently coming away in shreds, or as an entire membrane. The abortion is then comparatively easy. From the third to the sixth month, after the placenta is formed, the amnion is, as a rule, first ruptured by the uterine contractions, and the fœtus is expelled by itself. The placenta and membranes may then be shed as in ordinary labour. It often happens, however, that on account of the firmness of the placental adhesion at this period, the secundines are retained for a greater or less length of time. This subjects the patient to many risks, especially to those of profuse hæmorrhage, and of septicæmia.

Abortions are most dangerous between the third and sixth months.

For this reason, premature termination of the pregnancy is attended by much greater danger to the mother between the third and sixth months than at an earlier or later date. After the sixth month the course of events is not different from that attending ordinary labour. The prognosis to the child is more unfavourable in proportion to the distance from the full period of gestation at which premature labour takes place.

Causes.

The causes of abortion may conveniently be subdivided into the *predisposing* and *exciting*, the latter being often slight, and such as would have no effect in inducing uterine contractions in women unless associated with one or more of the former class of causes. The predisposition to abortion may depend on some condition interfering with the vitality of the ovum, or its relation to the maternal structures, or on certain conditions directly affecting the mother's health.

Causes referable to the fœtus.

One of the most common antecedents of abortion is the death of the fœtus, which leads to secondary changes, and ultimately produces the uterine contractions which end in its expulsion. The precise causes of death in any given case cannot always be accurately ascertained, as they sometimes depend on conditions which are traceable to the maternal structures, at others to the ovular, or, it may be, to a combination of the two.   Nor does it by any means follow that the death of the ovum immediately results in its expulsion. The mode in which death of the ovum produces abortion is not difficult to understand, for it necessarily leads to changes in the relations between the ovular and maternal structures ; these changes cause hæmorrhages—partly external, and partly into the membranes—which, in their turn, excite uterine contraction.   Extravasations of blood may take place in various positions.   One of the most common is into the decidual cavity, between the decidua vera and the decidua reflexa—or between the decidua vera and the uterine walls. If the hæmorrhage is only slight, and especially if it comes from that portion of the decidua near the internal os, and at a distance from the ovum, there need be no material

Extravasation of blood following the death of an ovum.

Fig. 93.

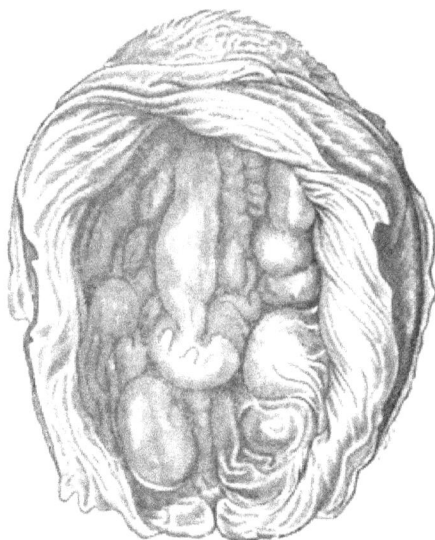

AN APOPLECTIC OVUM, WITH BLOOD EFFUSED IN MASSES UNDER THE FŒTAL SURFACE OF THE MEMBRANES.

separation, and pregnancy may continue. This explains the cases occasionally met with in which there is more or less hæmorrhage, without subsequent abortion. When the amount of extravasated blood is at all great, separation and

abortion necessarily result, and the decidua will be found
on expulsion to have coagula on its surface, and between its
various layers which are found to project into the cavity of
the amnion (fig. 93). In other cases hæmorrhage is still
more extensive, and, after breaking through the decidua reflexa, it forms clots between it and the chorion, and even in
the cavity of the amnion. Supposing expulsion to take place
shortly after coagula are deposited among the membranes,
the blood is little altered, and we have an ordinary abortion.
If, however, the ovum is retained, the coagulated fibrine,
and the placenta or membranes, undergo secondary changes
which lead to the formation of moles. The so-called *fleshy
mole* (fig. 94) is often retained for many weeks or months
after the death of the fœtus, and during this time there may
be but little modification of the usual symptoms of pregnancy; or, as is frequently the case, it gives rise to occasional
hæmorrhage, until at last uterine contractions come on, and
it is cast off in the form of a thick fleshy mass, having but
little resemblance to the ordinary products of conception.
The most probable explanation of its formation is, that when
hæmorrhage originally took place, the effusion of blood was
not sufficient to effect the entire separation and expulsion of
the ovum. Part of the membranes, or of the placenta—if
that organ had commenced to form—retained its organic
connection with the uterus, while the fœtus perished. The
attached portion of the placenta or membranes continues to
be nourished, although abnormally. The fœtus generally
entirely disappears, especially if it has perished at an early
period of utero-gestation, when it becomes dissolved in the
liquor amnii. Or it may become macerated, shrivelled, and
greatly altered in appearance. The effused blood becomes
decolorised from the absorption of the corpuscles; and,
according to Scanzoni, fresh vessels are developed in the
fibrine, which increase the vascular attachment of the mole
to the uterine walls. The placenta and membranes may go
on increasing in thickness, until they form a mass of considerable size. Careful microscopic examination will almost
always enable us to discover the villi of the chorion, altered
in appearance, often loaded with granular fatty molecules,
but sufficiently distinct to be readily recognisable.

Important as are the causes of abortion arising from some

morbid condition of the ovum, they are not more so than
those which depend on the maternal state, and it is to be
observed that the former are often indirect causes, produced
by primary maternal changes. Many of these maternal

Causes depending on the maternal state.

Fig. 91.

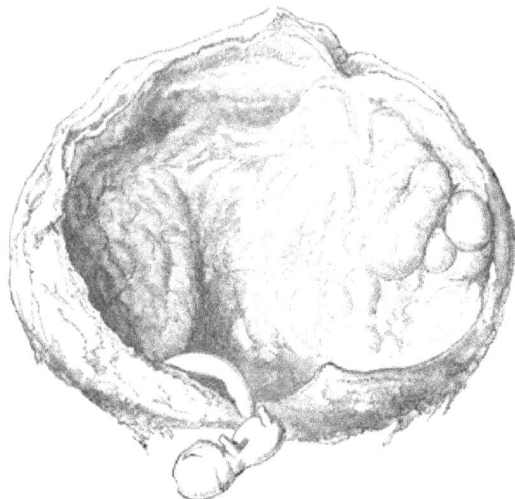

BLIGHTED OVUM, WITH FLESHY DEGENERATION OF THE MEMBRANES.

causes act by causing hyperæmia of the uterus, which leads
to extravasation of blood. Thus abortion is apt to occur in
women who lead unhealthy lives, such as those who occupy
over-heated and ill-ventilated rooms, or indulge to excess in
the fatigues and pleasures of society, in the use of alcoholic
drinks, and the like. Over-frequent coitus has been, for the
same reason, observed to produce a remarkable tendency to
abortion, and Parent-Duchatelet has noted that it is of very
frequent occurrence amongst women of loose life. Many
diseases strongly predispose to it, such as fevers, zymotic
diseases of all kinds, measles, scarlet fever, small-pox; and
diseases of the respiratory organs, such as bronchitis and
pneumonia. *Syphilis* is well known to be one of the most
frequent causes, and one that is likely to act in successive
pregnancies. It may act so that the pregnancy is brought
to a premature termination, time after time, until the constitutional disease is eradicated by appropriate treatment.
It acts in some cases through the influence of the father in

Influence of syphilis in pro-ducing abortion.

producing a diseased ovum ; and it is the only cause which can with certainty be traced to the state of the father's health. Many other morbid conditions of the blood also dispose to abortion. It has been observed to be a frequent result of lead-poisoning ; also of the presence of noxious gases in the atmosphere, such as an excess of carbonic acid.

Causes acting through the nervous system. Many causes act through the nervous system, such as fright, anxiety, sudden shock, and the like. Thus there are numerous instances on record in which women aborted suddenly after the receipt of some bad news, and it is said to have been of frequent occurrence in women immediately before execution. The influence of irritation propagated through the nervous system from a distance, tending to produce uterine contraction and abortion through the agency of reflex action, has been specially dwelt upon by Tyler Smith. Thus he points out that abortion not unfrequently occurs from the irritation of constant suckling in women who become pregnant during lactation. The effect of suckling in producing uterine contraction is, indeed, well known, and the application of the child to the breast, for this purpose, has long been recognised as a method of treatment in post-partum hæmorrhage. The irritation of the trifacial in severe toothache ; of the renal nerves in cases of gravel, in albuminuria, etc.; of the intestinal nerves in excessive vomiting, in diarrhœa, obstinate constipation, ascarides, etc., all act in the same way. We may, perhaps, also explain, by this hypothesis, the fact that women are more apt to abort at what would have been the menstrual epoch than at other times, as the ovarian nerves may then be subject to undue excitement. It is probable, however, that there may be also at these times more or less active congestion of the decidua, which may predispose to laceration of its capillaries and blood extravasation. Such congestion exists in those exceptional cases in which menstruation continues for one or more periods after conception, the blood probably escaping from the space between the decidua vera and reflexa ; and, therefore, there is no reason to question its also happening even when such abnormal menstruation is not present.

Tendency to abortion at the menstrual epochs.

Physical causes. Certain physical causes may produce abortion by separating the ovum. Thus it may follow a fall, a blow, or other accidents of a trivial character. On the other hand, women may be subjected to injuries of the severest kind without aborting. The probability, therefore, is that these apparently trivial causes only operate in women who, for some other reason, are predisposed to the accident. This is borne out by the fact—which is well known in these days, when the artificial production of abortion is, unhappily, far from a very rare event—that it is by no means easy to destroy the vitality of the fœtus. I myself know of a case, in which the uterine sound was passed several times into a pregnant uterus without producing abortion, the pregnancy proceeding to term. Oldham has related a similar case in which he in vain attempted to induce abortion by the sound in a case of contracted pelvis; and Duncan has mentioned an instance in which an intra-uterine stem pessary was unwittingly introduced, and worn for some time by a pregnant woman, without any bad effect. The fact that pregnancy is with difficulty interfered with when there is a healthy relation between the ovum and the uterus, no doubt, explains the disastrous effects of criminal abortion, which have been especially insisted on by many of our American brethren.

Occasional difficulty in producing abortion.

Morbid states of the uterus have an important influence in the production of abortion. Any condition which mechanically interferes with the proper development of the uterus is apt to operate in this way. Amongst these may be mentioned fibroid tumours; the presence of old peritoneal adhesions, rendering the womb a more or less fixed organ; but, above all, flexion and displacement of the uterus. Retroflexion of the uterus is, unquestionably, one of the most frequent factors in its production, not only on account of the irritation which the abnormal position sets up, but from interference with the uterine circulation, which leads to the effusion of blood, and the death of the ovum. An inflamed condition of the cervical and uterine mucous membranes will act in the same way, should pregnancy have occurred; although such a condition more often prevents conception taking place.

Causes depending on morbid states of the uterus.

One of the earliest indications of impending abortion is more or less hæmorrhage. This may at first be slight, and may last for a short time only, recurring after an interval of time; or it may commence with a sudden and profuse discharge. Occasionally it is very abundant, and its continuance ·

Symptoms.

and amount form one of the gravest symptoms of the accident. After the loss of blood has continued for a greater or less length of time—it may be even for some days—uterine contractions come on, recurring at regular intervals, and eventually lead to the expulsion of the ovum. More rarely the impending miscarriage commences with pains, which lead to laceration of vessels and hæmorrhage.

There is little chance of arresting abortion when pain and hæmorrhage co-exist.

As long as one or other of these symptoms exist alone, we may hope to avert the threatened miscarriage; but when both occur together there is little or no chance of its being arrested. Certain premonitory symptoms are described by authors as common in abortion, such as feverishness, shivering, a sensation of coldness; all of which are obscure and unreliable, and are certainly much more frequently absent than present.

If the pregnancy be early it is probable that the entire ovum will be shed with little trouble, and it often passes unperceived in the clots which surround it. It is, therefore, of importance that all the discharges should be very carefully examined. After the second month the rigid and undilated cervix presents a formidable obstacle to the escape of the ovum, and it may be a considerable time before there is sufficient dilatation to admit of its passage. This is gradually effected by the continuance of pains, but not without a severe loss of blood. It may be that the amnion is ruptured,

Occasional retention of the secundines.

and the fœtus expelled first. After a lapse of time the secundines are also shed, but there may be a considerable delay, amounting even to days, before this is effected. As long as any portions of the membranes are retained in utero, the patient is necessarily subjected to considerable risk, not only from the continuance of hæmorrhage, but also from septicæmia. Hence it may be laid down as a rule that we can never consider our patient out of danger until we have satisfied ourselves that the whole of the uterine contents have been expelled.

Treatment. Arrest of threatened miscarriage.

Our first endeavour in any case of impending miscarriage will be, of course, to avert the threatened accident. If hæmorrhage has not been excessive, and if, on vaginal examination, which should always be practised, we find no dilatation of the os, we may entertain a reasonable hope of success. If, on the contrary, we find the os beginning to open, if we are able to insert the finger through it so as to touch the ovum, especially if pains also exist, we are justified in considering abortion to be inevitable, and the indication will then be to have the ovum expelled, and the case terminated as soon as possible. In the former case the most absolute rest is the first thing to insist on. The patient should be placed in bed, not overburdened with clothes, in a cool temperature, and she should have a light and easily assimilated diet. All movements, even rising out of bed to empty the bladder or bowels, should be absolutely prohibited. To avert the tendency to the commencement of uterine contraction there is no remedy so useful as opium, which must be given freely, and frequently repeated. It may be administered either in the form of laudanum, or of Battley's sedative solution, which has the advantage of producing less general disturbance. It may be advantageously exhibited in doses of from 20 to 30 minims, and repeated after a few hours. A still better preparation is chlorodyne, which I have found of extreme value in arresting impending miscarriage, in doses of 10 minims, repeated every third or fourth hour. If, from any other cause, it is considered unadvisable to give the sedative by the mouth, it may be administered in a small starch enema *per rectum*. In all cases it will be necessary to keep the patient more or less under the influence of the drug for several days, and until all symptoms of miscarriage have passed away. Care should be taken that the bowels do not become locked up by the action of the opiates—as this might of itself be a cause of irritation—and their constipating effects ought to be obviated by small doses of castor-oil, or other gentle aperient. Various subsidiary methods of treatment have been recommended, such as bleeding from the arm, or the local application of leeches in supposed plethoric states of the system; revulsives, such as dry cupping to the loins; the application of ice, to check hæmorrhage; astringents, such as acetate of lead or gallic acid, for the same purpose. Most of these, if not hurtful, will be, at least, useless. The cases in which venesection would be beneficial are extremely rare, and the local applications, especially cold, are much more apt to favour than to prevent uterine action.

In cases of repeated miscarriage in successive pregnancies,

Prophy-
lactic
treatment.

a special course of prophylactic treatment is indicated, and is often attended with much success. In cases of this kind the first indication, and one which ought to be carefully attended to, is to seek for and, if possible, to remove or mitigate the cause which has given rise to the former abortions. Those causes which depend on constitutional states must first be carefully investigated, and treated according to the indications present. These may be obscure and not easily discovered; but it is certainly unwise to assume too readily the existence of what has been called 'a habit of abortion,' which further inquiry may prove to be only an indication of constitutional debility, degeneracy of the placental structures, or a latent and unsuspected syphilitic taint. If constitutional debility be present to a marked extent, a generous diet and a restorative course of treatment (preparations of iron, quinine, and other suitable tonics) may effect the desired object.

Treat-
ment in
cases
depending
on local
causes.

Local congestion of the uterus, or a general plethoric state of the patient, have often been supposed to be efficient causes of recurring abortion. Dr. Henry Bennet has especially dwelt on the influence of congestion and abrasions of the cervix in causing premature expulsion of the fœtus,[1] and recommends the topical application of nitrate of silver, or other caustics, to the inflammatory abrasions existing on the neck of the womb. Formerly venesection was a favourite remedy; and many authors have recommended the local abstraction of blood by leeches applied to the groin, or round the anus, or even to the cervix. The influence of general plethora is more than doubtful; and although local congestions are, probably, much more effective causes, still it would seem more judicious to treat them by rest and local sedatives rather than by topical applications, which, injudiciously applied, might produce the very accident they were intended to prevent.

The position of the uterus should be carefully investigated. If it be found to be retroflexed, a well-fitting Hodge's pessary should be applied, so as to support it until it has completely risen out of the pelvis.

Treat-
ment in
cases

The possibility of syphilitic infection should always be inquired into, for this poison may act on the product of con-

[1] On Inflammation of the Uterus, p. 432.

ception long after all appreciable traces of it have disappeared
from the infected parent.   Should there be recurrent abortions in a patient who had formerly suffered from syphilis,
or whose husband had at any time contracted the disease,
no time should be lost in using appropriate anti-syphilitic
remedies, which should invariably be administered both to
the husband and wife.   Diday especially insists that in such
cases it is not sufficient to submit the father and mother to
a mercurial course in the absence of pregnancy, but that, as
each successive impregnation occurs, the mother should again
commence anti-syphilitic treatment, even though she has no
visible traces of the disease.[1]   In this way there is reasonable
ground for hoping that infection of the ovum may be prevented.   I think, too, that we may be the more encouraged
to persevere in the treatment of these unfortunate cases,
from the fact that the syphilitic poison tends to wear itself
out.   I have seen several cases in which this taint at first
produced early abortion, then each successive pregnancy was
of longer duration, until eventually a living child was born.

In fatty degeneration of the chorion villi, and in other
morbid states of the placenta, which act by preventing the
proper nutrition of the fœtus, and the due aëration of its
blood, there is no reliable means of treatment except the
general improvement of the mother's health.   Simpson
strongly recommended the administration of chlorate of
potash in cases in which the child habitually dies in the
latter months of pregnancy, on the supposition that it supplied to the blood a large amount of oxygen, and thus made
up for any deficiency in the supply of that element through
the placental tufts.   The theory is, at best, a doubtful one,
although I believe the drug to be unquestionably beneficial
in cases of the kind.   It probably acts by its tonic properties
rather than in the manner Simpson supposed.   It may be
given in doses of 15 to 20 grains three times a day, and
may be advantageously combined with small doses of dilute
hydrochloric acid.   In frequently recurring premature labours with dead children, Simpson strongly recommended
the induction of premature labour a little before the time at
which we had reason to believe that the fœtus has usually
perished; or in other words, before the placental disease had

[1] Diday, *Infantile Syphilis, Syd. Soc. Trans.* p. 207.

advanced sufficiently far to interfere with its nutrition. The practice has constantly been adopted with success, and is perfectly legitimate, but the difficulty, of course, is to fix on the right time. Careful auscultation of the fœtal heart may be of some use in guiding us to a decision, as the death of the fœtus is generally preceded for some days by irregular, tumultuous, and intermittent action of the heart.

Treatment where no cause can be discovered.

There will always remain a certain number of cases in which no appreciable cause can be discovered. Under such circumstances prolonged rest, at least until the time has passed at which abortion formerly took place, will afford the best chance of avoiding a recurrence of the accident. There must always be some difficulty in carrying out this indication, inasmuch as the patient's health is apt to suffer in other ways from the confinement, and the want of fresh air and exercise which it entails. The strictness with which rest should be insisted on must vary in different cases, but it should be specially attended to at what would have been the menstrual periods. At these times the patient should remain in bed altogether; at others she may lie on a sofa, and, if circumstances permit, spend part of the day at least in the open air. Sexual intercourse should be prohibited.

Treatment when abortion is inevitable.

Should actual symptoms of abortion come on, the preventive treatment, already indicated, may be resorted to. Great care, however, should be used in prescribing opiates as preventives, and they should be given for a specified time only. I have seen, more than once, an incurable habit of opium-eating originate from the incautious and too long-continued exhibition of the drug in such cases.

When we have satisfied ourselves that abortion is inevitable, we must proceed to employ treatment that favours the expulsion of the ovum.

Removal of the ovum when within reach.

If the os be sufficiently dilated, and the pains strong, we may find the ovum separated and protruding from the os. We may then be able to detach it by the finger. For this purpose the uterus is depressed from without by the left hand, while an endeavour is made to scoop out the ovum with the examining finger. If it be out of reach, and yet appears detached, chloroform should be administered, the whole hand introduced into the vagina, and the finger into the uterine cavity. The complete detachment of the ovum can, in this way, be far more readily and safely effected than by using any of the many ovum forceps which have been invented for the purpose.

If the ovum be not sufficiently separated, or the os be undilated, means must be taken to control the hæmorrhage until the former can be removed or expelled. It is here that plugging of the vagina finds its most useful application. This may be done in various ways. That most usually employed is filling the vagina with a tolerably large sponge, in the interstices of which the blood coagulates. A better plan is to soak a number of pledgets of cotton wool in carbolised water and tie a string round each. The vagina can be completely and effectively packed with these; and this is best done through a speculum. Each pledget should be covered with glycerine, which completely prevents the offensive odour which otherwise always arises. The pledgets can be removed by traction on the strings, but if these are not used much pain is caused in getting them out of the vagina. The plug should never be left in for more than six or eight hours, after which a fresh one may be inserted if necessary. Two or three full doses of the liquid extract of ergot, of ʒss to ʒj each, or a subcutaneous injection of ergotine, may be given while the plug is in position. The plug itself is a strong excitant of uterine action, and the two combined often effect complete detachment, so that, on the removal of the tampon, the ovum may be found lying loose in the os uteri. If the os be undilated and the ovum entirely out of reach, the former may be opened by means of sponge or laminaria tents. I think a well-prepared sponge tent the most effectual, and it can be maintained *in situ* by a vaginal plug below it. It also acts as a most efficient plug, effectually controlling all hæmorrhage. In a few hours it opens up the os sufficiently to admit the finger.

The most troublesome cases are those in which the fœtus is first expelled, and the placenta and membranes remain in utero. As long as this is the case the patient can never be considered safe from the occurrence of septicæmia. Dr. Priestley has strongly insisted on the importance of removing the secundines as soon as possible. There can be no doubt that this should be done whenever it is feasible. Cases, however, are frequently met with in which any forcible

[margin: Plugging of the vagina.]

[margin: Retention of the membranes.]

attempt at removal would be likely to prove very hurtful, and in which it is better practice to control hæmorrhage by the plug or sponge tent, and wait until the placenta is detached, which it will generally be in a day or two at most. Under such circumstances fœtor and decomposition of the secundines may be prevented by intra-uterine injections of diluted Condy's fluid. Provided the os be sufficiently patulous to prevent the collection of the fluid in the uterine cavity, and not more than a drachm or two of fluid be injected at a time, so as simply to wash away and disinfect decomposing detritus, they can be used with perfect safety. Sometimes cases are met with in which the os has entirely closed, and in which we can only suspect the retention of the placenta by the history of the case, the continuance of hæmorrhage, or the presence of a fœtid discharge. Should we see reason to suspect this the os must be dilated with sponge or laminaria tents, and the uterine cavity thoroughly explored under chloroform. This condition of things is far from uncommon in women who have not had medical assistance from the first, and it often gives rise to very troublesome and anxious symptoms. It has been said that placentæ thus retained have been completely absorbed, and cases of the kind have been related by Naegele and Osiander. The spontaneous absorption, however, of so highly organised a body as the placenta would be a phenomenon of the most remarkable character; and it seems more natural to suppose that, in most cases of the kind, the placenta has been cast off without the knowledge of the patient. Sometimes the placenta never becomes entirely detached, and retaining organic connection with the uterine walls, forms what has been called a 'placental polypus.' This may produce secondary hæmorrhages, in the same way as an ordinary fibroid polypus. Barnes recommends the removal of these masses by means of the wire écraseur. Before their detection the os uteri must be opened up.

Retention in utero of a blighted ovum.

The cases, previously alluded to, in which an ovum has perished in early pregnancy and is retained in utero, are often puzzling, and may give rise to serious moral and medico-legal questions. The blighted ovum may be retained for many months, the outside limit according to McClintock,[1]

[1] Sydenham Society's edition of *Smellie's Midwifery*, vol. i. p. 169.

by whom the subject has been ably discussed, being nine
months. The appearance of the ovum when thrown off will
give no reliable clue to the length of time which has elapsed
since it perished.   The symptoms are often very obscure.
Generally there have been the usual indications of pregnancy
which, with or without signs of impending miscarriage, disappear or are modified, and then follows a period of ill-
health, with pelvic uneasiness, and irregular metrorrhagia,
which may be mistaken for menstruation.   Occasionally, but
by no means necessarily, there is a fœtid discharge, and this
probably exists only when the membranes have broken,
and air has access to the ovum.   In some cases obscure
septicæmic symptoms have been observed.   Such symptoms
are obviously too indefinite to lead to an accurate diagnosis.
In the course of time the ovum is generally thrown off,
with more or less hæmorrhage.   If the nature of the case
is detected, ergot may be given to promote the expulsion
of the uterine contents, and it may even be advisable to
dilate the cervix with sponge or laminaria tents, and remove
them artificially.

The frequency with which abortion leads to chronic
uterine disease should lead us to attach much more importance to the subsequent management of the patient than has
been customary. The usual practice is to confine the patient
to bed for two or three days only, and then to allow her to
resume her ordinary avocations, on the supposition that a
miscarriage requires less subsequent care than a confinement.
The contrary of this is, however, most probably the case ;
for the uterus has been emptied when it is unprepared for
involution, and that process is often very imperfectly performed. We should, therefore, insist on at least as much
attention being paid to rest as after labour at term.

Subsequent management.

PART III.

LABOUR.

CHAPTER I.

THE PHENOMENA OF LABOUR.

Delivery at term.
IN considering delivery at term we have to discuss two distinct classes of events.

One of these is the series of vital actions brought into play in order to effect the expulsion of the child ; and the other consists of the movements imparted to the child—the body to be expelled—in other words, the mechanism of delivery.

Causes of labour.
Before proceeding to the consideration of these important topics, a few words may be said as to the determining causes of labour. This subject has been from the earliest times a *quæstio vexata* among physiologists ; and many and various are the theories which have been broached to explain the curious fact that labour spontaneously commences, if not at a fixed epoch, at any rate approximately so. It must be admitted that, even yet, there is no explanation which can be implicitly accepted.

They are referred either to fœtal or maternal causes.
The explanations which have been given may be divided into two classes—those which attribute the advent of labour to the fœtus, and those which refer it to some change connected with the maternal generative organs.

The former is the opinion which was held by the older accoucheurs, who assigned to the fœtus some active influence in effecting its own expulsion. It need hardly be said that such fanciful views have no kind of physiological basis.

Changes in the fœtal circulation.
Others have supposed that there might be some change in the placental circulation, or in the vascular system of the

fœtus, which might solve the mystery. The latest hypothesis of this kind, which, however, is not fortified by any
evidence, is by Barnes, who says: 'I rather incline to the
opinion that when the fœtus has attained its full development, when its organs are prepared for external life, some
change takes place in its circulation, which involves a corre-
lative disturbance in the maternal circulation, which excites
the attempt at labour.'[1]

③ The majority of obstetricians, however, refer the advent
of labour to purely maternal causes. Among the more
favourite theories is one, which was originally started in this
country by Dr. Power, and adopted and illustrated by Depaul,
Dubois, and other writers. It is based on the assumption
that there is a sphincter action of the fibres of the cervix,
analogous to that of the sphincters of the bladder and rec- *The reflex sphincter theory.*
tum, and that when the cervix is taken up into the general
uterine cavity as pregnancy advances, the ovum presses upon
it, irritates its nerves, and so sets up reflex action, which
ends in the establishment of uterine contraction. This
theory was founded on erroneous conceptions of the changes
that occurred in the neck of the uterus; and, as it is
certain that obliteration of the cervix does not really take
place in the manner that Power believed when his theory
was broached, it is obvious that its supposed result cannot
follow.

④ Extreme distension of the uterus has been held to be *Disten-sion of the uterus.*
the determining cause of labour, a view lately revived by
Dr. King, of Washington,[2] who believes that contractions are
induced because the uterus ceases to augment in capacity,
while its contents still continue to increase. This hypothesis
is sufficiently disproved by a number of clinical facts which
show that the uterus may be subject to excessive and even
rapid distension—as in cases of hydramnios, multiple pregnancy, and hydatiform degeneration of the ovum—without
the supervention of uterine contractions.

⑤ Another inciter of uterine action has been supposed to *Fatty de-generation of the decidua.*
be the separation of the ovum from its connections to the
uterine parietes, in consequence of fatty degeneration of the
decidua occurring at the end of pregnancy. The supposed
result of this change, which undoubtedly occurs, is that the

[1] *Diseases of Women*, p. 434.   [2] *American Journal of Obstetrics*, vol. iii.

ovum becomes so detached from its organic adhesions as to
be somewhat in the position of a foreign body, and thus in-
cites the nerves so largely distributed over the interior of the
uterus. This theory, which has been widely accepted, was
originally started by Sir James Simpson, who pointed out
that some of the most efficient means of inducing labour
(such, for example, as the insertion of a gum-elastic catheter
between the ovum and the uterine walls) probably act in the
same way, viz., by effecting separation of the membranes and
detachment of the ovum.

Barnes instances, in opposition to this idea, the fact that
ineffectual attempts at labour come on at the natural term
of gestation in cases of extra-uterine pregnancy, when the
fœtus is altogether independent of the uterus, and therefore,
he argues, the cause cannot be situated in the uterus itself.
A fair answer to this argument would be that although,
in such cases, the womb does not contain the ovum, it does
contain a decidua, the degeneration and separation of which
might suffice to induce the abortive and partial attempts at
labour then witnessed.

Objections
to these
theories.

A serious objection to all these theories, which are based
on the assumption that some local irritation brings on contraction, is the fact, which has not been generally appreciated,
that uterine contractions are always present during pregnancy as a normal occurrence, and that they may be, and
often are, readily intensified at any time, so as to result in
premature delivery.

It is, indeed, most likely that, at or about the full term,
the nervous supply of the uterus is so highly developed, and
in so advanced a state of irritability, that it more readily
responds to stimuli than at other times. If by separation
of the decidua, or in some other way, stimulation of the
excitor nerves is then effected, more frequent and forcible
contractions than usual may result, and, as they become
stronger and more regular, terminate in labour. But, allowing this, it still remains quite unexplained why this should
occur with such regularity at a definite time.

Tyler
Smith's
ovarian
theory.

Tyler Smith tried, indeed, to prove that labour came on
naturally at what would have been a menstrual epoch, the
congestion attending the menstrual nisus acting as the ex-
citer of uterine contraction. He, therefore, refers the onset

of labour to ovarian, rather than to uterine, causes. Although this view is upheld with all its author's great talent, there are several objections to it difficult to overcome. Thus, it assumes that the periodic changes in the ovary continue during pregnancy, of which there is no proof. Indeed there is good reason to believe that ovulation is suspended during gestation, and with it, of course, the menstrual nisus. Besides, as has been well objected by Cazeaux, even if this theory were admitted, it would still leave the mystery unsolved, for it would not explain why the menstrual nisus should act in this way at the tenth menstrual epoch, rather than at the ninth or eleventh.

In spite, then, of many theories at our disposal, it is to be feared that we must admit ourselves to be still in entire ignorance of the reason why labour should come on at a fixed epoch. *The cause of labour at a fixed epoch is still unknown.*

The expulsion of the child is effected by the contractions of the muscular fibres of the uterus, aided by those of some of the abdominal muscles. These efforts are in the main entirely independent of volition. So far as regards the uterine contractions, this is absolutely true, for the mother has no power of originating, lessening, or increasing the action of the uterus. As regards the abdominal muscles, however, the mother is certainly able to bring them into action, and to increase their power by voluntary efforts; but, as labour advances, and as the head passes into the vagina and irritates the nerves supplying it, the abdominal muscles are often stimulated to contract, through the influence of reflex action, independently of volition on the part of the mother. *Mode in which the expulsion of the child is effected.*

There can be little doubt that the chief agent in the expulsion of the child is the contraction of the uterus itself. This opinion is almost unanimously held by accoucheurs, and the influence of the abdominal muscles is believed to be purely accessory. Dr. Haughton, however, maintains a view which is directly contrary to this. From an examination of the force of the uterine contractions, arrived at by measuring the amount of muscular fibre contained in the walls of the uterus, he arrives at the conclusion that the uterine contractions are chiefly influential in rupturing the membranes, and dilating the os uteri, bringing into action, if needful, a force equivalent to 54 lbs.; but when this is effected, and *The chief factor in expulsion is uterine contraction.*

the second stage of labour has commenced, he thinks the remainder of the labour is mainly completed by the contractions of the abdominal muscles, to which he attributes enormous powers, equivalent, if needful, to a pressure of 523·65 lbs. on the area of the pelvic canal.

These views bear on a topic of primary consequence in the physiology of labour. They have been fully criticised by Duncan, who has devoted much experimental research to the study of the powers brought into action in the expulsion of the child. His conclusions are that, so far from the enormous force being employed that Haughton estimated, in the large majority of cases the effective force brought to bear on the child by the combined action of both the uterine and abdominal muscles is less than 50 lbs.—that is, less than the force which Haughton attributed to the uterus alone. In extremely severe labours, when the resistance is excessive, he thinks that extra power may be employed; but he estimates the maximum as not above 60 lbs., including in this total the action of both the uterine and abdominal muscles. Joulin arrived at the conclusion that the uterine contractions were capable of resisting a maximum force of about one hundredweight. Both these estimates, it will be observed, are much under that of Haughton, which Duncan describes as representing 'a strain to which the maternal machinery could not be subjected without instantaneous and utter destruction.'

Reasons on which this conclusion is based.

There are many facts in the history of parturition which make it certain that the chief factor in the expulsion of the child is the uterus. Among these may be mentioned occasional cases in which the action of the abdominal muscles is materially lessened, if not annulled—as in profound anæsthesia, and in some cases of paraplegia—in which, nevertheless, uterine contractions suffice to effect delivery. The most familiar example of its influence, however, and one that is a matter of everyday observation in practice, is when inertia of the uterus exists. In such cases no effort on the part of the mother, no amount of voluntary action that she can bring to bear on the child, has any appreciable influence on the progress of the labour, which remains in abeyance until the defective uterine action is re-established, or until artificial aid is given.

The contraction of the uterus, then, being the main agent in delivery, it is important for us to appreciate its mode of action, and its effect on the ovum.

We have seen that intermittent and generally painless uterine contractions exist during pregnancy. As the period for delivery approaches, these become more frequent and intense, until labour actually commences, when they begin to be sufficiently developed to effect the opening up of the os uteri, with a view to the passage of the child. They are now accompanied by pain, which increases as labour advances, and is so characteristic that 'pains' are universally used as a descriptive term for the contractions themselves. It does not necessarily follow that uterine contractions are painless until they commence to effect dilatation of the os uteri. On the contrary, during the last days or even weeks of pregnancy, women constantly have irregular contractions, accompanied by severe suffering, which, however, pass off without producing any marked effect on the cervix. When labour has actually begun, if the hand is placed on the uterus, when a pain commences, the contraction of its muscular tissue is very apparent, and the whole organ is observed to become tense and hard, the rigidity increasing until the pain has reached its acme, the uterine walls then relaxing, and remaining soft until the next pain comes on. At the commencement of labour these pains are few, separated from each other by a considerable interval, and of short duration. In a perfectly typical labour the interval between the pains becomes shorter and shorter, while, at the same time, the duration of each pain is increased. At first they may occur only once in an hour or more, while eventually there may not be more than a few minutes' interval between them.

Uterine contractions at the commencement of labour.

If, when the pains are fairly established, a vaginal examination be made, the os uteri will be found to be thinned and dilated in proportion to the progress of the labour. During the contraction the bag of membranes will be felt to bulge, to become tense from the downward pressure of the liquor amnii within it, and to protrude through the os if it be sufficiently open. The membranes, with the contained liquor amnii, thus form a fluid wedge, which has a most important influence in dilating the os uteri (see Frontispiece). This

Mode in which dilatation of the cervix is effected.

does not, however, form the sole mechanism by which the os uteri is dilated, for it is also acted upon by the contractions of the muscular fibres of the uterus, which tend to pull it open. It is probable that the muscular dilatation of the os is effected chiefly by the longitudinal fibres, which, as they shorten, act upon the os uteri, the part where there is least resistance.

Partly then by muscular contraction, partly by mechanical pressure, the cervical canal is dilated, and as it opens up it becomes thinner and thinner, until it is entirely taken up into the uterine cavity.

Rapture of the membranes.
There is no longer any obstacle to the passage of the presenting part of the child into the cavity of the pelvis, and the force of the pains now generally effects the rupture of the membranes, and the escape of the liquor amnii. There is often observed, at this time, a temporary relaxation in the frequency of the pains, which had been steadily increasing; but they soon recommence with increased vigour. If the abdomen be now examined it will be observed to be much diminished in size, partly in consequence of the escape of the liquor amnii, partly from the descent of the fœtus into the pelvic cavity.

Change in the character of the pains.
The character of the pains soon changes. They become stronger, longer in duration, separated by a shorter interval, and accompanied by a distinct forcing effort, being generally described as 'the bearing down' pains. Now is the time at which the accessory muscles of parturition come into operation. The patient brings them into play in the manner which will be subsequently described, and the combined action of the uterine and abdominal muscles continues until the expulsion of the child is effected.

The precise mode of action of the uterus is somewhat doubtful.
The precise mode of uterine contraction is still somewhat a matter of dispute. It is generally described as commencing in the cervix, passing gradually upwards by peristaltic action, the wave then returning downwards towards the os uteri. This view was maintained by Wigand, and has been indorsed by Rigby, Tyler Smith, and many other writers. In support of it they instance the fact that, on the accession of a pain, the presenting part first recedes, the bag of membranes then becomes tense and protrudes through the os, and it is not until some time that the presenting part of the child itself

is pushed down. It is very doubtful if this view is correct; and a careful examination of the course of the pains would rather lead to the belief that the contractions commence at the fundus, where the muscular tissue is most largely developed, and gradually proceed downwards to the cervix; the waves of contraction being, however, so rapid that the whole organ seems to harden *en masse*. The apparent recession of the presenting part, and the bulging of the bag of membranes, are certainly no proof that the contractions begin at the cervix; for the commencing contraction would necessarily push down the fluid in front of the head, and cause the membranes to bulge, and the os to become tense, before its force was brought to bear on the fœtus itself. Indeed did the contraction commence at the lower part of the uterus, we should expect the opposite of what takes place to occur, and the waters to be pushed upwards, and away from the cervix. The fundal origin of the contraction is further illustrated by what is observed when the hand of the accoucheur is placed in the uterine cavity, as often happens in certain cases of hæmorrhage or turning; for if a pain then comes on, it will be felt to start at the fundus, and gradually compress the hand from above downwards.

The intermittent character of the contractions is of great practical importance. Were they continuous, not only would the muscular powers of the patient be rapidly exhausted, but by the obliteration of the vessels produced by the muscular contraction, the circulation through the placenta would be interfered with, and the life of the child imperilled. Hence one of the chief dangers of protracted labour, especially after the escape of the liquor amnii, is that the uterine fibres may enter into a state of tonic rigidity, a condition that cannot be long continued without serious risks both to the mother and child.

Value of the intermittent character of the pains.

The fact that the uterine contractions are altogether involuntary proves them to be excited—as indeed we would *à priori* infer from our knowledge of the anatomical arrangement of the nerves of the uterus—solely by the sympathetic system. Still it is a fact of every-day observation that they can be largely influenced by emotions. Various stimuli applied to the spinal system of nerves (as, for example, when the mammæ are irritated) have also a marked effect in

The contractions are incited through the sympathetic nerves.

inducing uterine contraction. The precise mode in which
such influence is conveyed to the uterus, in spite of the numerous experiments which have been made for the purpose of
determining how far labour is affected by destruction of the

In the second stage of labour the vaginal nerves act as inciters of reflex action.
spinal cord, is still a matter of doubt. After the fœtus has
passed through the cervix, the spinal nerves distributed to
the vagina and perinæum are excited by the pressure of the
presenting part, and through them the accessory powers of
parturition are chiefly brought into play. The contraction
of the muscles of the vagina itself is supposed to have some
influence in favouring the expulsion of the fœtus after the
birth of part of the body, and also in promoting the expulsion
of the placenta. In the lower animals the vagina has a very
marked contractile property, and is, in some of them, the
main agent by which the young are expelled. In the human
subject this influence is certainly of very secondary importance.

Character and source of pains during labour.
The amount of suffering experienced during labour varies
much in different cases, and is in direct proportion to the
nervous susceptibility of the patient. There are some women
who go through labour with little or no pain at all. This is
proved by the cases (of which there are numerous authentic
instances recorded) in which labour has commenced during
sleep, and the child has been actually born without the
mother awaking. I am acquainted with a lady, who has had
a large family, who assures me that, though labour is accompanied by a sense of pressure and discomfort, she experiences
nothing which can be called actual pain. Such a happy state
of affairs is, however, extremely exceptional, and, in the
vast majority of cases, parturition is accompanied by intense
suffering during its whole course, in some cases amounting
to anguish, which has probably no parallel under any other
condition.

The precise cause of the pain has been much discussed,
and is, no doubt, complex.

In the first stage.
In the early stage of labour, and before the dilatation of
the os, it is chiefly seated in the back, from whence it shoots
round the loins and down the thighs. It is then probably
produced, partly by pressure on the nerve-filaments caused
by contraction of the muscular fibres to which they are
distributed, and partly by stretching and dilatation of the

muscular tissue of the cervix. M. Beau believes that in this stage the pain is not produced, strictly speaking, in the uterus itself, but is rather a neuralgia of the lumbo-abdominal nerves. The pains at this time are generally described as 'acute' and 'grinding,' terms which sufficiently well express their nature. In highly nervous women these pains are often much less well borne than those of a later stage, and the suffering they undergo is indicated by their extreme restlessness and loud cries as each contraction supervenes. As the os dilates, and the labour advances into the expulsive stage, other sources of suffering are added.

The presenting part now passes into the vagina and presses on the vaginal nerves, as well as on the large nervous plexuses lying in the pelvis. As it descends lower it stretches the perinæum and vulva, and presses on the bladder and rectum. Hence cramps are produced in the muscles supplied by the nerve plexuses, as well as an intolerable sense of tearing and stretching in the vulva and perinæum, and often a distressing feeling of tenesmus in the bowels. By this time the accessory muscles of parturition are brought into action, and they, as well as the uterine muscles, are thrown into frequent and violent contractions, which, independently of the other causes mentioned, are sufficient of themselves to produce great pain, likened to that of colic, produced by involuntary and repeated contraction of the muscles of the intestines.

In the second stage.

Taking all these causes into consideration, there is no lack of sufficient explanation of the intolerable suffering which is so constant an accompaniment of childbirth.

The effect of the pains on the mother's circulation is well marked. The rapidity of the pulse increases distinctly with each contraction, and, as the pain passes off, it again declines to its former state. A similar observation has been made with regard to the sounds of the fœtal heart, especially after the expulsion of the liquor amnii. Hicks has pointed out that during a pain the muscular vibrations give rise to a sound which often resembles that of the fœtal heart, and which completely disappears when the muscular tissue relaxes. The effect of the pain in intensifying the uterine souffle has been already mentioned. The strong muscular efforts would naturally lead us to expect a marked elevation

Effect of the pains on the mother and fœtus.

of temperature during labour. Further observations on this point are required; but Squire asserts that there is generally only a very slight increase in temperature during delivery, rapidly passing off as soon as labour is over.

Division of labour into stages.

Such being the physiological facts in connection with the labour pains, we may now describe the ordinary progress of a natural labour—that is, one terminated by the natural powers, and with a head presenting.

For facility of description obstetricians have long been in the habit of dividing the course of labour into *stages*, which correspond pretty accurately with the natural sequence of events. For this purpose we generally talk of three stages: viz.① from the commencement of regular pains until the complete dilatation of the cervix; ② from the complete dilatation of the cervix until the expulsion of the child; ③, the concluding stage, comprising the permanent contraction of the uterus, and the separation and expulsion of the placenta. To these we may conveniently add a preparatory stage, antecedent to the regular commencement of the labour.

Preparatory stage.

For a short time before delivery, varying from a few days to a week or two, certain premonitory symptoms generally exist, which indicate the approaching advent of labour. Sometimes they are well marked, and cannot be mistaken; at others they are so slight as to escape observation. Amongst the most common is a sinking of the uterus into the pelvic cavity, resulting from the relaxation of the soft parts preceding delivery. The result is, that the upper edge of the uterine tumour is less high than before, and, in consequence, the pressure on the respiratory organs is diminished, and the woman often feels lighter, and altogether less unwieldy, than in the previous weeks. If a vaginal examination be made at this time, the lower segment of the uterus will be found to have sunk lower into the pelvic cavity; and the consequence of this is that, while the respiration is less embarrassed, and the patient feels less bulky, other accompaniments of pregnancy, such as hæmorrhoids, irritability of the bladder and bowels, and œdema of the limbs, become aggravated. The increased pressure on the bowels often induces a sort of temporary diarrhœa, which is so far advantageous that it empties the bowels of fæces which may have collected within them. As has already been pointed out, the contractions

Sinking of the uterus.

which have been going on at intervals during the latter months of pregnancy now get more and more marked, and they have the effect of producing a real shortening of the cervix, which is of great value preparatory to its dilatation. More marked mucous discharge from the cavity of the cervix also generally occurs a short time before labour, and it is not unfrequently tinged with blood from the laceration of minute capillary vessels. This discharge, popularly known as the '*shows*,' is a pretty sure sign that labour is not far off. It may, however, be entirely absent, even until the birth of the child. When copious it serves to lubricate the passages, and is generally coincident with rapid dilatation of the parts, and a speedy labour. Mucous discharge, or 'shows.'

During this time (premonitory stage) painful uterine contractions are often present, which, however, have no effect in dilating the cervix. In some cases they are frequent and severe, and are very apt to be mistaken for the commencement of real labour. Such '*false pains*,' as they are termed, are often excited and kept up by local irritations, such as a loaded or disordered state of the intestinal canal; and they frequently give rise to considerable distress, and much inconvenience both to the patient and practitioner. They are, it should be remembered, only the normal contractions of the uterus, intensified and accompanied with pain. False pains.

As labour actually commences, the uterine contractions become stronger, and the fact that they are '*true*' pains can be ascertained by their effect on the cervix. If a vaginal examination be made during one of these, the membranes will be felt to become tense and bulging during the pain, and the os uteri will be found partially dilated, and thinned at its edges. As labour advances this effect on the os becomes more and more marked. At first the dilatation is very slight, perhaps not more than enough to admit the tip of the examining finger, and both the upper and lower orifices of the cervix can be made out. As the pains get stronger and more frequent, dilatation proceeds in the way already described, and the cervix gets more thin and tense, until we can feel a thin circular ring (which is lax between the pains, but becomes rigid and tense during the contraction when the bag of waters bulges through it), without any distinction between the upper and lower orifices. During this time the First stage, or dilatation.

patient, although she may be suffering acutely, is generally able to sit up and walk about. The amount of pain experienced varies much according to the character of the patient. In emotional women of highly developed nervous susceptibilities it is generally very great. They are restless, irritable, and desponding, and when the pain comes on cry out loudly. The character of the cry is peculiar and well marked during the first stage, and has constantly been described by obstetric writers as characteristic. It is acute and high, and is certainly very different from the deep groans of the second stage, when the breath is involuntarily retained to assist the parturient effort. When dilatation is nearly completed various reflex nervous phenomena often show themselves. One of these is nausea and vomiting, another is uncontrollable shivering, which is not accompanied by a sense of coldness, the patient being often hot and perspiring. Both these symptoms indicate that the propulsive stage will shortly commence; and they may be regarded as favourable rather than otherwise, although they are apt to

Rupture of the membranes. alarm the patient and her friends. By this time the os is fully dilated, the membranes generally rupture spontaneously, and a considerable portion of the liquor amnii flows away. The head, if presenting, often acts as a sort of ball-valve, and, falling down on the aperture of the cervix, prevents the complete evacuation of the liquor amnii, which escapes by degrees during the rest of the labour, or may be retained in considerable quantity until the birth of the child.

It not unfrequently happens, if the membranes are somewhat tougher than usual, and the pains frequent and strong, that the fœtus is pushed through the pelvis, and even expelled, surrounded by the membranes. When this occurs the child is said to be born with a ' *caul*,' and this event would doubtless happen more frequently than it does were it not the custom of the accoucheur to rupture the membranes artificially as soon as the os is completely opened up, after which time their integrity is no longer of any value.

Second stage, or propulsion. The os is now entirely retracted over the presenting part, and is no longer to be felt, the vagina and the uterine cavity forming a single canal. Now the mucous discharge is generally abundant, so that the examining finger brings away long strings of glairy transparent mucus, tinged with

blood.   The pains, after a short interval of rest, become
entirely altered in character.   The uterus contracts tightly
round the fœtus, the presenting part descends into the pelvis,
and the true propulsive pains commence.   The accessory
muscles of parturition now come into play.   With each pain
the patient takes a deep inspiration, and thus fills the chest,
so as to give a *point d'appui* to the abdominal muscles.
For the same reason she involuntarily seizes hold of some
point of support, as the hand of a bystander or a towel tied
to the bed, and, at the same time, pushes with her feet
against the end of the bed, and so is able to bear down to
advantage.   The cries are no longer sharp and loud, but consist of a series of deep suppressed groans, which correspond
to a succession of short expirations made during the straining
effort.   In this way the abdominal muscles contract forcibly
on the uterus, which they further stimulate to action by
pressing upon it.   It is to be observed that these straining
efforts are, to a considerable extent, under the control of the
patient.   By encouraging her to hold her breath and bear
down they can be intensified; while if we wish to lessen
them we can advise her to call out, and when she does so
the abdominal muscles have no longer a fixed point of action.
Although the patient may thus lessen the effect of these
accessory muscles, it is entirely out of her power to stop their
action altogether.   As labour advances the head descends
lower and lower, receding somewhat in the intervals between
the pains, until eventually it comes down on the perinæum,
which it soon distends.

The pains now get stronger and more frequent, often
with scarcely a perceptible interval between them, until the
perinæum gets stretched by the advancing head.   In the interval between the pains the elasticity of the perinæal structures pushes the head upwards, so as to diminish the tension
to which the perinæum is subjected, the next pain again
putting it on the stretch, and protruding the head a little
further than before.   By this alternate advance and recession, the gradual yielding of the structures is favoured, and
risk of laceration greatly diminished.   During this time the
pressure of the head mechanically empties the bowel of its
contents.   During the last pains, when the perinæum is
stretched to the utmost, the anal aperture is dilated, some-

*Character
of the con-
traction of
the acces-
sory mus-
cles of par-
turition.*

*Distension
of the
perinæum
and birth
of the
child.*

times to the size of a five-shilling piece; and in this way the perinæum is relaxed, just as the distension, and consequent risk of laceration, are at their maximum. The apex of the head now protrudes more and more through the vulva, surrounded by the orifice of the vagina, and eventually it glides over the perinæum and is expelled. The intensity of the suffering at this moment generally causes the patient to call out loudly. The force of the abdominal muscles is thus lessened at the last moment, and this, in combination with the relaxation of the sphincter ani, forms an admirable contrivance for lessening the risk of perinæal injury. The rest of the body is generally expelled immediately by a single pain, and with it are discharged the remains of the liquor amnii, and some blood-clots from separation of the placenta; and so the second stage of labour terminates.

The third stage. Its importance. The third stage commences after the expulsion of the child. It is of paramount importance to the safety of the mother that it should be conducted in a natural and efficient manner; for it is now that the uterine sinuses are closed, and the frail barrier by which nature effects this may be very readily interfered with, and serious and even fatal loss of blood ensue. Unfortunately, it is too often the case that the practitioner's entire attention is fixed on the expulsion of the child, so that the natural history of the rest of delivery is very generally imperfectly studied and understood.

Contraction of the uterus and detachment of the placenta. As soon as the child is expelled, the uterine fibres contract in all directions, and the hand, following the uterus down, will find that it forms a firm rounded mass lying in the lower part of the abdominal cavity. By retraction of its internal surface, the placental attachments are generally separated, and the after-birth remains in the cavity of the uterus as a foreign body.

Mode in which hæmorrhage is prevented. The escape of blood from the open mouths of the uterine sinuses is now prevented in two ways, viz. (1) by the contractions of the uterine walls, and the more firm, persistent, and tonic this is, the more certain is the immunity from hæmorrhage; (2) by the formation of coagula in the mouths of the vessels. Any undue haste in promoting the expulsion of the placenta tends to prevent the latter of these two hæmostatic safeguards, and is apt to be followed by loss of blood. After a certain time, averaging from a quarter to half an hour, the uterus will be felt to harden, and, if the case be solely left to nature, what has been aptly called a miniature labour occurs. Pains come on, and the placenta is spontaneously expelled from the uterus, either into the canal of the vagina, or even externally. In most obstetric works it is stated that the after-birth may be separated either from its centre or edge, and that it is very generally expelled through the os in an inverted form, with its fœtal surface downwards, and folded transversely on itself. That this is the mode in which the placenta is often expelled, when traction on the cord is practised, is a matter of certainty. It then passes through the os very much in the shape of an inverted umbrella. It is certain, however, that this is not the natural mechanism of its delivery. What this is has been well illustrated by Duncan,[1] who has very clearly shown that, when this stage of labour is left entirely to nature, the separated placenta is expelled edgeways, its uterine and detached surface gliding along the inner surface of the uterus, the foldings of its structure being parallel to the long diameter of the uterine cavity (fig. 95). In this way it is expelled into the vagina, and during the process little or no hæmorrhage occurs. When the placenta is drawn out in the way too generally practised, it obstructs the aperture of the os, and, acting like the piston of a pump, tends to promote hæmorrhage. The corollaries as to treatment drawn from these facts will be subsequently considered. I am anxious, however, here to direct attention to nature's mechanism, because I believe there is no part of labour about the management of which erroneous views are more prevalent than that of this stage, and none in which they are more apt to lead to serious consequences ; and unless the mode in which nature effects the expulsion of the placenta, and prevents hæmorrhage, is thoroughly understood, we shall

Margin notes: Spontaneous expulsion of the placenta. Its mechanism when left to nature.

Fig. 95.

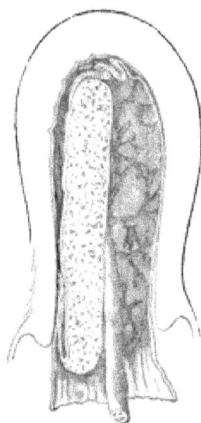

MODE IN WHICH THE PLACENTA IS NATURALLY EXPELLED. (After Duncan.)

[1] *Edin. Med. Journ.* April 1871.

certainly fail in assisting her in a proper manner. In the large proportion of cases, when left entirely to themselves, the placenta would be retained, if not in the uterus, at any rate in the vagina, for a considerable time—possibly for several hours—and such delay would very unnecessarily tire the patience of the practitioner, and be prejudicial to the patient. It is, therefore, our duty in the majority of cases to promote the expulsion of the after-birth; and when this is properly and scientifically done, we increase rather than diminish the patient's safety and comfort. But, in order to do this, we must assist nature, and not act in opposition to her method, as is so often the case.

When once the placenta is expelled, the uterus contracts still more firmly, and in a typical case is felt just within the pelvic brim, hard and firm, and about the size of a cricket ball. Generally for several hours, or even for one or two days, it occasionally relaxes and contracts, and these contractions give rise to the '*after-pains*,' from which women often suffer much. The object of these pains is, no doubt, to expel any coagula that may remain in the uterus, and therefore, however unpleasant they may be to the patient, they must be considered, unless very excessive, to be salutary rather than otherwise.

The length of labour varies extremely in different cases, and it is quite impossible to lay down any definite rules with regard to it. Subject to exceptions, labour is longer in primiparæ than in multiparæ, on account of the greater resistance of the soft parts in the former, especially of the structures about the vagina and vulva. It is also generally stated that the difficulty of labour increases with the age of the patient, and that in elderly primiparæ it is likely to be unusually tedious, from rigidity of the soft parts. It is very doubtful if this opinion has any real basis, and in such cases the practitioner often finds himself agreeably disappointed in the result. Mr. Roper,[1] indeed, argues that the wasting of the tissues which occurs after forty years of age diminishes their resistance, and that first labours, after that age, are easier, as a rule, than in early life. The habits and mode of life of patients have, no doubt, a considerable influence on the duration of labour, but we are not in possession of any very

[1] *Obst. Trans.* v. 7.

reliable facts with regard to this subject. It is reasonable to suppose that the tissues of large, muscular, strongly developed women will offer more resistance than those of slighter build. On the other hand, women of the latter class, especially in the upper ranks of life, more often develop nervous susceptibilities, which may be expected to influence the length of their labours. The average duration of labour, calculated from a large number of cases, is from eight to ten hours; even in primiparæ, however, it is constantly terminated in one or two hours from its commencement, and may be extended to twenty-four hours without any symptoms of urgency arising. In multiparæ it is frequently over in even a shorter time. Indications calling for interference may arise at any time during the progress of labour, independently of its length. The proportion between the length of the first and second stages also varies considerably. The first stage is generally the longest; and it is stated by Cazeaux to be normally about twice the length of the second. This is probably under the mark, and I believe Joulin to be nearer the truth in stating that the first stage should be to the second as four or five to one, rather than as two to one. Often when the first stage has been very prolonged, the second is terminated rapidly.

Proportion of length of second stage.

The practitioner is constantly asked as to the probable length of labour, and the uncertainty of this should always lead him to give a most guarded opinion. Even when labour is progressing apparently in the most satisfactory manner, the pains frequently die away, and delivery may be delayed for many hours. In the first stage a cervix that is apparently rigid and unyielding may rapidly and unexpectedly dilate, and delivery soon follow. In either case, if the practitioner has committed himself to a positive opinion he is apt to incur blame, and it is far better always to be extremely cautious in our predictions on this point.

Necessity of caution in expressing an opinion as to the possible duration of labour.

A somewhat larger proportion of deliveries occur in the early hours of the morning than at other times. Thus West[1] found that out of 2,019 deliveries, 780 took place from 11 P.M. to 7 A.M., 662 from 7 A.M. to 3 P.M., and 577 from 3 P.M. to 11 P.M.

Period of the day at which labour occurs.

* [1] *Amer. Med. Journ.* 1854.

CHAPTER II

MECHANISM OF DELIVERY IN HEAD PRESENTATIONS.

Importance of the subject.

IT is quite impossible to over-estimate the importance of thoroughly understanding the mechanism of the passage of the fœtus through the pelvis. This dominates the whole scientific practice of midwifery, and the practitioner cannot acquire more than a merely empirical knowledge, such as may be possessed by an uneducated midwife, or conduct the more difficult cases requiring operative interference, with safety to the patient or satisfaction to himself, unless he thoroughly masters the subject.

In treating of the physiological phenomena of labour, it was assumed that we had to do with an ordinary case of head presentation, the description being applicable, with slight variations, to presentations of other parts of the fœtus. So in discussing the mechanical phenomena of delivery, I shall describe more in detail the mechanism of head presentations, reserving any account of the mechanism of

Frequency of head presentations.

other presentations until they are separately studied. Head presentation is so much more frequent than that of any other part—amounting to 95 per cent. of all cases—that this mode of studying the subject is fully justified; and, when once the student has mastered the phenomena of delivery in head presentations, he will have little difficulty in understanding the mechanism of labour when other parts of the fœtus present, based, as it always is, on the same general plan.

Mode of recognising the position of the head by its sutures

In entering on this study we come to appreciate the importance of the sutures and fontanelles in enabling us to detect the position of the fœtal head, and to watch its progress through the pelvis; and unless the 'tactus eruditus' by which these can be distinguished from each other has been acquired, the practitioner will be unable to satisfy himself of the exact progress of the labour. Nor is this always easy. Indeed, it requires considerable experience and practice before it is possible to make out the position of the head with absolute certainty; but this knowledge should always be aimed at, and the student will never regret the time and trouble he spends in acquiring it.

At the commencement of labour the long diameter of the head lies in almost any diameter of the pelvic brim, except in the antero-posterior, where there is not space for it. In the large majority of cases, however, it enters the pelvis in one or other of the oblique diameters, or in one between the oblique and transverse; but until it has fairly passed through the brim, it more frequently lies directly in the transverse diameter than has been generally supposed. Hence obstetricians are in the habit of describing the head as lying in four positions, according to the parts of the pelvis to which the occiput points; the first and third positions being those in which the long diameter of the head occupies the right oblique diameter of the pelvis, the second and fourth those in which it lies in the left oblique. Many subdivisions of these positions have been made, which only complicate the subject, and render it more difficult to understand.

The positions, then, of the fœtal head after it has entered the brim, which it is of importance to be able to distinguish in practice, are :—

First (or *left occipito-cotyloid*).—The occiput points to the left foramen ovale, the sinciput to the right sacro-iliac synchondrosis, and the long diameter of the head lies in the right oblique diameter of the pelvis.

Second (or *right occipito-cotyloid*).—The occiput points to the right foramen ovale, the forehead to the left sacro-iliac synchondrosis, and the long diameter of the head lies in the left oblique diameter of the pelvis.

Third (or *right occipito-sacro-iliac*).—The occiput points to the right sacro-iliac synchondrosis, the forehead to the left foramen ovale, and the long diameter of the head lies in the right oblique diameter of the pelvis. This position is the reverse of the first.

Fourth (or *left occipito-sacro-iliac*).—The occiput points to the left sacro-iliac synchondrosis, the forehead to the right

[marginal notes:] and fontanelles.

Position of the head at the commencement of labour.

There are four generally described.

foramen ovale, and the long diameter of the head lies in the left oblique diameter of the pelvis. This position is the reverse of the second.

The rela-
tive fre-
quency
of these
positions.
The relative frequency of these positions has long been, and still is, a matter of discussion among obstetricians. According to Naegele, to whose classical essay we owe the greater part of our knowledge of the subject, the head lies in the right oblique diameter in 99 per cent. of all cases. More recent researches have thrown some doubt on the accuracy of these figures, and many modern obstetricians believe that the second position, which Naegele believed only to be observed as a transitional stage in the natural progress of the third position, is much more common than he supposed. This question will be more fully discussed when we treat of the mechanism of occipito-posterior delivery, and, in the meantime, it may serve to show the discrepancy which exists in the opinions of modern writers, if we append the following table of the relative frequency of the various positions,[1] copied from Leishman's work :—

	First Position	Second Position	Third Position	Fourth Position	Not Classified
Naegele . . .	70·	——	29·	——	1·
Naegele, jun. . .	64·64	—	32·88	——	2·47
Simpson and Barry .	76·45	·29	22·68	·58	—
Dubois . . .	70·83	2·87	25·66	·62	——
Murphy . . .	63·23	16·18	16·18	4·42	——
Swayne . . .	86·36	9·79	1·04	2·8	——

Here it will be seen that all obstetricians are agreed as to the immensely greater frequency of the first position—the only point at issue being the relative frequency of the second and third.

Explana-
tion of the
frequency
with which
the head
lies in
the right
oblique
diameter.
Various explanations have been given of the greater frequency with which the head lies in the right oblique diameter. By some it is referred to the natural tendency of the back of the fœtus, as shown by the experimental researches of Höning and other writers, to be directed, in consequence of gravitation, forwards and to the left side of the mother in the erect attitude, and backwards and to her right side in the recumbent. The explanation given by Simpson was that

[1] Leishman's *System of Midwifery*, p. 341.

the head lay in the right oblique diameter in consequence of the measurement of the left oblique being more or less lessened by the presence of the rectum. When the rectum is collapsed, indeed, the narrowing of the diameter is slight ; but it is so often distended by fæcal matter—sometimes, when constipation exists, to a very great extent—that it may really have a very important influence in determining the position of the fœtal head.

In describing the mechanism of delivery, it will be well for us to concentrate our attention on the first, or most common, position, dwelling subsequently more briefly on the differences between it and the less common ones.

In this position, when the head commences to descend, the occiput lies in the brim pointing to the left ileo-pectineal eminence, the forehead is directed to the right sacro-iliac synchondrosis, and the sagittal suture runs obliquely across the pelvis in the right-oblique diameter. The back of the

Descrip-
tion of
the first
position.

Fig. 96.

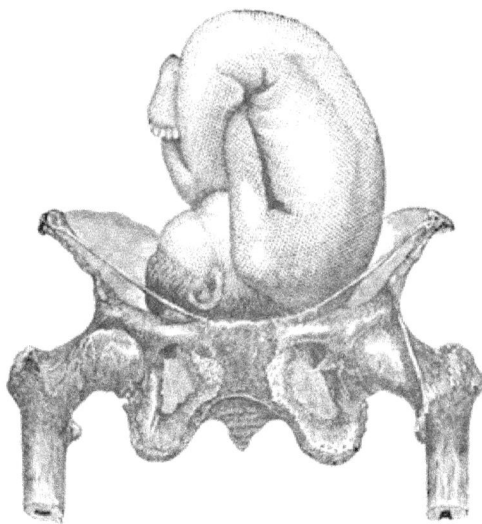

ATTITUDE OF CHILD IN FIRST POSITION. (After Hodge.)

child is turning towards the left side of the mother's abdo-men, the right shoulder to her right side, the left to her left side (fig. 96). If a vaginal examination be now made (the patient lying in the ordinary obstetric position), and the os

be sufficiently open, the finger will impinge upon the protuberance of the right parietal bone, which is described as
the 'presenting part,' a term which has received various
definitions, the best of which is probably that adopted by
Tyler Smith, viz. 'that portion of the fœtal head felt most
prominently within the circle of the os uteri, the vagina,
and the os tincæ, in the successive stages of labour.' If the
tip of the examining finger be passed slightly upwards, it
will feel the sagittal suture running obliquely across the
pelvis, and, if this be traced downward and to the left, it
will come upon the triangular posterior fontanelle, with the
lambdoidal sutures diverging from it. If the finger could
be passed sufficiently high in the opposite direction, upwards
and to the right, it would come upon the large anterior
fontanelle; but, at this time, that is too high up to be within
reach. The chin is slightly flexed upon the sternum, this
flexion, as we shall presently see, being greatly increased as
the head begins to descend.

The head, at the commencement of labour, generally lies
within the pelvic brim, especially in primiparæ. In multiparæ, owing to the relaxation of the abdominal parietes, the
uterus is apt to fall somewhat forwards, and the head consequently is more entirely above the brim, but is pushed
within it as soon as labour actually commences.

Naegele's
views as to
obliquity
of the
head at
the brim.
Naegele—and his description has been adopted by most
subsequent writers—describes the head, at this period, as
lying obliquely in relation to the brim, the right parietal
bone, on which the examining finger impinges, being supposed
by him to be much lower than the left. The accuracy of
this view has, of late years, been contested, and it is now
pretty generally admitted that this obliquity does not exist,
and that the head enters the brim of the pelvis with both
parietal bones on the same level, and with its bi-parietal
diameter parallel to the plane of the inlet (fig. 97). Naegele's view was adopted, partly because the finger always
felt the right parietal protuberance lowest, and partly because it was at that point that the 'caput succedaneum,' or
swelling observed on the head after delivery, was always
formed. Both arguments are, however, fallacious; for the
right parietal bone is the part which would naturally be felt
lowest, on account of the oblique position of the pelvis to

the trunk; while, with regard to the caput succedaneum, it has been conclusively proved by Duncan that it does not form on the point most exposed to pressure, as Naegele

Fig. 97.

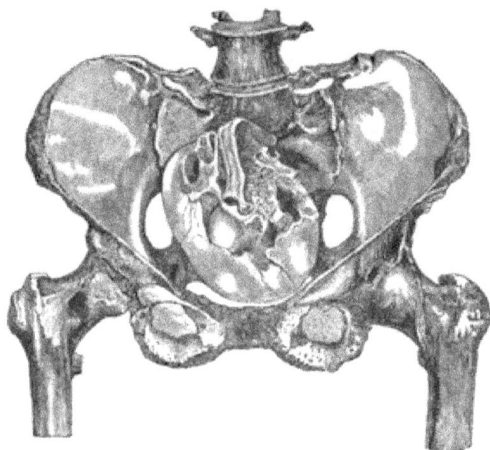

FIRST POSITION : MOVEMENT OF FLEXION.

assumed, but on the part of the head where there is least pressure, that is, the part lying over the axis of the vaginal canal.

In tracing the progress of the head from the position just described, obstetricians have been in the habit of dividing the movements it undergoes into various stages, which are convenient for the purpose of facilitating description. It must be borne in mind that these are not evident and distinct stages, which can always be made out in practice, but that they run insensibly into one another, and often occur simultaneously, or nearly so, in rapid labour. They may be described as: 1. *Flexion.* 2. *First movement of descent.* 3. *Levelling or adjusting movement.* 4. *Rotation.* 5. *Second movement of descent and extension.* 6. *External rotation.* *Division of mechanical movements into stages.*

1. *Flexion.*—The first movement of the head consists of a rotation on its bi-parietal diameter, by which the chin of the child becomes bent on the sternum, and the occiput descends lower than the forehead. By this there is a clear gain of at least a half-inch, for the occipito-bregmatic diameter (3¼ inches) becomes substituted for the occipito-frontal (4½ inches) (fig. 97). *Flexion.*

The movement is most marked when the pelvis is narrow, and, in some cases of pelvic deformity, it takes place to an extreme degree; while, in unusually large and roomy pelves, it occurs to a very slight extent, or not at all. The reason of this flexion is twofold. Solayres and the majority of obstetricians explain it by saying that the expulsive force is communicated to the head through the vertebral column, and inasmuch as the head is articulated much nearer the occiput than the sinciput, the resistance being equal, the former must be pushed down. This is doubtless the correct explanation of the flexion *after* the membranes are ruptured; but, before that happens, the ovum is practically a bag of water, which is equally compressed at all points by the uterine contraction, and is pushed downwards through the os *en masse*, the expulsive force not being transmitted through the vertebral column at all. Under such circumstances flexion is probably effected in the following way: the head being articulated nearer the occiput than the forehead, and being equally pressed upon from below by the resisting structures, the pressure is more effectual on the forehead—consequently that is forced upwards, and the occiput descends. This explanation would also hold good after the rupture of the membranes, and probably both causes assist in affecting the movement.

Descent and levelling movement.

2 and 3. *Descent* and *Levelling Movement.*—The movements of *descent* and *levelling* may be described together. As soon as the head is liberated from the os uteri, it descends pretty rapidly through the pelvis, until the occiput reaches a point nearly opposite the lower part of the foramen ovale (fig. 98), and the sinciput is opposite the second bone of the sacrum. A levelling movement now occurs, the anterior fontanelle comes to be more easily within reach, more on a level with the posterior, and the chin is no longer so much flexed on the sternum. This change is due to the fact that the anterior end of the ovoid experiences greater resistance than the posterior, and as soon as this resistance counterbalances and exceeds that applied to the latter, the sinciput must descend. The right side of the head also descends more than the left from a similar cause, so that the head becomes, as it were, slightly flexed on the right shoulder. This obliquity of the head on its transverse diameter in the

lower part of the pelvis has been denied by Küneke,[1] who
maintains that the head passes through the entire pelvis in

Fig. 98.

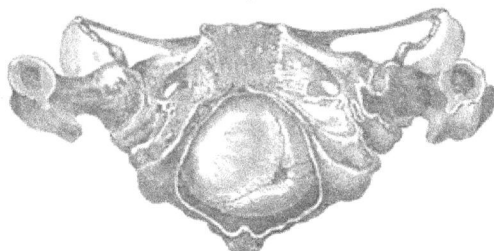

FIRST POSITION : OCCIPUT IN THE CAVITY OF THE PELVIS.   (After Hodge.)

the same position as it enters the brim, that is, with both
parietal bones on a level, so that the point of intersection
of the transverse and antero-posterior diameters of the pelvis
would correspond with the sagittal suture.   There is, however, good reason to believe that in the lower half of the
pelvic cavity the head is not truly synclitic, as Küneke describes, but that the right parietal bone is on a somewhat
lower level than the left.

    4. *Rotation.*—The movement of *rotation* is very impor-   Rotation.
tant.   By it the long diameter of the head is changed from

Fig. 99.

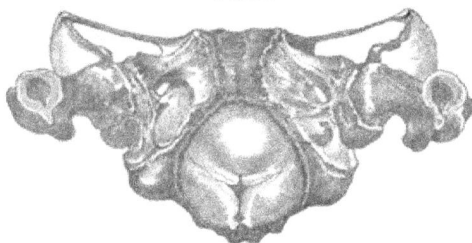

FIRST POSITION : OCCIPUT AT OUTLET OF THE PELVIS.   (After Hodge.)

the oblique diameter of the pelvic cavity to the antero-posterior diameter of the outlet (fig. 99), or to a diameter nearly
corresponding to it, so that the long diameter of the head is
brought into relation with the longest diameter of the pelvic

    [1] *Die Vier Factoren der Geburt*, Berlin, 1869.

outlet. This alteration almost always takes place, and may
be readily observed by the accoucheur who carefully watches
the progress of labour. Various explanations have been
given of its causes. The one most generally adopted is, that
it is due to the projection inwards of the ischial spines, which
narrow the transverse diameter of the pelvic outlet. As the
pains force the occiput downwards, its rotation backwards is
prevented by the projection of the left ischial spine, while its
rotation forwards is favoured by the smooth bevelled surface
of the ascending ramus of the ischium. Similarly the ischial
spine on the opposite side prevents the rotation forwards of
the forehead, which is guided backwards to the cavity of the
sacrum by the smooth surface of the sacro-ischiatic ligaments.
These arrangements, therefore, give a screw-like form to the
interior of the pelvis; and as the pains force the head downwards they are effectual in imparting to it the rotatory movement which is of such importance in adapting it to the
longest measurement of the outlet.

By most of the German obstetricians the influence of the
ischial spines and of the smooth pelvic planes in producing
rotation is not admitted. They rather refer the change of
direction to the increased resistance the head meets from the
posterior wall of the pelvis, and from the perinæal structures.
Whichever part of the head first meets this resistance, which
is much greater than that of the interior part of the pelvis,
must necessarily be pressed forwards; and as, in the large
majority of cases, the posterior fontanelle descends first, it is
thus pressed forwards until rotation is effected. This view
has the advantage of accounting equally well for the rotation in occipito-posterior as in occipito-anterior positions, the
former of which, on the more ordinarily received theory, are
not quite satisfactorily explicable. It does not follow that the
smooth surfaces of the pelvic planes are without influence
in favouring the rotation. On the contrary, they probably
greatly facilitate it; but it is more simply and effectually
explained by the latter theory than by that which attributes
so important an action to the ischial spines.

In some rare cases the head escapes rotation and reaches
the perinæum still lying in the oblique diameter. Even here,
however, rotation is generally effected, often suddenly, just
as the head is about to pass the vulva, and it is very rarely

expelled in the oblique position. The movement at this
stage may be explained by the perinæum, which is attached
at its sides, and grooved in its centre; to the hollow so
formed the long diameter of the head accommodates itself,
and is thus rotated into the antero-posterior diameter of the
outlet.

5. *Extension.*—By the process just described the face is   Extension
turned back into the hollow of the sacrum; but the head
does not lie absolutely in the antero-posterior diameter of the
pelvic outlet, but rather in one between it and the oblique.
The occiput is still forced down by the pains, and, in consequence of its altered position, is enabled to pass between

Fig. 100.

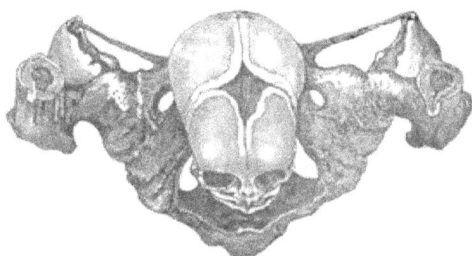

FIRST POSITION : HEAD DELIVERED. (After Hodge.)

the rami of the pubis, and advances until its further descent
is checked by the nape of the neck, which is pressed under
and against the arch of the pubes. By this means the occiput is fixed, and, the pains continuing, the uterine force no
longer acts on the occiput, but on the anterior part of the
head, which is now pushed down and separated from the
sternum. This constitutes *extension.* As the head descends,
the soft structures of the perinæum are stretched, and the
coccyx pushed back so as to enlarge the outlet. The pains
continue to distend the perinæum more and more, the head
advancing and receding with each pain. As the forehead
descends, the sub-occipito-bregmatic, the sub-occipito-frontal,
and the sub-occipito-mental diameters successively present;
the occiput turns more and more upwards in front of the
pubes (fig. 100), and, at last, the face sweeps over the perinæum and is born.

Y 2

The mechanical cause of this movement may be readily explained. As soon as the occiput has passed under the arch of the pubis, and is no longer resisted by the anterior pelvic walls, the head is subjected to the action of two forces: that of the uterine pressure acting downwards and backwards; and that of the resistance of the posterior walls of the pelvis and the soft parts acting almost directly forwards. The necessary result is that the head is pushed in a direction intermediate between these two opposing forces— that is, downwards and forwards in the axis of the pelvic outlet.

In addition to the slight obliquity which exists as regards the direct relation of the long diameter of the head to the antero-posterior diameter of the outlet at the moment of its expulsion, the head also lies somewhat obliquely in relation to its own transverse diameter, so that, in the majority of cases, the right parietal bone is expelled before the left.

External rotation.

6. *External Rotation.*—Shortly after the head is expelled, as soon as renewed uterine action commences, it may be observed to make a distinct rotatory movement, the occiput turning to the left thigh of the mother, and the face turning upwards to the right thigh (fig. 101). The reason of this is evident. When the head descends in the right oblique diameter the shoulders lie in the opposite or left oblique diameter, and as the head rotates into the antero-posterior diameter, they are necessarily placed more nearly in the transverse. As soon as the head is expelled the shoulders are subjected to the same uterine force and pelvic resistance as the head has just been, and they are acted on in precisely the same way. Consequently they too rotate, but in the opposite direction, into the antero-posterior diameter of the outlet, or nearly so, just as the head did, and as they do so they necessarily carry the head with them, and cause its external rotation.

The two shoulders are soon expelled, the left shoulder generally the first, sweeping over the perinæum in the same manner as the face. This is, however, not always the case, and they are often expelled simultaneously, or the right shoulder may come first. The body soon follows, and the second stage of labour is completed.

In the second position (right occipito-cotyloid) the long
diameter of the head lies in the left oblique diameter of the
pelvis.   On making a vaginal examination, in the ordinary

Fig. 101.

EXTERNAL ROTATION OF HEAD IN FIRST POSITION.   (After Hodge.)

obstetric position, the finger, passing upwards and to the
right, feels the small posterior fontanelle; downwards and
to the left, it feels the anterior.   The sagittal suture lies
obliquely across the pelvis in the left oblique diameter.
The description of the mechanism of delivery is precisely
the same as in the first position, substituting the word left
for right.   Thus the finger impinges on the left parietal
bone, the occiput turns from right to left during rotation.
After the birth of the head the occiput turns to the right
thigh of the mother, the face to the left thigh.

In the third position the head enters the pelvic brim
with the occiput directed backwards to the right sacro-iliac
synchondrosis, and the sinciput forwards to the left foramen
ovale (fig. 102).   The posterior fontanelle is directed backwards, the anterior fontanelle forwards, while the examining
finger impinges on the left parietal bone.   The mechanism
of delivery in these cases is of much interest.   In the large
majority of cases, during the progress of delivery the occiput
rotates forwards along the right side of the pelvis, until it
comes to lie almost in the antero-posterior diameter of the
outlet, and passes under the pubic arch, the forehead passing
over the perinæum.   It will be seen that during part of this
extensive rotation the head must lie in the second position,

and the case terminates just as if it had been in the second position from the commencement of labour.

Manner in which the occiput is rotated forwards.

How is it that this rotation is effected, and that the sinciput, occupying the position of the occiput in the first position, should not be rotated forwards to the pubes as that is? This, no doubt, may be explained by the fact that the uterine force transmitted through the vertebral column causes the occiput to descend lower than the sinciput, so that in most cases, in making a vaginal examination, the posterior fontanelle can be readily felt, while the anterior is high up and out of reach. The head is, therefore, extremely

Fig. 102.

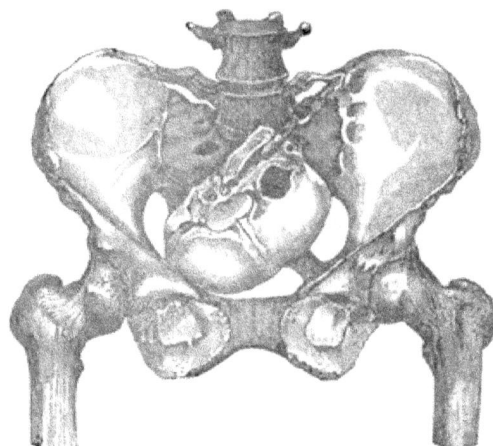

THIRD POSITION OF OCCIPUT, AT BRIM OF PELVIS.

flexed, and so descends into the pelvic cavity, until the occiput, being now below the right ischial spine, experiences the resistance of the pelvic floor, opposite the right sacro-ischiatic ligament, by which it is directed forwards. The forehead is, at this time, supposing flexion to be marked, too high to be influenced by the anterior pelvic plane. Pressure continuing, the occiput rotates forwards, the forehead passes round the left side of the pelvis, and labour is terminated as in the second position.

The period of labour at which rotation takes place varies. In the majority of cases it does not occur until the head is on the floor of the pelvis, for it is then that resistance is most felt; but the greater the resistance, the sooner will rotation be produced. Hence it is more likely to occur early when the head is large and the pelvis comparatively small.

The facility with which this movement is effected obviously depends upon the complete flexion of the chin on the sternum, by which the anterior fontanelle is so elevated that its rotation backwards is not resisted by the inward projection of the left ischial spine, and the occiput is correspondingly depressed. If, however, this flexion is not complete, and the anterior fontanelle is so low as to be readily within reach of the finger, considerable difficulty is likely to be experienced. In many such cases rotation is still eventually effected, but in others it is not; and the labour is then terminated with the face to the pubes, but at the expense of considerable delay and difficulty. According to Dr. Uvedale West, of Alford, who devoted much careful study to the subject, this termination occurs in about 4 per cent. of occipito-posterior positions. When it is about to happen the anterior fontanelle may be felt very low down, and sometimes even the forehead and superciliary ridges. The uterine force pushes down the occiput, the sinciput being fixed behind the pubes, which it obviously cannot pass under, as does the occiput in the first position. The sinciput, therefore, becomes more flexed and pushed upwards, while the resistance of the pelvic floor directs the occiput forwards. The perinæum now becomes enormously distended by the back part of the head, and is in great danger of laceration. The occiput is eventually, but not without much difficulty, expelled. A process of extension now occurs, the nape of the neck being fixed, as it were, against the centre of the perinæum, the expelling force now acting on the forehead, and producing rotation of the head on its transverse axis. The forehead and face are thus protruded, and the body follows without difficulty.

It is said that, in a few exceptional cases, where the anterior fontanelle is much depressed, the labour may terminate by the conversion of the presentation into one of the face, the head rotating on its transverse axis, the forehead passing to the posterior part of the pelvis, and the chin emerging under the perinæum. It is obvious, however, that this change can only occur when the head is unusually small, and it must of necessity be extremely rare.

Termination with face to pubes.

Relative
frequency
of second
and third
positions.Reference has already been made to Naegele's views as
to the rarity of the second position, and to his opinion that
cases in which the occiput was found to point to the right
foramen ovale were only transitional stages in the rotation
of occipito-posterior positions. Such an assumption, however, is unwarrantable, unless the case has been watched from
the very commencement of labour. Many perfectly qualified
observers have arrived at the conclusion that second positions
are far more common than Naegele supposed ; and in the
table already quoted it will be seen that while Murphy estimates the second and third as being equally frequent, Swayne
believes the second to be much more common than the third.
It is probable that the weight of Naegele's authority has
induced many observers to classify second positions as third
positions in which partial rotation has already been accomplished. My own experience would certainly lead me to
think that second positions are very far from uncommon.
The question, however, must be considered to be in abeyance,
until further observations by competent authorities enable
us to decide it conclusively.

Fig. 103.

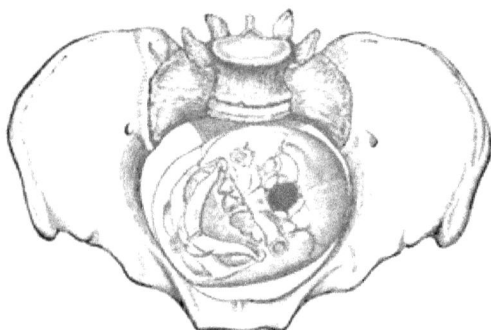

FOURTH POSITION OF OCCIPUT AT PELVIC BRIM.

Fourth or
left occi-
pito-sacro-
iliac.
The fourth position is just as much the reverse of the
second as the third is of the first. The occiput points to
the left (fig. 103) sacro-iliac synchondrosis, and the finger
impinges on the right parietal bone. The mechanism is
precisely the same as in the third position, the rotation
taking place from left to right.

The formation of the caput succedaneum has been already
alluded to. This term is applied to the œdematous swelling
which forms on the head, and is produced by effusion from
the obstruction of the venous circulation caused by the pressure to which the head is subjected. It follows that the
size of the swelling is in direct proportion to the length of
the labour. In rapid deliveries, in which the head is forced
through the pelvis quickly, it is scarcely, if at all, developed ;
while after protracted labours it is large and distinct, and
may obscure the diagnosis of the position, by preventing the
sutures and fontanelles being felt. Its situation varies according to the position of the head : thus, in the first and
fourth positions it forms on the right parietal bone, in the
second and third on the left ; and we may therefore verify,
by inspection of its site, the accuracy of our diagnosis.

An ordinary mistake which has been made by obstetricians is to regard the caput succedaneum as formed at the
point where the head has been most subjected to pressure ;
while, in fact, it forms on that part which is most unsup-
ported by the maternal structures, and where the swelling
may consequently most readily occur. Therefore, in the
early stages of the labour, it always forms on the part of the
head which lies in the circle of the os uteri ; while, in subsequent stages, it forms on that which lies in the axis of the
vaginal canal, and eventually is most prominent on the part
that is first expelled from the vulva.

A few words may be said as to the alteration in the form
of the fœtal head which occurs in tedious labours, and results
from the moulding which it has undergone in its passage
through the pelvis. The smaller the pelvis, and the greater
the pressure applied to the head during delivery, the more
marked this is. The result is, that in vertex presentations
the occipito-mental and occipito-frontal diameters are elongated to the extent of an inch, or even more, while the
transverse diameters are lessened, from compression of the
parietal bones. This moulding is of unquestionable value
in facilitating the birth of the child. The amount of
apparent deformity is very considerable, and may even give
rise to some anxiety. It is well to remember, therefore,
that it is always transient, and that in a few hours, or days

Formation
of the
caput suc-
cedaneum.

It is not
formed
where the
pressure is
greatest.

Alteration
in the
shape of
the head
from
moulding.

at most, the elasticity of the soft cranial bones causes them to resume their natural form. The caput succedaneum also disappears rapidly; therefore no amount of deformity from either of these causes need give rise to anxiety, or call for any treatment.

CHAPTER III.

MANAGEMENT OF NATURAL LABOUR.

ALTHOUGH labour is a strictly physiological function, and in a large majority of cases might, no doubt, be safely accomplished without assistance from the accoucheur, still medical aid, properly **given, is** always of value in facilitating the **process, and** is often absolutely essential for the safety of the mother and child.

The management of the pregnant woman before delivery is a point which should always receive the attention of the medical attendant, since it is of consequence that **the labour** should come on when she is in as good a state of **health as** possible. For this purpose ordinary **hygienic precautions** should **never be** neglected **in the latter months of gestation.** The patient should **take regular and gentle exercise, short of** fatigue, and, **if** the weather **permit, should spend as much** of her time as possible in **the open air.** Hot rooms, late hours, and excitement of all kinds should be strictly avoided. The diet should be simple, nutritious, and unstimulating. The state of the bowels should be particularly attended to. During the few days preceding labour the descent of the uterus often causes pressure on the rectum, and prevents its evacuation. **Hence it is customary to** prescribe occasional **gentle aperients, such as** small doses of castor-oil, for a few **days before the** expected period of delivery. Some caution, **however, is** necessary, as **it is** certainly not very uncommon **for labour** to be determined rather sooner than was anticipated, in consequence of the irritation of too large **a** purgative dose. The state of the bowels should **always be** inquired into at the commencement of labour, and, **if there be any reason to** suspect that they are loaded, a copious enema **should** be administered. This is always a proper precaution

Preparatory treatment.

to take, for a loaded rectum is a common cause of irregular and ineffective uterine action ; and even when it does not produce this result, the escape of the fæces, in consequence of pressure on the bowel during the propulsive stage, is always disagreeable both to the patient and practitioner.

Dress of patient during pregnancy. The dress of the patient during pregnancy may be here adverted to ; for much discomfort may arise, and the satisfactory progress of labour may even be interfered with, from errors in this respect.

After the uterus has risen out of the pelvis the ordinary corset which most women wear is apt to produce very injurious pressure ; still more so when attempts are made to conceal the increased size by tight lacing. After the fourth or fifth month, therefore, the comfort of the patient is much increased by wearing a specially constructed pair of stays with elastic let into the sides and front, so that they accommodate themselves to the gradual increase of the figure. Such are made by all stay-makers, and should be worn whenever the circumstances of the patient permit. Failing this, it is better to avoid the use of the corset altogether, and to have as little pressure on the uterus as possible ; although many women cannot do without the support to which they are accustomed. To multiparæ, especially if there be much laxity of the abdominal parietes, a well-fitting elastic abdominal belt is often a great comfort. This is constructed so that it can be tightened when the patient is walking and in the erect position, when such support is most required, and readily loosened when desired.

Necessity of attending to the first summons. It is hardly necessary to insist on the necessity of the practitioner attending immediately to the first summons to the patient. It is true that he may very often be sent for long before he is actually required. But on the other hand, it is quite impossible to foresee what may be the state of any individual case. By prompt attention he may be able to rectify a malposition, or prevent some impending catastrophe, and thus save his patient from consequences of the utmost gravity.

Articles to be taken by the accoucheur. The practitioner should always be provided with the articles which he may require. The ordinary obstetric cases, containing one or two bottles and a catheter, such as are sold by most instrument-makers, are cumbrous and useless :

while 'obstetric bags' are expensive luxuries not within the reach of all. Everyone can manufacture an excellent obstetric bag for himself, at a small expense, by having compartments for holding bottles stitched on to the sides of an ordinary leather bag, such as is sold for a few shillings at any portmanteau-maker's. It is a great comfort to have at hand all that may be required, and the bag should contain chloroform or other anæsthetic, chloral, laudanum, the liquor ferri perchloridi of the Pharmacopœia, the liquid extract of ergot, and a hypodermic syringe, with bottles containing ether and a solution of ergotine for subcutaneous injection. If it also contain a Higginson's syringe, a small elastic catheter, a good pair of forceps, and one or two suture needles, with some silver wire or carbolised catgut, the practitioner is provided against any ordinary contingency. Other articles that may be required, such as thread, scissors, and the like, are generally provided by the nurse or patient.

On arriving at the house the practitioner should have his visit announced to the patient, and he will very often find that the first effect of his presence is to arrest the pains that have been hitherto progressing rapidly; thereby affording a very conclusive proof of the influence of mental impressions on the progress of labour. If the pains be not already propulsive, it is well that he should occupy himself at first in general inquiries from the attendants as to the progress of the labour, and in seeing that all the necessary arrangements are satisfactorily carried out, so as to allow the patient time to get accustomed to his presence. If he have any choice in the matter, he should endeavour to secure a large, airy, and well-ventilated apartment for the lying-in room, as far removed as possible from without. He may also see to the bed, which should be without curtains, and prepared for the labour by having a water-proof sheeting laid under a folded blanket or sheet, on which the patient lies. These receive the discharges during labour, and can be pulled from under the patient after delivery, so as to leave the dry clothes beneath. Among the lower classes, the lying-in chamber is considered a legitimate meeting-place for numerous female friends to gossip, whose conversation is often distressing, and is certainly injurious, to a woman in

Duties on first visiting the patient.

the excitable condition associated with labour. The medical
attendant should, therefore, insist on as much quiet as possible, and should allow no one in the room except the nurse
and some one friend whose presence the patient may desire.
The husband's presence must be left to the wishes of the
patient. Some women like their husbands to be with them,
while others prefer to be without them, and the medical attendant is bound to act in accordance with the patient's desire.

Vaginal examination. If pains be actually present a vaginal examination is
essential, and should not be delayed. It enables us to ascertain whether the labour has commenced or not, and
whether the presentation is natural or otherwise. The pains,
although apparently severe, may be altogether spurious, and
labour may not have actually commenced. It is of much
importance, both for our own credit and comfort, that we
should be able to diagnose the true character of the pains ;
for if they be so-called 'false' pains, we might wait hours
in fruitless expectation of progress, while delivery is still far
off. The necessity of ascertaining, therefore, the actual state
of affairs need not further be insisted on.

Character of false pains. *False* pains are chiefly characterised by their irregularity,
sometimes coming on at short intervals, sometimes with
many hours between them ; they also vary much in intensity, some being very sharp and painful, while others are
slight and transient. In these respects they differ from the
true pains of the first stage, which are at first slight and
short, and gradually recur with increased force and regularity.
The situation of the two kinds of pains also varies, the false
pains being chiefly situated in front, while the true pains
are felt most in the back, and gradually shoot round towards
the abdomen. Nothing short of a vaginal examination will
enable us to clear up the diagnosis satisfactorily. If the
labour have actually commenced, the os will be more or less
dilated, and its edges thinned ; while with each pain the
cervix will become rigid, and the membranes tense and
prominent. The false pains, on the contrary, have no effect
on the cervix, which remains flaccid and undilated : or, if
the os be sufficiently open to admit the tip of the finger, the
membranes will not become prominent during the contraction. Under such circumstances we may confidently assure
the patient that the pains are false, and measures should be

taken to remove the irritation which produces them. In the
large majority of cases the cause of the spurious pains will
be found to be some disordered state of the intestinal tract ;
and they will be best remedied by a gentle aperient—such *Their treatment.*
as castor-oil, or the compound colocynth pill with hyoscyamus—followed by, or combined with, a sedative, such as
twenty minims of laudanum or chlorodyne. Shortly after
this has been administered the false pains will die away, and
not recur until true labour commences.

Fig. 104.

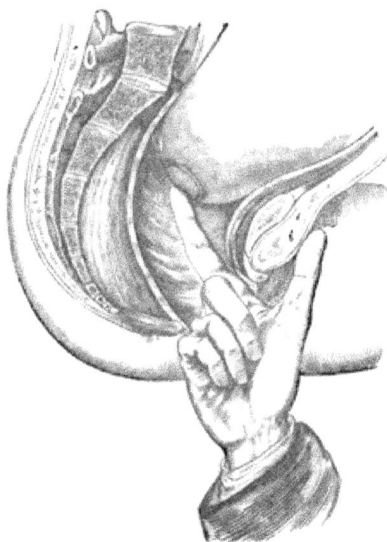

EXAMINATION DURING THE FIRST STAGE.

For a vaginal examination the patient is placed by the *Mode of conducting a vaginal examina- tion.*
nurse on her left side, close to the edge of the bed, with the
legs flexed on the abdomen. The practitioner being seated
by the edge of the bed, passes the index finger of the right
hand, previously lubricated with carbolised oil or cold cream,
up to the vulva, and gently insinuates it into the orifice of the
vagina, then pushes it backwards in the axis of the vaginal
outlet, and finally turns it upwards and forwards so as to
more readily reach the cervix. This it may not always be
easy to do, for at the commencement of labour the cervix
may be so high as to be reached with difficulty, or it may be
directed backwards so as to point towards the cavity of the

sacrum. The exploration is often much facilitated by depressing the uterus from without, by the left hand placed
on the abdomen. Our object is not only to ascertain the
state of the cervix as to softness and dilatation, but also the
presentation, the condition of the vagina, and the capacity
of the pelvis. The examination is generally commenced
during a pain, at which time it is less depressing to the
patient; but in order to be satisfactory the finger must remain in the vagina until the pain is over, the examination being concluded in the interval between this pain and the next.

Object of
the exami-
nation.

In head presentations the round mass of the cranium is
generally at once felt through the lower part of the uterus,
and then we have the satisfaction of being able to assure
the patient that all is right. If the os be sufficiently dilated,
we can also feel through it the occiput covered by the membranes. It is impossible at this time to make out the exact
position of the head by means of the sutures and fontanelles,
which are too high up to be within reach. Nor should any
attempt be made to do so, for fear of prematurely rupturing
the membranes. The fact that the head is presenting is all
that we require to know at this stage of the labour.

No attempt
should be
made, at
this time,
to ascer-
tain the
position of
the head.

The condition of the os itself, as to rigidity and dilatation,
will materially assist us in forming an opinion as to the
progress and probable duration of the labour; but, although
the friends will certainly press for an opinion on this point,
the cautious practitioner will be careful not to commit himself to a positive statement, which may so easily be falsified.
It will suffice to assure the friends that everything is satisfactory, but that it is impossible to say with any certainty
how rapidly, or the reverse, the case may progress.

The condi-
tion of the
os as indi-
cating the
progress of
labour.

If the pains be not very frequent or strong, and the os
not dilated to more than the size of a shilling, a considerable
delay may be anticipated, and the presence of the medical
attendant is useless. He may, therefore, safely leave the
patient for an hour or more, provided he be within easy
reach. It is needless to say that this should never be done
unless the exact presentation be made out. If some part
other than the head be presenting, it will probably be
impossible to make it out until dilatation has progressed
further; and the practitioner must be incessantly on the
watch until the nature of the case be made out, so as to be

able to seize the most favourable moment for interference, should that be necessary.

The position of the patient is a matter of some moment in the first stage. It is a decided advantage that she should not be then in a recumbent position on her side, as is usual in the second stage; for it is of importance that the expulsive force should act in such a way as to favour the descent of the head into the pelvis, i.e. perpendicularly to the plane of its brim, and also that the weight of the child should operate in the same way. Therefore the ordinary custom of allowing the patient to walk about, or to recline in a chair, is decidedly advantageous; and it will often be observed that the pains are more lingering and ineffective if she lie in bed. If the patient be a multipara, or if the abdomen be somewhat pendulous, an abdominal bandage, by supporting the uterus, will greatly favour the progress of this stage. Keeping the patient out of bed has the further advantage of preventing her being unduly anxious for the termination of the labour; and a little cheerful conversation will keep up her spirits, and obviate the mental depression which is so common. Good beef-tea may be freely administered, with a little brandy and water occasionally, if the patient be weak, and will be useful in supporting her strength.

Position of patient during first stage.

Over-frequent vaginal examinations at this period should be avoided, for they serve no useful purpose, and are apt to irritate the cervix. It will be necessary, however, to ascertain the progress of the dilatation at intervals.

Vaginal examinations.

When once the os is fully dilated the membranes may be artificially ruptured if they have not broken spontaneously, for they no longer serve any useful purpose, and only retard the advent of the propulsive stage. This can be easily done by pressing on them, when they are rendered tense during a pain, by some pointed instrument, such as the end of a hair-pin, which is always at hand. In some cases, indeed, it is even expedient to rupture the membranes before the os is fully dilated. Thus it not unfrequently happens, when the amount of liquor amnii is at all excessive, that the os dilates to the size of a five-shilling-piece or more; but, although it is perfectly soft and flaccid, it opens up no further until the liquor amnii is evacuated, when the propulsive pains rapidly complete its dilatation. Some experience and judgment are

Artificial rupture of the membranes.

It is sometimes advisable to rupture the membranes before the os is fully dilated.

required in the detection of such cases, for if we evacuated
the liquor amnii prematurely the pressure of the head on
the cervix might produce irritation, and seriously prolong the
labour. This manœuvre is most likely to be useful when the
pains are strong and the os perfectly flaccid, but when the
membranes do not protrude through the os and effect further
dilatation.

It is sometimes not easy to ascertain whether the membranes are ruptured or not. This is most likely to be the
case when the head is low down, and the amount of liquor
amnii is so small that the pouch does not become prominent
during the pains. A little care, however, will enable us, if
the membranes be ruptured, to feel the rugosities of the scalp
covered with hair, and to distinguish it from the smooth
polished surface of the membranes.

Treatment of the propulsive stage.
After the evacuation of the liquor amnii there is generally
a lull in the progress of the labour, the pains, however, soon
recurring with increased force and frequency, and propelling
the head through the pelvic cavity. The change in the
character of the pains is soon appreciated by the bearing-
down efforts by which they are accompanied, as well as by
their increased length and intensity.

Position of the patient during the second stage.
It is now advisable that the patient be placed in bed; and
in this country it is usual for her to lie on her left side, with
her nates parallel to the edge of the bed, and her body lying
across it. This is the established obstetric position in England,
and it would be useless to attempt to insist on any other, even
if it were advisable. Although the dorsal position is preferred
on the Continent, it is difficult to see wherein its advantages
consist. It certainly leads to unnecessary exposure of the
person, and it is, on the whole, less easy to reach the patient,
so placed, for the necessary manipulations. Moreover, the
dorsal position increases the risk of laceration of the perinæum, by bringing the weight of the child's head to bear
more directly upon it. Thus Schroeder found that lacerations occurred in 37·6 per cent. of cases delivered on the
back, as against 24·4 per cent. in other positions.

The patient usually remains in bed during the whole of
this stage, and it is customary for the nurse to tie to the foot
of the bed a jack-towel, which is laid hold of and used as a
support in making bearing-down efforts. If the pains be

few and far between, and the patient finds it more comfortable to get up occasionally, there is no reason why she should not do so. On the contrary, we shall subsequently see in treating of lingering labour, the pains under such circumstances are often increased in the sitting posture, in consequence of the weight of the child producing increased pressure on the nerves of the vagina.

At this time vaginal examination, which should be more frequently repeated than in the first stage, enables us to ascertain precisely the position of the head, by means of the sutures and fontanelles, as well as to watch its progress.

Detection of the position of head.

It not unfrequently happens that the head descends into the pelvis, even to its floor, without the os having entirely disappeared. The anterior lip especially is apt to get caught between the head and pubes, to become swollen by the pressure to which it is subjected, and then to retard the progress of the labour. There can be no reasonable objection to attempting to prevent this cause of delay by pressing on the incarcerated lip during the interval of the pains, so as to push it above the head, and maintain it there during the pains until the head descends below it. This manœuvre, if done judiciously, and without any undue roughness or force, is certainly not liable to be attended by any of the evil consequences which many obstetricians have attributed to it; it is indeed a matter of common sense that the injury to the cervix is likely to be less if it be pushed gently out of the way than if it be left to be tightly jammed for hours between the presenting part and the bony pelvis. This mode of assistance is very different from the digital dilatation of a rigid cervix, which was formerly much practised, especially in Edinburgh, in consequence of the recommendation of Hamilton, and which was properly objected to by the great majority of obstetricians.

Management of the anterior lip of cervix when impacted between the head and pelvis.

If the pains be producing satisfactory progress, no further interference is required. The medical attendant should, however, see that the bladder is evacuated; and if it have not been so for some hours, it may be necessary to draw off the urine by the catheter. Whenever the labour is lengthy, he should occasionally practise auscultation, so as to satisfy himself that the fœtal circulation is being satisfactorily carried on.

Regulation of the voluntary bearing down efforts.

The regulation of the bearing-down efforts at this time is of importance. It is common for the nurse to urge the patient to help herself by straining, and it is certain that by voluntary action of this kind she can materially increase the action of the accessory muscles of parturition. If the pains be strong, and the labour promise to be rapid, such voluntary exertions are not likely to be prejudicial. On the other hand, if the case be progressing slowly, they only unnecessarily fatigue the patient, and should be discouraged. When the perinæum is distended we may even find it advisable to urge the patient to cease all voluntary effort, and to cry out, for the express purpose of lessening the tension to which the perinæum is subjected. This is the stage in which anæsthesia is most serviceable, but its employment must be separately discussed.

Distension of the perinæum.

As the head descends more and more the perinæum becomes distended, and there is considerable difference of opinion amongst accoucheurs as to the management of the case at this time. In most obstetric works the practitioner is advised to endeavour to prevent laceration by the manœuvre that is described as 'supporting the perinæum.' By this is meant, laying the palm of the hand on the distended structures, and pressing firmly upon them during the acme of the pain, with the view of mechanically preventing their tearing. There can be little doubt that this, or some modification of it, is the practice now followed by the large majority of practitioners. Of late years the evil effects likely to follow it have been specially dwelt upon by Graily Hewitt, Leishman, Goodell, and other writers, who maintain that by pressure exerted in this fashion we not only fail to prevent, but actually favour, laceration, in consequence of the pressure producing increased uterine action, just at the time when forcible distension of the perinæum is likely to be hurtful. Therefore some hold that the perinæum ought to be left entirely alone, and that the head should be allowed gradually to distend it, without any assistance on the part of the practitioner.

Evil effects of pressure on the perinæum.

Much error may be traced to a misconception of what is required. The term 'supporting the perinæum' conveys an unquestionably erroneous idea, and it is certain that no one can prevent laceration by mechanical support. If the term

The object aimed at should be relaxation

'relaxation of the perinæum' was employed, we should have a far more accurate idea of what should be aimed at, and, if this be borne in mind, I think it cannot be questioned that nature may be most usefully assisted at this stage. of the perinæal structures.

Dr. Goodell, of Philadelphia, has specially studied this subject, and has recommended a method the object of which is to relax the perinæum. His advice is, that one or two fingers of the left hand should be inserted into the rectum, by which the perinæum should be hooked up and pulled forward over the head, towards the pubis, the thumb of the same hand being placed on the advancing head, so as to restrain its progress if needful. I have adopted this plan frequently, and believe that it admirably answers its purpose, especially when the perinæum is greatly distended, and laceration is threatened. It must be admitted that the insertion of the fingers into the anal orifice, in the manner recommended, is repugnant both to the practitioner and Dr. Goodell's method.

Fig. 105.

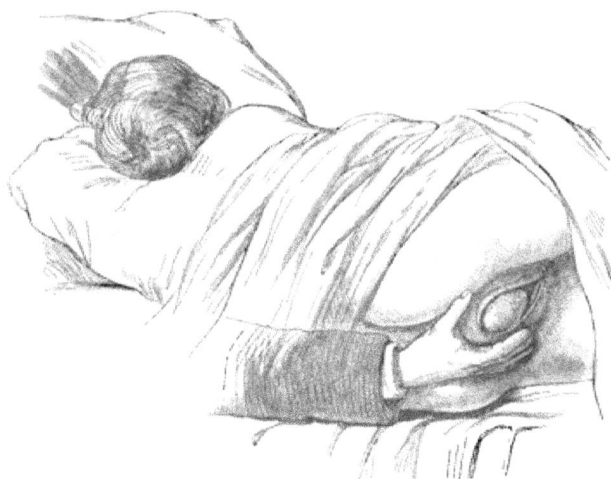

MODE OF EFFECTING RELAXATION OF THE PERINÆUM.

patient, and the same result can be obtained in a less unpleasant way. I mention it, however, to show what it is that the practitioner must aim at. If, when the head is distending the perinæum greatly, the thumb and forefinger of the right

hand are placed along its sides, it can be pushed gently
forward over the head at the height of the pain, while the
tips of the fingers may, at the same time, press upon the
advancing vertex, so as to retard its progress if advisable
(fig. 105). By this means the sudden and forcible stretching
of the perinæal structures is prevented, and the chance of
laceration reduced to a minimum, while nature's mode of relaxing the tissues, by dilatation of the anal orifice, is favoured.
This is very different from the mechanical support that is
usually recommended, and the less pressure that is applied
directly to the perinæum the better. Nor is it either needful
or advisable to sit by the patient with the hand applied to
the perinæum for hours, as is so often practised. Time should
be given for the gradual distension of the tissues by the
alternate advance and recession of the head, and we need
only intervene to assist relaxation when the stretching has
reached its height, and the head is about to be expelled. A
napkin may be interposed between the hand and the skin,
for the purpose of cleanliness. Should the perinæum be
excessively tough and resistant, assiduous fomentation with
a hot sponge may be resorted to, and will be of some service
in promoting relaxation.

When the tension is so great that laceration seems inevitable, it is generally recommended that a slight incision
should be made on each side of the central raphé, with the
view of preventing spontaneous laceration. This may no
doubt be done with perfect safety, but I question if it is
likely to be of use. The idea is that an incised wound is
likely to heal more readily than a lacerated one. When,
however, a distended perinæum ruptures, its structures are so
thinned that the tear is always linear ; and, as a matter of
fact, the edges of the tear are always as clean, and as closely
in apposition, as if the cut had been made with a knife.
Moreover, the laceration invariably heals perfectly, if only
the edges be brought into contact at once with one or two
metallic sutures. I believe, therefore, that Goodell is right
in stating that incision of the perinæum is rarely, if ever,
necessary, unless it is hardened by previous cicatrisation. In
almost all first labours, the fourchette is torn, but requires
no treatment of any kind. In some cases, do what we will,
more or less laceration occurs, and the perinæum should

always be examined after the expulsion of the child, to see if any tear has taken place.

If it has given way to any extent, I believe that it is good practice to insert one or two interrupted sutures of silver wire or carbolised gut at once. Immediately after delivery the sensibility of the tissues is deadened by the distension to which they have been subjected, and the sutures can be inserted with little or no pain. It is quite true that lacerations of an inch or less will generally heal perfectly well of themselves; but this is not invariably the case, while healing almost certainly follows if the edges be brought together at once. In the severer forms of laceration, extending back to, or even through, the sphincter, the precaution is all the more necessary, and a subsequent operation of gravity may in this way be avoided. The sutures can be removed without difficulty in a week or so, when complete adhesion has taken place. *Treatment of lacerations.*

The head, when expelled, should be received in the palm of the right hand, while the left hand is placed upon the abdomen to follow down the uterus as it contracts and expels the body. There is generally some little delay after the expulsion of the head, and we should now see if the cord surround the neck, and, if it does so, it should be drawn over the head, and, if this is not possible, it may be tied and divided between the ligatures. The expulsion of the body should be left entirely to the uterine contractions. If there be undue delay we may endeavour to excite uterine action by friction on the fundus, and it will rarely happen that sufficient contraction does not now come on. If we display undue haste in withdrawing the body, we run the risk of emptying the uterus while its tissues are relaxed, and so favour hæmorrhage. If, however, there seem serious danger of the child being asphyxiated, its expulsion may be favoured by gently passing the forefinger of each hand within the axillæ, and using traction; but it is only very exceptionally that such interference is required. *Expulsion of the child.*

As the uterus contracts, it should be carefully followed down through the abdominal parietes by the left hand, which should grasp it as the body is expelled, with the view of seeing that it is efficiently contracted. This is a point of vital importance in preventing hæmorrhage, which will presently be more especially considered. *Promotion of uterine contraction after the birth of the child.*

Ligature of the cord.

As soon as the child cries we may proceed to tie and separate the cord. For this purpose the nurse usually provides ligatures composed of several strands of whitey-brown thread; but tape, or any other suitable material, may be employed. It is important, especially if the cord be very thick and gelatinous, to see that it is thoroughly compressed, so that the vessels are obliterated, otherwise secondary hæmorrhage might occur. The cord is tied about an inch and a half from the child, and it is usual, though of course not essential, to place a second ligature about two inches nearer the placental extremity of the cord. The latter is, perhaps, of some use by retaining the blood, and thus increasing the size of the placenta, and favouring its more ready expulsion by uterine contraction. The cord is then divided with scissors between the ligatures, the child wrapped up in flannel, and given to the nurse, or a bystander, to hold, while the attention of the practitioner is concentrated on the mother, with a view to the proper management of the third stage of labour. The researches of Budin,[1] Ribemont,[2] and others show that there is a distinct advantage in not tying the cord until the child has cried lustily, as the act of respiration tends to withdraw the placental blood, and thus increases the entire amount of blood in the fœtus. It is said that after late ligature of the cord the child is more vigorous and active than when it is tied too early.

Treatment of the cord by laceration.

The cord may, if preferred, be treated with perfect safety by laceration. This method was first brought under my notice by my friend Dr. Stephen, who has employed it for many years, and in several hundred cases. The cord is twisted round the index fingers of both hands, and torn through, the lacerated vessels retracting without any hæmorrhage. It is a close imitation of the method instinctively adopted by the lower animals, who gnaw the cord asunder, and has the advantage of dispensing with ligatures altogether. I have used it myself in a large number of cases, but prefer, on the whole, the plan usually adopted.

Importance of proper manage-

There is unquestionably no period of labour where skilled management is more important, and none in which mistakes are more frequently made. By proper care at this time the

---

[1] Budin, *Progrès Médicale.* 1876.
[2] *Archiv. de Toxolog.* Oct. 1879.

risk of post-partum hæmorrhage is reduced to a minimum, ment of
the efficient contraction of the uterus is secured, the amount third stage.
and intensity of after-pains are lessened, and the safety and
comfort of the patient greatly promoted. Moreover, the
general practice, as to the management of this stage, is
opposed to the natural mechanism of placental expulsion,
and is far from being well adapted to secure the important
objects which we ought to have in view. Let us see what is
the practice usually recommended and followed, and then we
shall be in a position to understand in what respects it is
erroneous. For this purpose I cannot do better than copy
the directions contained in one of our most deservedly popular
obstetric text-books, which undoubtedly expresses the usual
practice in the management of this stage. 'When the binder
is applied, the patient may be allowed to rest a while, if there
is no flooding; after which, *when the uterus contracts,* gentle
traction may be made by the funis, to ascertain if the placenta
be detached. If so, and especially if it be in the vagina, it
may be removed by continuing the traction steadily in the
axis of the upper outlet at first, at the same time making
pressure on the uterus.'[1]

This may fairly be taken as a sufficiently accurate descrip- Objections
tion of the practice usually followed.[2] The objections I have to ordin-
ary prac-
to make are: (1) That it inculcates the common error of tice.
relying on the binder as a means of promoting uterine contraction, advising its application before the expulsion of the
placenta; while I hold that the binder should never be applied until after the placenta is expelled, and not even then,
unless the uterus is perfectly and permanently contracted.
(2) That it teaches that traction on the cord should be used
as a means of withdrawing the placenta; whereas the uterus
itself should be made to expel the after-birth, and, in nineteen cases out of twenty, the finger need never be introduced
into the vagina after the birth of the child, nor the cord
touched. This may seem an exaggerated statement to those
who have accustomed themselves to the usual method of
dealing with the placenta: but I feel confident that all who
have learnt the method of expression of the placenta would
testify to its accuracy.

---

[1] Churchill's *Theory and Practice of Midwifery,* p. 162.

[2] This practice is further illustrated by the annexed diagram, contained

Expression of the placenta— its object.

The cardinal point to bear in mind is, that the placenta should be expelled from the uterus by a *vis a tergo*, not drawn out by a *vis a fronte*. That uterine pressure after the birth of the child has been recommended by many English writers is certain, and the Dublin school especially have dwelt on its importance as a preventive of post-partum hæmorrhage; but the distinct enunciation of the doctrine that the placenta should be pressed, and not drawn, out of the uterus, we owe to Crede and other German writers; and it is only of late years that this practice has become at all common. Those who have not seen placental expression practised find it difficult to understand that, in the large majority of cases, the uterus may be made to expel the placenta out of the vagina; but such is unquestionably the fact. A little practice is no doubt necessary to effect this satisfactorily: but when once the knack has been learnt, there is little difficulty likely to be experienced.

Importance of not removing the pla-

Before describing the method of placental expression, a word of caution may be said against undue haste in attempting expression of the placenta, a mistake that is often made,

in most obstetric works, which represents the accoucheur as withdrawing the placenta by traction, and which I insert as an illustration of what ought *not* to be done (fig. 106).

Fig. 106.

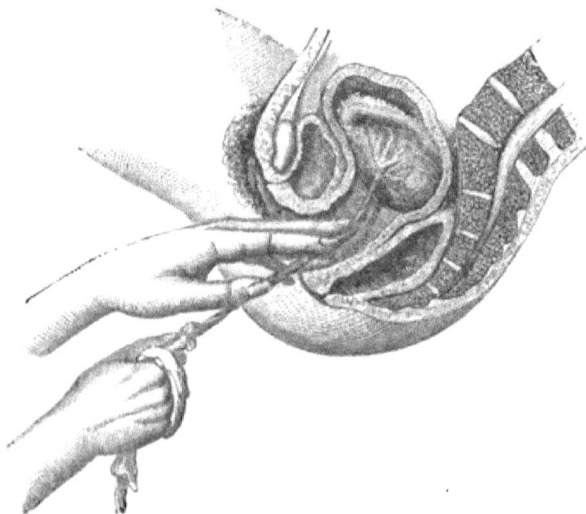

USUAL METHOD OF REMOVING THE PLACENTA BY TRACTION ON THE CORD.

and which, I believe, tends to increase the risk of post-partum hæmorrhage. So long as we satisfy ourselves that the uterus is fairly contracted, so as to avoid the possibility of its distension with blood, a certain delay after the birth of the child is useful, from its giving time for coagula to form within the uterine sinuses, by which their open mouths are closed up. The importance of this point has been specially dwelt upon by McClintock, who lays down the rule that 15 or 20 minutes should be allowed to elapse, after the birth of the child, before any attempt to remove the after-birth is made. This is a good and safe practical rule, as it gives ample time for the complete detachment of the placenta, and the coagulation of the blood in the uterine sinuses.

centa hurriedly.

During this interval the practitioner or nurse should sit by the bedside, with the hand on the uterus to secure contraction and prevent distension ; but not kneading or forcibly compressing it. When we judge that a sufficient time has elapsed, we may proceed to effect expulsion. For this purpose the fundus should be grasped in the hollow of the left

Mode of effecting expression of the placenta.

Fig. 107.

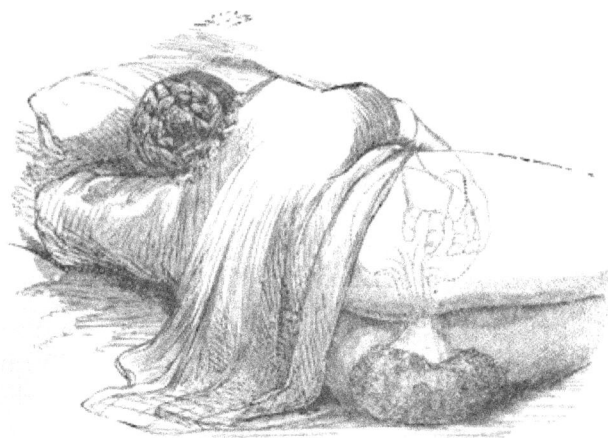

ILLUSTRATING EXPRESSION OF THE PLACENTA.

hand, the ulnar edge of the hand being well pressed down behind the fundus, and *when the uterus is felt to harden,* strong and firm pressure should be made downwards and backwards in the axis of the pelvic brim. If this manœuvre be properly carried out, and sufficiently firm pressure made, in almost every case the uterus may be made to expel the placenta into the bed, along with any coagula that may be in its cavity (fig. 107). The uterine surface of the placenta is generally expelled first, as is represented in the diagram, the cord being within the membranes; whereas the fœtal surface, and root of the cord, are the parts which appear first when the placenta is removed by traction (fig. 106). If we do not succeed at the first effort, which is rarely the case if extrusion be not attempted too soon after the birth of the child, we may wait until another contraction takes place, and then re-apply the pressure. I repeat that, after a little practice, the placenta may be entirely expelled in this way, in nineteen cases out of twenty, without even touching the cord, and the bugbear of retained placenta will cease to be a source of dread.

Should we fail in causing the uterus to expel the placenta, a vaginal examination may be made, and, if the placenta be found lying entirely in the vagina, it may be carefully withdrawn. If, however, the cord can be traced up through the os, showing that the placenta is still within the uterine cavity, we must again resort to pressure to effect its expulsion, and not attempt to withdraw it by traction. Such cases may fairly be classed as retained placenta, but they should be very rarely met with, and are discussed elsewhere. When they do occur often in the hands of the same practitioner, it is fair to conclude that he has not properly acquired the art of managing this stage of labour. Generally speaking, the placenta should be expelled within twenty minutes after the birth of the child; but no doubt, in the large majority of cases, expulsion might be effected sooner were it advisable to attempt it.

Management of the membranes.

When the mass of the placenta is expelled, the membranes generally still remain in the vagina, and they should be twisted into a rope, and very gently withdrawn, so as not to leave any portion behind. This is a precaution the importance of which I would strongly urge, for I believe that the chance of part of the membranes being torn off and left in utero is the one objection to the method recommended. With due care, however, this accident may be avoided, and the risk will be lessened if the placenta is received into the palm of the right hand, on expression, so as to avoid any strain on the membranes.

The duties of the medical attendant are not even now over. For at least ten minutes after the extrusion of the placenta, he should keep his hand on the firmly contracted uterus, gently kneading it, without any force, for the purpose of promoting firm and equable contraction, and causing it to throw off the coagula that may form in its cavity. Compression of the uterus should be continued some time after the expulsion of the placenta.

The subsequent comfort and safety of the patient may be promoted by administering, at this time, a full dose of ergot of rye, such as a drachm, or more, of the liquid extract. The property possessed by this drug of producing tonic and persistent contraction of the uterine fibres, which renders it of doubtful utility as an oxytocic during labour, is of special value after delivery, when such contraction is precisely what we desire. I have long been in the habit of administering the drug at this period, and believe it to be of great value, not only as a prophylactic against hæmorrhage, but as a means of lessening after-pains. Administration of ergot of rye.

When we are satisfied that the uterus is permanently contracted we may apply the binder, but this should rarely be done until at least half an hour after the birth of the child. The soiled clothes should be gently withdrawn from under the patient, moving her as little as possible, and the binder should be, at the same time, slipped under the body, taking care that it is passed well below the hips, so as to secure a firm hold. No kind of bandage is better than a piece of stout jean, of sufficient breadth to extend from the trochanters to the ensiform cartilage ; a jack-towel or bolster slip answers the purpose very well. These are preferable, at any rate at first, to the shaped binders that are often used. One or two folded napkins are generally placed over the uterus, so as to form a pad to keep up pressure. Once in position, the binder is pulled tight, and fastened by pins. The utility of careful bandaging after delivery can scarcely be doubted, although some years ago it became the fashion to dispense with it. It gives a comfortable support to the lax abdominal walls, keeps up a certain amount of pressure on the uterus, and tends to restore the figure of the patient. After the bandage is applied, a warm napkin should be placed Application of the binder.

on the vulva, as a means of estimating the quantity of the discharge, and the patient may be allowed to rest.

After-treatment.

Unless the labour have been very long and fatiguing, an opiate, often exhibited as a matter of routine, is unadvisable; although it may be well to leave one with the nurse, to be given if the patient cannot sleep, or if the after-pains be very troublesome. The practitioner may now leave the room, but not the house, and at least an hour should elapse after delivery before he takes his departure. Before doing so he should visit the patient, inspect the napkin to see that there is not too much discharge, and satisfy himself that the uterus is contracted, and not distended with coagula. He should also count the pulse, which, if the patient be progressing satisfactorily will be found at its normal average. If, however, it be beating over 100 per minute, he should on no account leave, for such a rapidity of the circulation renders it extremely probable that hæmorrhage is impending. This is a good practical rule, laid down by McClintock in his excellent paper 'On the Pulse in Child-bed,' attention to which may often save the patient from disastrous consequences.

Before leaving, the practitioner should see that the room is darkened, all bystanders excluded, and the patient left as quiet as possible to recover from the shock of labour.

CHAPTER IV.

ANÆSTHESIA IN LABOUR.

A FEW words may be said as to the use of anæsthetics during labour, a practice which has become so universal that no argument is required to establish its being a perfectly legitimate means of assuaging the sufferings of childbirth. Indeed, the tendency in the present day is in the opposite direction; and a common error is the administration of chloroform to an extent which materially interferes with the uterine contractions, and predisposes to subsequent post-partum hæmorrhage. Anæsthesia during labour.

Practically speaking the only agent hitherto employed in this country is chloroform, although the bi-chloride of methylene, and ether, have been occasionally tried. Of late years, chloral has been extensively used by some; and as I believe it to be an agent of very great value, I shall first indicate the circumstances under which it may be employed. Agents employed.

The peculiar value of chloral in labour is, that it may be safely administered at a time when chloroform cannot be generally employed. The latter, while it annuls suffering, very frequently tends, in a marked degree, to diminish uterine action. This is a familiar observation to all who have employed it much during labour, as the diminution of the force and intensity of the pains, and the consequent retardation of the labour, often oblige us to suspend its inhalation, at least temporarily. Indeed, this very property of annulling uterine action is one of its most valuable qualities in obstetrics, as in certain cases of turning. For such purposes it is necessary to give it to the surgical extent, which we endeavour to avoid when it is used simply to lessen the suffering of ordinary labour. Still it is not always easy to limit its action in this way, and thus it very frequently does Chloral may be administered when chloroform is inadmissible. Objections to the use of chloroform.

more than we wish. Such diminution in the intensity of uterine contraction is comparatively of less consequence in the propulsive stage, and it is generally more than counter-balanced by the relief it affords. In the first stage it is otherwise, and, practically speaking, chloroform is generally not admissible until the head is in the pelvic cavity.

Chloral is especially the anæs-thetic of the first stage.

Chloral, on the other hand, has no such relaxing effects on uterine contraction. It cannot, it is true, compete with chloroform in its power of relieving pain, but it produces a drowsy state in which the pain is not felt nearly so acutely as before. It is, therefore, in the first stage of labour, while the pains are cutting and grinding, and during the dilatation of the cervix, that it finds its most useful application. It is especially valuable in those cases, so frequently met with in the upper classes, in which the pains produce intolerably acute suffering, but with little effect on the progress of the labour. In them the os is often thin and rigid, and the pains very frequent and acute, but little or no dilatation is effected. When the patient is brought under the influence of chloral, however, the pains become less frequent but stronger, nervous excitement is calmed, and the dilatation of the cervix often proceeds rapidly and satisfactorily. Indeed I know of nothing which answers so well in cases of rigid, undilatable cervix, and I believe its administration to be far more effective, under such circumstances, than any of the remedies usually employed.

Object and mode of adminis-tration.

The object is to produce a somnolent condition, which shall be protracted as long as possible. For this purpose 15 grains of chloral may be administered every twenty minutes, until three doses are given. This generally suffices to pro-duce the desired effect. The patient becomes very drowsy, dozes between the pains, and wakes up as each contraction commences. It may be necessary to give a fourth dose, at a longer interval, say an hour after the third dose, to keep up and prolong the soporific action, but this is seldom necessary, and I have rarely given more than a drachm of chloral during the entire progress of labour. Another advantage of this treatment is that, while it does not interfere with the use of chloroform in the second stage, it renders it necessary to give less than otherwise would be called for, and thus its action can be more easily kept within bounds. On the whole, therefore, I am inclined to consider chloral a very valuable aid in the management of labour, and believe that it is destined to be much more extensively used than is at present the case. So far as my experience has yet gone I have not met with any symptoms which have led me to think that it has produced bad effects; and I have known many patients sleep quietly through labour, without expressing any excessive suffering, or asking for chloroform, who, under ordinary circumstances, would have been most urgently calling for relief. It occasionally happens that the patient cannot retain the chloral from its tendency to produce sickness; it may then be readily given per rectum in the form of enema.

Generally speaking, we do not think of giving chloroform until the os is fully dilated, the head descending, and the pains becoming propulsive. It has often, indeed, been administered earlier, for the purpose of aiding the dilatation of a rigid cervix, and there is no doubt that it often succeeds well when employed in this way; but I have already stated my belief that chloral answers this purpose better.

Chloroform is generally not given until the first stage is completed.

There is one cardinal rule to be remembered in giving chloroform during the propulsive stage, and that is, that it should be administered intermittently, and never continuously. When the pain comes on a few drops may be scattered over a Skinner's inhaler, which affords one of the best means of administering it in labour, or placed within the folds of a handkerchief twisted into the form of a cone. During the acme of the pain the patient inhales it freely, and at once experiences a sense of great relief; and, as soon as the pain dies away, the inhaler should be removed. In the interval between the pains the effect of the drug passes off, so that the higher degree of anæsthesia should never be produced. Indeed, when properly given, consciousness should not be entirely abolished, and the patient, between the pains, should be able to speak, and understand what is said to her. This intermittent administration constitutes the peculiar safety of chloroform administered in labour, and it is a fortunate circumstance that, as yet, there is, I believe, no case on record of death during the inhalation of chloroform for obstetric purposes. This is obviously due to the effect of each inhalation passing off before a fresh dose is administered.

Chloroform should only be given during the pains, and withdrawn in the intervals.

The effect on the pains should be carefully watched. If

Its effects on the pains should be carefully watched.

they become very materially lessened in force and frequency, it may be necessary to stop the inhalation for a short time, commencing again when the pains get stronger, which effect may be often completely and easily prevented by mixing the chloroform with about one third of absolute alcohol, which, originally recommended, I believe, by Dr. Sansom, increases the stimulating effects of chloroform, and thus diminishes its tendency to produce undue relaxation. The amount administered must vary, of course, with the peculiarities of each individual case and the effect produced, but it need never be large. As the head distends the perinæum, and the pains get very strong and forcing, it may be given more freely and to the extent of inducing even complete insensibility just before the child is born.

Ether as a substitute for chloroform.

In cases in which chloroform has lessened the force of the pains ether may be given instead with great advantage. It certainly often acts well when chloroform is inadmissible on account of its effects on the pains, and, so far as my experience goes, it has not the property of relaxing the uterus, but, on the contrary, has sometimes seemed to me distinctly to intensify the pains. Of late I have used a mixture of one part of absolute alcohol, two of chloroform, and three of ether. This is less disagreeable than ether, and has not the over-relaxing effects of chloroform.

Precautions against hæmorrhage should always be taken when chloroform has been freely given.

Bearing in mind the tendency of chloroform to produce uterine relaxation, more than ordinary precautions should always be taken against post-partum hæmorrhage in all cases in which it has been freely administered.

When administered to the surgical degree it should be given by another medical man.

In cases of operative midwifery it is often given to the extent of producing complete anæsthesia. In all such cases it should be administered, when possible, by another medical man, and not by the operator, because the giving of chloroform to the surgical degree requires the undivided attention of the administrator, and no man can do this and operate at the same time. I once learnt an important lesson on this point. I had occasion to apply the forceps in the case of a lady who insisted on having chloroform. When commencing the operation I noticed some suspicious appearances about the patient, who was a large stout woman, with a feeble circulation. I therefore stopped, allowed her to regain consciousness, and delivered her without anæsthesia, much to her own annoyance.   Just one month after labour she went to a dentist to have a tooth extracted, and took chloroform, during the inhalation of which she died.   This impressed on my mind the lesson that no man can do two things at the same time.   The partial unconsciousness of incomplete anæsthesia, in which the patient is restless and tossing about, renders the application of forceps, as well as all other operations, very difficult.   Therefore, unless the patient can be completely and fully anæsthetised, it is better to operate without chloroform being given at all.

CHAPTER V.

PELVIC PRESENTATIONS.

UNDER the head of *pelvic* presentations it is customary to include all cases in which any part of the lower extremities of the child presents. By some these are further subdivided into *breech, footling,* and *knee* presentations; but, although it is of consequence to be able to recognise the feet and the knee when they present, so far as the mechanism and management of delivery are concerned, the cases are identical, and, therefore, may be most conveniently considered together.

Frequency. Presentations coming under this head are far from uncommon; those in which the breech alone occupies the pelvis are met with, according to Churchill, once in 52 labours, while Ramsbotham estimates that it presents more frequently, viz. once in 38·8 labours. Footling presentations occur only once in 92 cases. They are probably often the mere conversion of original breech presentations, the feet having come down during the labour, either in consequence of the sudden escape of the liquor amnii, when the breech was still freely movable above the brim, or from some other cause. Knee presentations are extremely rare, as may be readily understood if it be borne in mind that to admit them the thighs must be extended, hence the vertical measurement of the child must be greatly increased, and therefore it could not be readily accommodated within the uterine cavity, unless of unusually small size. As a matter of fact, Mme. La Chapelle found only one knee presentation in upwards of 3,000 cases.

Causes. The causes of pelvic presentations are not known. They are probably the same as those which produce other varieties of malpresentations; and it is not unlikely that, in certain women, there may be some peculiarity in the shape of the uterine cavity which favours their production. It would be difficult otherwise to explain such a case as that mentioned by Velpeau, in which the breech presented in six labours.

The results, as regards the mother, are in no way more **Prognosis.** unfavourable than in vertex presentation. The first stage of the labour is generally tedious, since the large rounded mass of the breech does not adapt itself so well as the head to the lower segment of the uterus, and dilatation of the cervix is consequently apt to be retarded. The second stage is, however, if anything, more rapid than in vertex cases; and even when it is protracted, the soft breech does not produce such injurious pressure on the maternal structures as the hard and unyielding head.

The result is very different as regards the child. Dubois **The infan-** calculated that one out of 11 children was still-born. **tile mor-** **tality in** Churchill estimates the mortality as much higher, viz. one **pelvic pre-** **sentations** in 3½. The latter certainly indicates a larger number of **is great.** still-births than is consistent with the experience of most practitioners, and more than should occur if the cases be properly managed; but there can be no doubt that the risk to the child is, even under the most favourable circumstances, very great. Even when the child is not lost it may be seriously injured. Dr. Rugè has tabulated a series of 29 cases in which there were found to be fractures of bones or other injuries.[1]

The chief source of danger is pressure on the umbilical **Causes of** cord, in the interval elapsing between the birth of the body **fœtal** **mortality.** and the head. At this time the cord is very generally compressed between the head of the child and the pelvic walls, so that circulation in its vessels is arrested. Hence the aëration of the fœtal blood cannot take place; and, pulmonary respiration not having been yet established, the child dies asphyxiated. There are other conditions present which tend, although in a minor degree, to produce the same result. One of these is that the placenta is probably often separated by the uterine contractions when the bulk of the body is being expelled, as, indeed, takes place, under analogous circumstances, when the vertex presents; the necessary result being the arrest of placental respiration. Joulin thinks that the same effect may be produced by the compression

[1] *Bul. Gén. de Thérap.* August 1875.

of the placenta between the contracted uterus and the hard
mass of the fœtal skull. Probably all these causes combine
to arrest the functions of the placenta ; and, if the delivery
of the head, and consequently the establishment of pulmonary respiration, be delayed, the death of the child is almost
inevitable. The corollary is that the danger to the child is
in direct proportion to the length of time that elapses between the birth of the body and that of the head.

The risk to the child is greater in footling than in breech
cases, because in the former the maternal structures are less
perfectly dilated, in consequence of the small size of the feet
and thighs, and, therefore, the birth of the head is more apt
to be delayed.

Diagnosis.   Inasmuch as the long axis of the child corresponds with
the long axis of the uterus, in pelvic as in vertex presentations, there is nothing in the shape of the uterus to arouse
By abdo-   suspicion as to the character of the case. Still, it is often
minal pal- sufficiently easy to recognise a pelvic presentation by ab-
pation.   dominal examination, if we have occasion to make one. The
facility with which it may be done depends a good deal on
the individual patient. If she be not very stout, and if the
abdominal parietes be lax and non-resistant, we shall generally be able to feel the round head at the upper part of the
uterus, much firmer and more defined in outline than the
breech. The conclusion will be fortified if we hear the fœtal
heart beating on a level with, or above, the umbilicus. The
greater resistance on one side of the abdomen will also enable us to decide, with tolerable accuracy, to which side the
back of the child is placed. Information thus acquired is, at
the best, uncertain ; and we can never be quite sure of the
existence of a pelvic presentation until we can corroborate
the diagnosis by vaginal examination.

Results of   The first circumstance to excite suspicion on examination
vaginal   *per vaginam*, even when the os is undilated, is the absence
examina-   of the hard globular mass felt through the lower segment of
tion.   the uterus, so characteristic of vertex presentations. When
the os is sufficiently open to allow the membranes to pro-
Peculi-   trude, although the presenting part is too high up to be
arity in   within reach, we may be struck with the peculiar shape
shape of   of the bag of membranes, which, instead of being rounded,
the mem-   projects a considerable distance through the os, like the
branes.

finger of a glove. This is a peculiarity met with in all mal-presentations alike, and is, indeed, much less distinct in breech than in footling presentations, because in the former the membranes are more stretched, just as they are in vertex cases. When the membranes rupture, instead of the waters dribbling away by degrees, they often escape with a rush, in consequence of the pelvic extremity not filling up the lower part of the uterus so accurately as the head, which acts as a sort of ball-valve, and prevents the sudden and complete discharge of the waters.

Often, on first examining, even when the membranes are ruptured, the presentation is too high up to be made out accurately. All that we can be certain of is, that it is not the head; and the case must be carefully watched, and examinations frequently repeated, until the precise nature of the presentation can be established. If the breech present, the finger first impinges on a round, soft prominence, on depressing which a bony protuberance, the trochanter major, can be felt. On passing the finger upwards it reaches a groove, beyond which a similar fleshy mass, the other but-tock, can be felt. In this groove various characteristic points, diagnostic of the presentation, can be made out. Towards one end we can feel the movable tip of the coccyx, and above it the hard sacrum, with its rough projecting promi-nences. These points, if accurately made out, are quite characteristic, and resemble nothing in any other presenta-tion. In front there is the anus, in which it is sometimes, but by no means always, possible to insert the tip of the finger. If this can be done it is easy to distinguish it from the mouth, with which it might be confounded, by observing that the hard alveolar ridges are not contained within it. Still more in front we may find the genital organs, the scrotum in male children being often much swollen if the labour has been protracted. Thus it is often possible to recognise the sex of the child before birth.

Diagnosis of the breech.

The breech might be mistaken for the face, especially if the latter be much swollen; but this mistake can readily be avoided by feeling the spinous processes of the sacrum.

Differen-tial dia-gnosis.

The knee is recognised by its having two tuberosities with a depression between them. It might be confounded with the heel, the elbow, or the shoulder. From the heel it is distinguished by having two tuberosities instead of one ; from the elbow, by the latter having one sharp tuberosity, with a depression on one side, instead of a central depression and two lateral prominences ; and from the shoulder, by the latter being more rounded, having only one prominence, running from which the acromion and clavicle can be traced.

Diagnosis of the foot.

The foot may be mistaken for the hand. This error will be avoided by remembering that all the toes are in the same line, and that the great toe cannot be brought into apposition with the others, as the thumb can with the fingers. The internal border of the foot is much thicker than the external, whereas the two borders of the hand are of the same thickness. Moreover, the foot is articulated at right angles to the leg, and cannot be brought into a line with it, as the hand can with the arm. Finally, the projection of the calcaneum is characteristic, and resembles nothing in the hand.

Mechanism.

As is the case in other presentations, obstetricians have very variously subdivided breech presentations, with the effect of needlessly complicating the subject. The simplest division, and that which will most readily impress itself on the memory of the student, is to describe the breech as presenting in four positions, analogous to those of the vertex, the sacrum being taken as representing the occiput, and the positions being numbered according to the part of the pelvis to which it points. Thus we have—

Division of breech presentations into four positions, corresponding to those of the vertex.

First, or *left sacro-anterior* (corresponding to the first position of the vertex). The sacrum of the child points to the left foramen ovale of the mother.

Second, or *right sacro-anterior* (corresponding to the second vertex position). The sacrum of the child points to the right foramen ovale of the mother.

Third, or *right sacro-posterior* (corresponding to the third vertex position). The sacrum of the child points to the right sacro-iliac synchondrosis of the mother.

Fourth, or *left sacro-posterior* (corresponding to the fourth vertex position). The sacrum of the child points to the left sacro-iliac synchondrosis of the mother.

Of these, as with the corresponding vertex positions, the first and third are the most common, their comparative

frequency, no doubt, depending on the same causes.   The
mechanical conditions to which the presenting part is subjected are also identical, but the alterations of position of
the breech in its progress are by no means so uniform as
those of the head, on account of its less perfect adaptation
to the pelvic cavity.   The mechanism of the delivery of the
shoulders and head in breech presentations, moreover, is of
much greater practical importance than that of the body in
vertex presentations, inasmuch as the safety of the child
depends on its speedy and satisfactory accomplishment.
Bearing these facts in mind, it will suffice to describe briefly
the phenomena of delivery in the first and third breech
positions.

In the first position (fig. 108) the sacrum of the child   Position of
points to the left foramen ovale ; its back is consequently   the child
at brim.

Fig. 108.

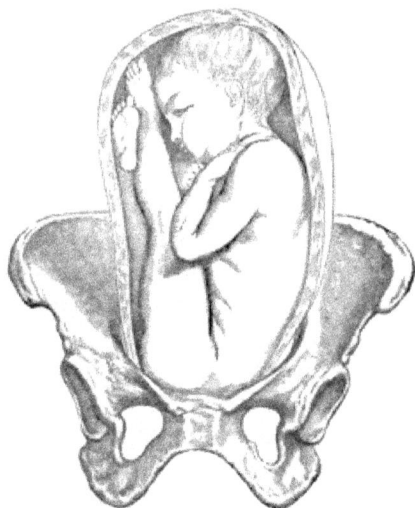

FIRST, OR LEFT SACRO-ANTERIOR POSITION OF THE BREECH.

placed to the left side of the uterus and anteriorly, and its
abdomen looks to the right side of the uterus and posteriorly.
The sulcus between the buttocks lies in the right oblique
diameter of the pelvis, while the transverse diameter of the
buttocks lies in the left oblique diameter, the left buttock
being most easily within reach.   As in vertex presentations,

the hips of the child lie on the same level at the pelvic brim, although Naegele describes the left hip as placed lower than the right.

Descent.   As the pains act on the body of the child, the breech is gradually forced through the pelvic cavity, retaining the same relations as at the brim, its progress being generally more slow than that of the head, until it reaches the lower pelvic strait, when the same mechanism which produces rotation of the occiput comes to operate upon it. The result is a rotation of the child's pelvis, so that its transverse diameter comes to lie approximately in the antero-posterior diameter of the outlet, its antero-posterior diameter corresponds to the transverse diameter of the mother's pelvis, the left hip lies behind the pubis, and the right towards the

Rotation is not so general or complete as in vertex presentations.   sacrum. This rotation, which is admitted by the majority of obstetricians, is altogether denied by Naegele. There can be no doubt, however, that it does generally take place, but by no means so constantly as the corresponding rotation of the vertex; and it is not uncommon for it to be entirely absent, and for the hips to be born in the oblique diameter of the outlet. The body of the child is said frequently not to follow the movement imparted to the hips, so that there is more or less of a twist in the vertebral column.

Expulsion of the hips and body.   The left hip now becomes firmly fixed behind the pubis, and a movement of extension, analogous to that of the head in vertex presentations, takes place. The right, or posterior, hip revolves round the fixed one, gradually distends the perinæum, and is expelled first, the left hip rapidly following. As soon as both hips are born, the feet slip out, unless the legs are completely extended upon the child's abdomen. The shoulders soon follow, lying in the left oblique diameter of the pelvis (fig. 109). The left shoulder rotates forwards behind the pubis, where it becomes fixed, the right shoulder sweeping over the perinæum, and being born first. The

Birth of the arms — evils of undue interference.   arms of the child are generally found placed upon its thorax, and are born before the shoulders. Sometimes they are extended over the child's head, thus causing considerable delay, and greatly increasing the risk to the child. It is now generally admitted that such extension is most apt to occur when traction has been made on the child's body with

the view of hastening delivery, and that it is rarely met with
when the expulsion of the body is left entirely to the natural
powers.

When the shoulders are expelled the head enters the   Delivery of
pelvis in the opposite, or right oblique diameter, the face   the head.

Fig. 109.

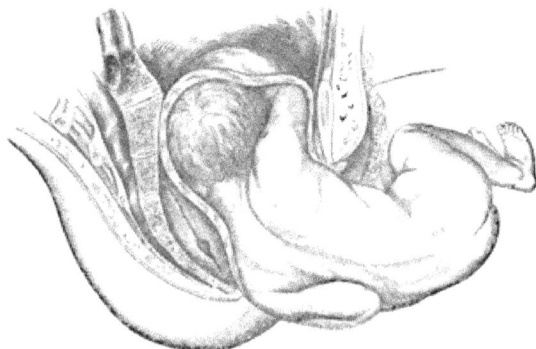

PASSAGE OF THE SHOULDERS AND PARTIAL ROTATION OF THE THORAX.

looking to the right sacro-iliac synchondrosis. As the greater
part of the child is now expelled, and as the head has entered
the vagina, the uterus, having a comparatively small mass to
contract upon, must obviously act at a mechanical disadvan-
tage. Still the pressure of the head on the vagina is a
powerful inciter, the accessory muscles of parturition are
brought into strong action, and there may be sufficient force
to insure expulsion of the head without artificial aid. On
account of the great resistance to the descent of the occiput
from its articulation with the spinal column, the pains
have the effect of forcing down the anterior portion of the
head, and this insures the complete flexion of the chin upon
the sternum (fig. 110). This is a great advantage from a
mechanical point of view, as it causes the short occipito-
mental diameter of the head to enter the pelvis in the axis
of the uterus and the brim. If the head should be in a
state of partial extension—as sometimes happens when the
pelvis is unusually roomy—the occipito-frontal diameter is
placed in a similar relation to the brim, a position certainly
less favourable to the easy birth of the head. As the head
descends it experiences a movement of rotation, the occiput

passing forwards and to the right, behind the pubic arch, the face turning backwards into the hollow of the sacrum. The body of the child will be observed to follow this movement, so that its back is turned towards the mother's abdomen, its anterior surface to the perinæum. The nape of the neck now becomes firmly fixed under the arch of the pubis, the pains act chiefly on the anterior portion of the head, and cause it to sweep over the perinæum, the chin being first born, then the mouth and forehead, and lastly the occiput.

Sacro-posterior positions. It is needless to describe the differences between the mechanism of the second and first positions, which the student, who has mastered the subject of vertex presentations, will readily understand. It is necessary, however, to

Delivery in the third position of the breech. say a few words as to sacro-posterior positions, choosing for that purpose the third, which is the more common of the two. This is exactly the opposite of the first position. The sacrum of the child points to the right sacro-iliac synchondrosis, its abdomen looks forward and to the left side of the mother. The transverse diameter of the child's pelvis lies in the left oblique diameter, the right hip being anterior.

The birth of the body is effected in the same way as in sacro-anterior positions. The birth of the body generally takes place exactly in the way that has been already described, the right hip being towards the pubis.

Rotation forwards of the hips. As the head descends into the pelvis the occiput most usually rotates along its right side—the rotation having been often already partially effected when that of the hips had been made—until it comes to rest behind the pubis, the face passing backwards along the left side of the pelvis into the hollow of the sacrum. This change corresponds exactly to the anterior rotation of the occiput in occipito-posterior positions, and is the natural and favourable termination.

Sometimes rotation does not occur. Sometimes, forward rotation does not take place, and the occiput then turns backwards into the hollow of the sacrum. What then generally occurs is, that the pains con-

Usual termination of such cases. tinue, for the reason already mentioned, to depress the chin and produce strong flexion of the face on the sternum, the occiput becoming fixed on the anterior border of the perinæum. The pains continuing to act chiefly on the anterior part of the head, the face is born first behind the pubis, the occiput only slipping over the perinæum after the forehead has been expelled.

A second mode of termination of such positions is mentioned in most works, on the authority of one or two recorded cases; but although mechanically possible, it is certainly an event of extreme rarity. The chin, instead of being flexed on the sternum, is greatly extended, so that the

Second mode in which such cases occasionally end.

Fig. 110.

DESCENT OF THE HEAD.

face of the child looks upwards towards the pelvic brim. The child then hitches over the upper edge of the pubis and becomes fixed there, while the force of the uterine contractions is expended on the posterior part of the head, which descends through the pelvis, distending the perinæum, and is born first, the face subsequently following.

The mechanism of the delivery of the body and head in cases in which the feet originally present does not differ, in any important respect, from that which has been already described, and requires no separate notice.

Mechanism of feet presentations.

From what has been said of the natural mechanism, it is evident that one of the most fruitful causes of difficulty and complication is undue interference on the part of the practitioner. It is, no doubt, tempting to use traction on the partially born trunk in the hope of expediting delivery; but when it is remembered that this is almost certain to produce extension of the arms above the head, and subsequently extension of the occiput on the spine, both of which seriously increase the difficulty of delivery, the necessity

Treatment.

Importance of avoiding

of leaving the case as much as possible to nature will be apparent.

Having once, therefore, determined the existence of a pelvic presentation, nothing more should be done until the birth of the breech. The membranes should be even more carefully prevented from prematurely rupturing than in vertex presentations, since they serve to dilate the genital passages better than the presenting part. Hence they should be preserved intact, if possible, until they reach the floor of the pelvis, instead of being punctured as soon as the os is fully dilated. The breech when born should be received and supported in the palm of the hand.

When the body is expelled as far as the umbilicus, the dangers to the child commence ; for now the cord is apt to be pressed between the body of the child and the pelvic walls. To obviate this risk as much as possible, a loop of the cord should be pulled down, and carried to that part of the pelvis where there is most room, which will generally be opposite one or the other sacro-iliac synchondrosis. As long as the cord is freely pulsating we may be satisfied that the life of the child is not gravely imperilled, although delay is fraught with danger, from other sources which have been already indicated. In most cases the arms now slip out ; but it may

happen, even without any fault on the part of the accoucheur, that they are extended above the head, and it is of great importance that we should be thoroughly acquainted with the best means of liberating them from their abnormal position.

They must, of course, never be drawn directly downwards, or the almost certain result would be fracture of the fragile bones. We should endeavour to make the arm sweep over the face and chest of the child, so that the natural movements of its joints should not be opposed. If the shoulders be within easy reach, the finger of the accoucheur should be slipped over that which is posterior—because there is likely to be more space for this manœuvre towards the sacrum—and gently carried downwards towards the elbow, which is drawn over the face, and then onwards, so as to liberate the forearm. The same manœuvre should then be applied to the opposite arm. It may be that the shoulders are not easily reached, and then they may be depressed by altering the position of the child's body. If this be carried well up to the mother's abdomen, the posterior shoulder will be brought lower down ; and, by reversing this procedure and carrying the body back over the perinæum, the anterior shoulder may be similarly depressed. It is only very exceptionally, however, that these expedients are required.

The arms being extracted, some degree of artificial as- **Birth of** sistance is, at this time, almost always required. If there **the head.** be much delay, the child will almost certainly perish. Attempts have been made, in cases in which delivery of the head could not be rapidly effected, to establish pulmonary respiration by passing one or two fingers into the vagina, so as to press it back and admit air to the child's mouth, or by passing a catheter or tube into the mouth. Neither of these expedients is reliable, and we should rather seek to aid nature in completing the birth of the head as rapidly as possible. The first thing to do, supposing the face to have rotated into the cavity of the sacrum, is to carry the body of the child well up towards the pubis and abdomen of the mother without applying any traction, for fear of interfering with the all-important flexion of the chin on the sternum. If now the patient bear down strongly, the natural powers may be sufficient to complete delivery. If there be any de- **Manage-** lay, traction must be resorted to, and we must endeavour to **ment when** apply it in such way as to insure flexion. For this purpose, **of the head** while the body of the child is grasped by the left hand, and **is delayed.** drawn upwards towards the mother's abdomen, the index and middle fingers of the right hand are placed on the back of the child's neck, so that their tips press on either side of the **Import-** base of the occiput, and push the head into a state of flexion. **ance of** In most works we are advised to pass the index and middle **flexion of** fingers of the left hand at the same time over the child's **the chin.** face, so as to depress the superior maxilla. Dr. Barnes insists that this is quite unnecessary, and that extraction in the manner indicated, by pressure on the occiput, is quite suffi- cient. Should it not prove so, flexion of the chin may be very effectually assisted by downward pressure on the forehead through the rectum. One or two fingers of the left hand can readily be inserted into the bowel, and the expulsion of the head is thus materially facilitated.

Value of
pressure
through
the abdo-
men.

By far the most powerful aid, however, in hastening delivery of the head, should delay occur, is pressure from above.
This has been, strangely enough, almost altogether omitted
by writers on the subject. It has been strongly recommended by Professor Penrose, and there can be no question
of its utility. Indeed, as the uterus contracts tightly round
the head, uterine expression can be applied almost directly
to the head itself, and without any fear of deranging its
proper relation to the maternal passages. It is very seldom,
indeed, that a judicious combination of traction on the part
of the accoucheur, with firm pressure through the abdomen
applied by an assistant, will fail in effecting delivery of the
head before the delay has had time to prove injurious to the
child.

Applica-
tion of the
forceps to
the after-
coming
head.

Many accoucheurs—among others, Meigs and Rigby—ad-
vocate the application of the forceps when there is delay in
the birth of the after-coming head. If the delay be due to
want of expulsive force in a pelvis of normal size, manual
extraction, in the manner just described, will be found to be
sufficient in almost every case, and preferable, as being more
rapid, easier of execution, and safer to the child. The forceps may be quite properly tried, if other means have failed ;
especially if there be some disproportion between the size of
the head and the pelvis.

Manage-
ment of
sacro-
posterior
positions.
Rotation
forward of
the hips
may be
favoured.

Difficulties in delivery may also occur in sacro-posterior positions. Up to the time of the birth of the head
the labour usually progresses as readily as in sacro-anterior
positions. If the forward rotation of the hips do not take
place, much subsequent difficulty may be prevented by gently
favouring it by traction applied to the breech during the
pains, the finger being passed for this purpose into the fold
of the groin.

Some have
advised
twisting
round of
the body.

It is after the birth of the shoulders that the absence of
rotation is most likely to prove troublesome. It has been
recommended that the body should then be grasped, in the
interval between the pains, and twisted round so as to bring
the occiput forward. It is by no means certain, however,
that the head would follow the movement imparted to the
body, and there must be a serious danger of giving a fatal
twist of the neck by such a manœuvre. The better plan

A better
plan is to

is to direct the face backwards, towards the cavity of the

sacrum, by pressing on the anterior temple during the continuance of a pain. In this way the proper rotation will generally be effected without much difficulty, and the case will terminate in the usual way.

press back the face during the pains.

If rotation of the occiput forwards do not occur, it is necessary for the practitioner to bear in mind the natural mechanism of delivery under such circumstances. In the majority of cases the proper plan is to favour flexion of the chin by upward pressure on the occiput, and to exert traction directly backwards, remembering that the nape of the neck should be fixed against the anterior margin of the perinæum. If this be not remembered, and traction be made in the axis of the pelvic outlet, the delivery of the head will be seriously impeded. In the rare cases in which the head becomes extended, and the chin hitches on the upper margin of the pubis, traction directly forwards and upwards may be required to deliver the head; but before resorting to it care should be taken to ascertain that backward extension of the head has really taken place.

Management of cases in which forward rotation does not occur.

It remains for us to consider the measures which may be adopted in those very troublesome cases in which the breech refuses to descend, and becomes impacted in the pelvic cavity, either from uterine inertia, or from disproportion between the breech and the pelvis. Here, unfortunately, the peculiar shape of the presenting part, which is unadapted for the application of the forceps, renders such cases very difficult to manage.

Management of impacted breech presentations.

Two measures have been chiefly employed: 1st, bringing down one or both feet, so as to break up the presenting part, and convert it into a footling case; 2nd, traction on the breech, either by the fingers, a blunt hook, or fillet passed over the groin.

Barnes insists on the superiority of the former plan, and there can be no question that, if a foot can be got down, the accoucheur has a complete control over the progress of the labour, which he can gain in no other way. If the breech be arrested at or near the brim, there will generally be no great difficulty in effecting the desired object. It will be necessary to give chloroform to the extent of complete anæsthesia, and to pass the hand over the child's abdomen in the same manner, and with the same precautions, as in

Bringing down a foot.

performing podalic version, until a foot is reached, which is seized and pulled down. If the feet be placed in the usual way close to the buttocks, no great difficulty is likely to be experienced. If, however, the legs be extended on the abdomen, it will be necessary to introduce the hand and arm very deeply, even up to the fundus of the uterus, a procedure which is always difficult, and which may be very hazardous. Nor do I think that the attempt to bring down the feet can be safe when the breech is low down and fixed in the pelvic cavity. A certain amount of repression of the breech is possible, but it is evident that this cannot be safely attempted when the breech is at all low down.

Traction on the groin.

Under such circumstances traction is our only resource, and this is always difficult and often unsatisfactory. Of all contrivances for this purpose none is better than the hand of the accoucheur. The index finger can generally be slipped over the groin without difficulty, and traction can be applied during the pains. Failing this, or when it proves insufficient, an attempt should be made to pass a fillet over the groins. A soft silk handkerchief, or a skein of worsted, answers best, but it is by no means easy to apply. The simplest plan, and one which is far better than the expensive instruments contrived for the purpose, is to take a stout piece of copper wire and bend it double into the form of a hook. The extremity of this can generally be guided over the hips, and through its looped end the fillet is passed. The wire is now withdrawn, and carries the fillet over the groins. I have found this simple contrivance, which can be manufactured in a few moments, very useful, and by means of such a fillet very considerable tractive force can be employed. The use of a soft fillet is in every way preferable to the blunt hook which is contained in most obstetric bags. A hard instrument of this kind is quite as difficult to apply, and any strong traction employed by it is almost certain to seriously injure the delicate fœtal structures over which it is placed. As an auxiliary the employment of uterine expression should not be forgotten, since it may give material aid when the difficulty is only due to uterine inertia. After a

Examination of the child.

difficult breech labour is completed the child should be carefully examined to see that the bones of the thighs and arms have not been injured. Fractures of the thigh are far from

uncommon in such cases, and the soft bones of the newly
born child will readily and rapidly unite if placed at once in
proper splints.

Failing all endeavours to deliver by these expedients, Embryo-
there is no resource left but to break up the presenting part tomy.
by scissors, or by craniotomy instruments; but fortunately
so extreme a measure is but rarely necessary.

CHAPTER VI.

PRESENTATIONS OF THE FACE.

Face presentations.

PRESENTATIONS of the face are by no means rare; and, although in the great majority of cases they terminate satisfactorily by the unassisted powers of nature, yet every now and again they give rise to much difficulty, and then they may be justly said to be amongst the most formidable of obstetric complications. It is, therefore essential that the practitioner should thoroughly understand the natural history of this variety of presentation, with the view of enabling him to intervene with the best prospect of success.

Erroneous views formerly held on the subject.

The older accoucheurs had very erroneous views as to the mechanism and treatment of these cases, most of them believing that delivery was impossible by the natural efforts, and that it was necessary to intervene by version in order to effect delivery. Smellie recognised the fact that spontaneous delivery is possible, and that the chin turns forwards and under the pubis; but it was not until long after his time, and chiefly after the appearance of Mme. La Chapelle's essay on the subject, that the fact that most cases could be naturally delivered was fully admitted and acted upon.

Frequency.

The frequency of face presentation varies curiously in different countries. Thus, Collins found that in the Rotunda Hospital there was only 1 case in 497 labours, although Churchill gives 1 in 249 as the average frequency in British practice; while in Germany this presentation is met with once in 169 labours. The only reasonable explanation of this remarkable difference is, that the dorsal decubitus, generally followed abroad, favours the transformation of vertex presentations into those of the face.

The mode in which this change is effected—for it can hardly be doubted that, in the large majority of cases, face presentation is due to a backward displacement of the occiput

after labour has actually commenced, but before the head
has engaged in the brim—has been made the subject of
various explanations.

It has generally been supposed that the change is induced
by a hitching of the occiput on the brim of the pelvis, so as
to produce extension of the head, and descent of the face ;
the occurrence being favoured by the oblique position of
the uterus so frequently met with in pregnancy. Hecker
attaches considerable importance to a peculiarity in the shape
of the fœtal head generally observed in face presentations,
the cranium having the dolicho-cephalous form, prominent
posteriorly, with the occiput projecting, which has the effect
of increasing the length of the posterior cranial lever arm,
and facilitating extension when circumstances favouring it
are in action. Dr. Duncan [1] thinks that uterine obliquity
has much influence in the production of face presentation,
but in a different way to that above referred to. He points
out that, when obliquity is very marked, a curve in the
genital passages is produced, the convexity of which is
directed to the side towards which the uterus is deflected.
When uterine contraction commences, the fœtus is propelled
downwards, and the part corresponding to the concavity of
the curve is acted on to the greatest advantage by the propelling force, and tends to descend. Should the occiput
happen to lie in the convexity of the curve so formed, the
tendency will be for the forehead to descend. In the
majority of cases its descent will be prevented by the increased resistance it meets with, in consequence of the
greater length of the anterior cranial lever arm ; but if the
uterine obliquity be extreme, this may be counterbalanced,
and a face presentation ensues. The influence of this obliquity is corroborated by the observation of Baudelocque,
that the occiput in face presentations almost invariably
corresponds to the side of the uterine obliquity. A further
corroboration is afforded by the fact that in face presentation
the occiput is much more frequently directed to the right
than to the left ; while right lateral obliquity of the uterus
is also much more common.

These theories assume that face presentations are produced during labour. In a few cases they certainly exist

Mode in
which face
presenta-
tions are
produced.

[1] *Edin. Med. Jour.* vol. xv.

before labour has commenced. It is possible, however, as we know that uterine contractions exist independently of actual labour, that similar causes may also be in operation, although less distinctly, before the commencement of labour.

Diagnosis.   The diagnosis is often a matter of considerable difficulty at an early period of labour, before the os is fully dilated and the membranes ruptured, and when the face has not entered the pelvic cavity. The finger then impinges on the rounded mass of the forehead, which may very readily be mistaken for the vertex. At this stage the diagnosis may be facilitated by abdominal palpation in the way suggested by Hecker. If the face is presenting at the brim, palpation will enable us to distinguish a hard, firm, and rounded body, immediately above the pubes, which is the forehead and sinciput; on the other side will be felt an indistinct, soft substance, corresponding to the thorax and neck. When labour is advanced, and the head has somewhat descended, or when the membranes are ruptured, we should be able to make out the nature of the presentation with certainty. The diagnostic marks to be relied on are the edges of the orbits, the prominence of the nose, the nostrils (their orifices showing to which part of the pelvis the chin is turned), and the cavity of the mouth, with the alveolar ridges. If these be made out satisfactorily, no mistake should occur. The most difficult cases are those in which the face has been a considerable time in the pelvis. Under such circumstances the cheeks become greatly swollen and pressed together, so as to resemble the nates. The nose might then be mistaken for the genital organs, and the mouth for the anus. The orbits, however, and the alveolar ridges, resemble nothing in the breech, and should be sufficient to prevent error. Considerable care should be taken not to examine too frequently and roughly, otherwise serious injury to the delicate structures of the face might be inflicted. When once the presentation has been satisfactorily diagnosed, examinations should be made as seldom as possible, and only to assure ourselves that the case is progressing satisfactorily.

Necessity of care in examination.

Mechanism.   If we regard face presentations, as we are fully justified in doing, as being generally produced by the extension of the occiput in what were originally vertex presentations, we can readily understand that the position of the face in relation to the pelvis must correspond to that of the vertex. This is, in fact, what is found to be the case, the forehead occupying the position in which the occiput would have been placed had extension not occurred.

The face, then, like the head, may be placed with its long diameter corresponding to almost any of the diameters of the brim, but most generally it lies either in the transverse diameter, or between this and the oblique, while, as it descends in the pelvis, it more generally occupies one or other of the oblique diameters. It is common in obstetric works to describe two principal varieties of face presentation, viz. the right and left mento-iliac, according as the chin is turned to one or other side of the pelvis. It is better, however, to classify the positions in accordance with the part of the pelvis to which the chin points. We may, therefore, describe four positions of the face, each being analogous to one of the ordinary vertex presentations, of which it is the transformation.

First position.—The chin points to the right sacro-iliac synchondrosis, the forehead to the left foramen ovale, and the long diameter of the face lies in the right oblique diameter of the pelvis. This corresponds to the first position of the vertex, and, as in that, the back of the child lies to the left side of the mother.

Second position.—The chin points to the left sacro-iliac synchondrosis, the forehead to the right foramen ovale, and the long diameter of the face lies in the left oblique diameter of the pelvis. This is the conversion of the second vertex position.

Third position.—The forehead (fig. 111) points to the right sacro-iliac synchondrosis, the chin to the left foramen ovale, and the long diameter of the face lies in the right oblique diameter of the pelvis. This is the conversion of the third vertex position.

Fourth position.—The forehead points to the left sacro-iliac synchondrosis, the chin to the right foramen ovale, and the long diameter of the face lies in the left oblique diameter of the pelvis. This is the conversion of the fourth vertex position.

The relative frequency of these presentations is not yet positively ascertained. It is certain that there is not the

Marginal notes:

The positions of the face correspond to those of the vertex.

The four positions generally met with.

The relative frequency

of these positions is not certainly known. preponderance of first facial that there is of first vertex positions, and this may, no doubt, be explained by the supposition that an unusual vertex position may of itself facilitate the transformation into a face presentation. Winckel con-

Fig. 111.

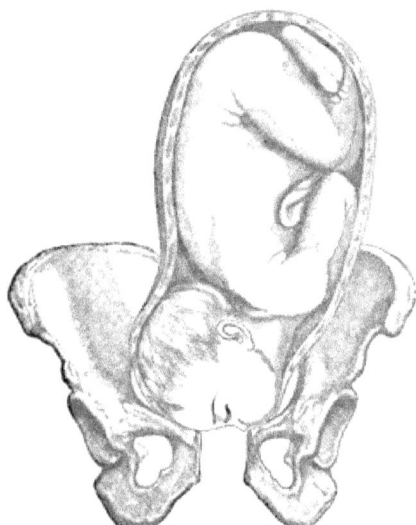

THIRD POSITION IN FACE PRESENTATIONS.

cludes that, *cæteris paribus*, a face presentation is more readily produced when the back of the child lies to the right than when it lies to the left side of the mother; the reason for this being probably the frequency of right lateral obliquity of the uterus. We shall presently see that, with very rare exceptions, it is absolutely essential that the chin should rotate forwards under the pubes before delivery can be accomplished; and, therefore, we may regard the third and fourth face positions, in which the chin from the first points anteriorly, as more favourable than the first and second.

The mechanism is practically the same as in vertex presentations. The mechanism of delivery in face is practically the same as in vertex presentations; and we shall have no difficulty in understanding it if we bear in mind that in face cases the forehead takes the place of, and represents the occiput in, vertex presentations. For the purpose of description we will take the first position of the face—

1. The first step consists in the *extension* of the head,

which is effected by the uterine contractions as soon as the
membranes are ruptured. By this the occiput is still more
completely pressed back on the nape of the neck, and the
fronto-mental, rather than the mento-bregmatic, diameter is
placed in relation to the pelvic brim. This corresponds to
the stage of flexion in vertex presentations.

Description of delivery in the first position of the face. Extension.

The chin descends below the forehead, from precisely
the same cause as the occiput in vertex presentations. On
account of the extended position of the head the presenting
face is divided into portions of unequal length in relation to
the vertebral column, through which the force is applied, the
longer lever arm being towards the forehead. The resistance
is, therefore, greatest towards the forehead, which remains
behind while the chin descends.

2. *Descent.*—As the pains continue, the head (the chin
being still in advance) is propelled through the pelvis. It is
generally said that the face cannot descend, like the occiput,
down to the floor of the pelvis, its descent being limited by
the length of the neck. There is here, however, an obvious
misapprehension. The neck, from the chin to the sternum,
when the head is forcibly extended, measures from 3½ to 4
inches, a length that is more than sufficient to admit of the
face descending to the lower pelvic strait. As a matter of fact,
the chin is frequently observed in mento-posterior positions
to descend so far that it is apparently endeavouring to pass
the perinæum before rotation occurs. At the brim the two
sides of the face are on a level, but as labour advances the
right cheek descends somewhat, the caput succedaneum forms
on the malar bone, and, if a secondary caput succedaneum
form, on the cheek.

Descent.

3. *Rotation* is by far the most important point in the
mechanism of face presentations; for unless it occurs, delivery, with a full-sized head and an average pelvis, is
practically impossible. There are, no doubt, exceptions to
this rule, which must be separately considered, but it is
certain that the absence of rotation is always a grave and
formidable complication of face presentation. Fortunately it
is only very rarely that this is not effected. The mechanical
causes are precisely those which produce rotation of the
occiput forwards in vertex presentations. As it is accomplished, the chin passes under the arch of the pubes, and the

Rotation.

occiput rotates into the hollow of the sacrum (fig. 112) ; and
then commences—

Fig. 112.

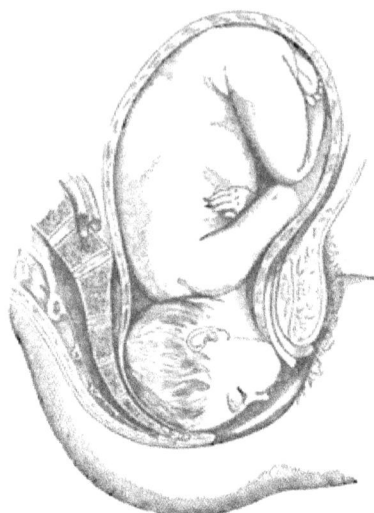

ROTATION FORWARDS OF CHIN.

Fig. 113.

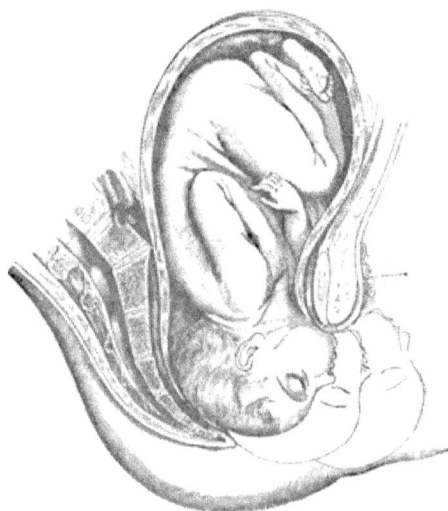

PASSAGE OF THE HEAD THROUGH THE EXTERNAL PARTS IN FACE PRESENTATION.

Flexion.   4. *Flexion*, a movement which corresponds to extension
in vertex cases.   The chin passes as far as it can under the

pubic arch, and there becomes fixed. The uterine force is now expended on the occiput, which revolves, as it were, on its transverse axis (fig. 113), the under surface of the chin resting on the pubes as a fixed point. This movement goes on until, at last, the face and occiput sweep over the distended perinæum.

5. *External rotation* is precisely similar to that which takes place in head presentations, and, like it, depends on the movements imparted to the shoulders.

Such is the natural course of delivery in the vast majority of cases; but, in order fully to understand the subject, it is necessary to study those rare cases in which the chin points backwards, and forward rotation does not occur. These may be taken to correspond to the occipito-posterior positions, in which the face is born looking to the pubes; but, unlike them, it is only very exceptionally that delivery can be naturally completed. The reason of this is obvious, for the occiput

External rotation.

Mento-posterior positions in which rotation does not take place.

Fig. 114.

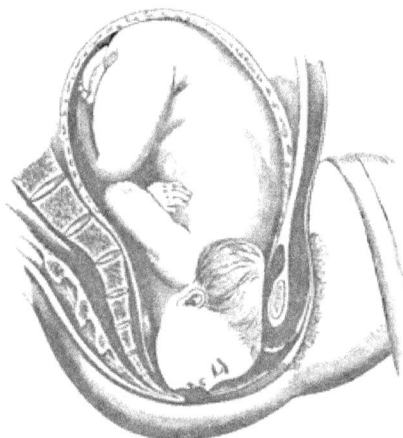

ILLUSTRATING THE POSITION OF THE HEAD WHEN FORWARD ROTATION OF THE CHIN
DOES NOT TAKE PLACE.

gets jammed behind the pubes, and there is no space for the fronto-mental diameter to pass the antero-posterior diameter of the outlet (fig. 114). Cases are indeed recorded in which delivery has been effected with the chin looking posteriorly; but there is every reason to believe that this can only happen when the head is either unusually small, or the pelvis unusually large. In such cases the forehead is pressed down until a portion appears at the ostium vaginæ, when it becomes firmly fixed behind the pubes, and the chin, after many efforts, slips over the perinæum. When this is effected, flexion occurs, and the occiput is expelled without difficulty. The forehead is probably always on a lower level than the chin.

Dr. Hicks[1] has published a paper, in which he attempts to show that this termination of face presentations is not so rare as is generally supposed, and he gives a single instance in which he effected delivery with the forceps; but he practically admits that special conditions are necessary, such as the 'antero-posterior diameter of the outlet particularly ample,' and a diminished size of the head. When delivery is effected it is probable, as Cazeaux has pointed out, that the face lies in the oblique diameter of the outlet, and that the chin depresses the soft structures at the side of the sacro-ischiatic notch, which yield to the extent of a quarter of an inch or more, and thereby permit the passage of the occipito-mental diameter of the head. It must, however, be borne well in mind, that spontaneous delivery in mento-posterior positions is the rare exception, and that supposing rotation does not occur—and it often does so at the last moment—artificial aid in one form or another will be almost certainly required.

Prognosis of face presentations.

As regards the mother, in the great majority of cases the prognosis is favourable, but the labour is apt to be prolonged, and she is, therefore, more exposed to the risks attending tedious delivery As regards the child, the prognosis is much more unfavourable than in vertex presentations. Even when the anterior rotation of the chin takes place in the natural way, it is estimated that 1 out of 10 children is stillborn; while if not, the death of the child is almost certain. This increased infantile mortality is evidently due to the serious amount of pressure to which the child is subjected, and probably depends in many cases on cerebral congestion, produced by pressure on the jugular veins, as the neck lies in the pelvic cavity. Even when the child is born alive, the face is always greatly swollen and disfigured. In some cases the

---

[1] *Obst. Trans.* vol. vii.

deformity produced in this way is excessive, and the features are often scarcely recognisable. This disfiguration passes away in a few days; but the practitioner should be aware of the probability of its occurrence, and should warn the friends, or they might be unnecessarily alarmed, and possibly might lay the blame on him.

After what has been said as to the mechanism of delivery in face presentation, it is obvious that the proper course is to leave the case alone, in the expectation of the natural efforts being sufficient to complete delivery. Fortunately, in the large majority of cases, this course is attended by a successful result.

Treatment. Most cases should be left to the natural efforts.

The older accoucheurs, as has been stated, thought active interference absolutely essential, and recommended either podalic version, or the attempt to convert the case into a vertex presentation, by inserting the hand and bringing down the occiput. The latter plan was recommended by Baudelocque, and is even yet followed by some accoucheurs. Thus Dr. Hodge[1] advises it in all cases in which face presentation is detected at the brim; but although it might not have been attended with evil consequences in his experienced hands, it is certainly altogether unnecessary, and would infallibly lead to most serious results if generally adopted. It may, however, be allowable in certain cases in which the face remains above the brim, and refuses to descend into the pelvic cavity. Even then it is questionable whether podalic version should not be preferred, as being easier of performance, giving, when once effected, a much more complete control over delivery, and being less painful to the mother. Version is certainly preferable to the application of the forceps, which are introduced with difficulty in so high a position of the face, and do not take a secure hold.

Management of cases in which the vertex does not descend.

Schatz[2] has more recently suggested the rectification of face presentations at an early stage, before the rupture of the membranes by manipulation through the abdomen. He raises the fœtal body by pressure on the shoulder and breast through the abdominal wall by one hand, while the breech is raised and steadied by the other. By this means the occiput is elevated, and then the breech is pressed downwards, when head flexion is produced by the resistance of the pelvic walls.

Rectification by abdominal palpation.

[1] *System of Obstetrics*, p. 335.   [2] *Arch. f. Gyn.* B. v. 313.

Of this method I have had no practical experience, but it obviously requires an unusual amount of skill and practice in abdominal palpation.

Difficulties from arrest in the pelvic cavity.

When once the face has descended into the pelvis, difficulties may arise from two chief causes: uterine inertia, and non-rotation forwards of the chin.

The treatment of the former class must be based on precisely the same general principles as in dealing with protracted labour in vertex presentations. The forceps may be applied with advantage, bearing in mind the necessity of getting the chin under the pubes, and, when this has been effected, of directing the traction forwards, so as to make the occiput slowly and gradually distend and sweep over the perinæum.

Difficulties arising from non-rotation of chin forwards.

The second class of difficult face cases are much more important, and may try the resources of the accoucheur to the utmost. Our first endeavour must be, if possible, to secure the anterior rotation of the chin. For this purpose various manœuvres are recommended. By some, we are advised to introduce the finger cautiously into the mouth of the child, and draw the chin forwards during a pain; by others, to pass the finger up behind the occiput and press it backwards during the pain. Schroeder points out that the difficulty often depends on the fact of the head not being sufficiently extended, so that the chin is not on a lower level than the forehead; and that rotation is best promoted by pressing the forehead upwards with the finger during a pain, so as to cause the chin to descend. Penrose[1] believes that non-rotation is generally caused by the want of a *point d'appui* below, on account of the face being unable to descend to the floor of the pelvis, and that, if this is supplied, rotation will take place. In such cases he applies the hand, or the blade of the forceps, so as to press on the posterior cheek. By this means the necessary *point d'appui* is given; and he relates several interesting cases in which this simple manœuvre was effectual in rapidly terminating a previously lengthy labour. Any, or all, of these plans may be tried. We must bear in mind, in using them, that rotation is often delayed until the face is quite at the lower pelvic strait, so that we need not too soon despair of its occurring. If, how-

[1] *Amer. Supplement to Obst. Journ.* April, 1876.

ever, in spite of these manœuvres, it do not take place, what
is to be done? If the head be not too low down in the
pelvis to admit of version, that would be the simplest and
most effectual plan. I have succeeded in delivering in this
way, when all attempts at producing rotation had failed; but
generally the face will be too decidedly engaged to render it
possible. An attempt might be made to bring down the
occiput by the vectis, or by a fillet; but if the face be in the
pelvic cavity, it is hardly possible for this plan to succeed.
An endeavour may be made to produce rotation by the
forceps; but it should be remembered that rotation of the
face mechanically in this way is very difficult, and much
more likely to be attended with fatal consequences to the
child than when it is effected by the natural efforts. In
using forceps for this purpose, the second or pelvic curve is
likely to prove injurious, and a short straight instrument is
to be preferred. If rotation be found to be impossible, an
endeavour may be made to draw the face downwards, so as to
get the chin over the perinæum, and deliver in the mento-
posterior position; but, unless the child be small, or the
pelvis very capacious, the attempt is unlikely to succeed.
Finally, if all these means fail, there is no resource left but
lessening the size of the head by craniotomy, a *dernier
ressort* which, fortunately, is very rarely required.

It sometimes happens that the head is partially extended,
so as to bring the os frontis into the brim of the pelvis, and
form what is described as a '*brow presentation.*' Should the
head descend in this manner, the difficulties, although not
insuperable, are apt to be very great, from the fact that the
long cervico-frontal diameter of the head is engaged in the
pelvic cavity. The diagnosis is not difficult, for the os
frontis will be detected by its rounded surface; while the
anterior fontanelle is within reach in one direction, the orbit
and root of the nose in another.

Brow presentations.

Fortunately, in the large majority of cases, brow presentations are spontaneously converted into either vertex or face
presentations, according as flexion or extension of the head
occurs; and these must be regarded as the desirable terminations and the ones to be favoured. For this purpose
upward pressure must be made on one or other extremity of
the presenting part during a pain, so as to favour flexion, or

In most cases they are spontaneously converted into either face or vertex presentations.

extension ; or, if the parts be sufficiently dilated, an attempt may be made to pass the hand over the occiput and draw it down, thus performing cephalic version. The latter is the plan recommended by Hodge, who describes the operation as easy. It is questionable, however, if a well-marked brow presentation be distinctly made out while the head is still at the brim, whether podalic version would not be the easiest and best operation. If the forehead have descended too low for this, and if the endeavour to convert it into either a face or vertex presentation fail, the forceps will, probably, be required. In such cases the face generally turns towards the pubes, the superior maxilla becomes fixed behind the pubic arch, and the occiput sweeps over the perinæum. Very great difficulty is likely to be experienced, and, if conversion into either a vertex or face presentation cannot be effected, craniotomy is not unlikely to be required.

The forceps or craniotomy may be required.

CHAPTER VII.

DIFFICULT OCCIPITO-POSTERIOR POSITIONS.

A FEW words may be said in this place as to the management of occipito-posterior positions of the head, especially of those in which forward rotation of the occiput does not take place. It has already been pointed out that, in the large majority of these cases, the occiput rotates forward without any particular difficulty, and the labour terminates in the usual way, with the occiput emerging under the arch of the pubes.

Difficult occipito-posterior positions.

In a certain number of cases such rotation does not occur, and difficulty and delay are apt to follow. The proportion of cases in which face to pubes terminations of occipito-posterior positions occurs has been variously estimated, and they are certainly more common than most of our text-books lead us to expect. Dr. Uvedale West,[1] who studied the subject with great care, found that labour ended in this way in 79 out of 2,585 births, all these deliveries being exceptionally difficult.

Rotation forwards of the occiput does not always occur.

He believed that forward rotation of the head is prevented by the absence of flexion of the chin on the sternum, so that the long occipito-frontal, instead of the short sub-occipito-bregmatic, diameter of the head is brought into contact with the pelvic diameter; hence the occiput is no longer the lowest point, and is not subjected to the action of those causes which produce forward rotation. Dr. Macdonald, who has written a thoughtful paper on the subject,[2] believes that the non-rotation forward of the occiput is chiefly due to the large size of the head, in consequence of which 'the forehead gets so wedged into the pelvis anteriorly that its tendency to slacken and rotate forward does not come into play.' Dr. West's explanation, which has an important bearing on the management of these cases, seems to explain most correctly the non-occurrence of the natural rotation.

Causes of face to pubes delivery.

[1] *Cranial Presentations*, p. 33.   [2] *Edin. Med. Jour.* Oct. 1874.

The important question for us to decide is, How can we best assist in the management of cases of this kind when difficulties arise, and labour is seriously retarded?

Mode of treatment of such cases. Upward pressure on the forehead. Dr. West, insisting strongly on the necessity of complete flexion of the chin on the sternum, advises that this should be favoured by upward pressure on the frontal bone, with the view of causing the chin to approach the sternum, and the occiput to descend, and thus to come within the action of the agencies which favour rotation. Supposing the pains to be strong, and the fontanelle to be readily within reach, we may, in this way, very possibly favour the descent of the occiput, and without injuring the mother, or increasing the difficulties of the case in the event of the manœuvre failing. The beneficial effects of this simple expedient are sometimes very remarkable. In two cases in which I recently adopted it, labour, previously delayed for a length of time without any apparent progress, although the pains were strong and effective, was in each instance rapidly finished almost immediately after the upward pressure was applied. The rotation of the face backwards may at the same time be favoured by pressure on the pubic side of the forehead during the pains.

Traction on the occiput. Others have advised that the descent of the occiput should be promoted by downward traction, applied by the vectis or fillet. The latter is the plan specially advocated by Hodge;[1] and the fillet certainly finds one of its most useful applications in cases of this kind, as being simpler of application, and probably more effective, than the vectis.

Over-active endeavours at assistance should be avoided. Although any of these methods may be adopted, a word of caution is necessary against prolonged and over-active endeavours at producing flexion and rotation when that seems delayed. All who have watched such cases must have observed that rotation often occurs spontaneously at a very advanced period of labour, long after the head has been pressed down for a considerable time to the very outlet of the pelvis, and when it seems to have been making fruitless endeavours to emerge; so that a little patience will often be sufficient to overcome the difficulty.

When necessary, the forceps In the event of assistance being absolutely required, there is no reason why the forceps should not be used. The

---

[1] *System of Obstetrics*, p. 308.

instrument is not more difficult to apply than under ordinary *may be used.* circumstances, nor, as a rule, is much more traction necessary. Dr. Macdonald, indeed, in the paper already alluded to, maintains that in persistent occipito-posterior positions there is almost always a want of proportion between the head and the pelvis, and that, therefore, the forceps will be generally required, and he prefers them to any artificial attempts at rectification. Some peculiarities in the mode of delivery are necessary to bear in mind. In most works it is taught that the operator should pay special attention to the rotation of the head, and should endeavour to impart this movement by turning the occiput forward during extraction. Thus Tyler Smith says, 'In delivery with the forceps in occipito-posterior presentations, the head should be slowly rotated during the process of extraction so as to bring the vertex towards the pubic arch, and thus convert them into occipito-anterior presentations.' The danger accompanying any for-*Danger of* cible attempt at artificial rotation will, however, be evident *ing artifi-* on slight consideration. It is true that in many cases, when *cial rota-* simple traction is applied, the occiput will, of itself, rotate *occiput.* forwards, carrying the instrument with it. But that is a very different thing from forcibly twisting round the head with the blades of the forceps, without any assurance that the body of the child will follow the movement. It is impossible to conceive that such violent interference should not be attended with serious risk of injury to the neck of the child. If rotation do not occur, the fair inference is, that the head is so placed as to render delivery with the face to the pubes *Delivery* the best termination, and no endeavour should be made to *to pubes* prevent it. This rule of leaving the rotation entirely to *when rota-* nature, and using traction only, has received the approval of *not take* Barnes and most modern authorities, and is the one which *place.* recommends itself as the most scientific and reasonable.

These are cases in which the pelvic curve of the forceps *Objection* is of doubtful utility. When applied in the usual way the *to curved* convexity of the blades points backwards. If rotation accom-*ments in* pany extraction, the blades necessarily follow the movement *such cases.* of the head, and their convex edges will turn forwards. It certainly seems probable that such a movement would subject the maternal soft parts to considerable risk. I have, however, more than once seen such rotation of the instrument

happen without any apparent bad result; but the dangers
are obvious. Hence it would be a wise precaution, either to
use a pair of straight forceps for this particular operation, or
to remove the blades and leave the case to be terminated by
the natural powers, when the head is at the lower strait, and
rotation seems about to occur. When there is no rotation,
more than usual care should be taken with the perinæum,
which is necessarily much stretched by the rounded occiput.
Indeed the risk to the perinæum is very considerable, and,
even with the greatest care, it may be impossible to avoid
laceration.

Necessity of guarding the perinæum in occipito-posterior delivery.

Bearing these precautions in mind, delivery with the
forceps in occipito-posterior positions offers no special difficulties or dangers.

CHAPTER VIII.

PRESENTATIONS OF THE SHOULDER, ARM, OR TRUNK—COMPLEX
PRESENTATIONS—PROLAPSE OF THE FUNIS.

In the presentations already considered the long diameter of the fœtus corresponded with that of the uterine cavity, and in all of them, the birth of the child by the maternal efforts was the general and normal termination of labour. We have now to discuss those important cases in which the long diameter of the fœtus and uterus do not correspond, but in which the long fœtal diameter lies obliquely across the uterine cavity. In the large majority of these it is either the shoulder or some part of the upper extremity that presents; for it is an admitted fact that, although other parts of the body, such as the back or abdomen, may, in exceptional cases, lie over the os at an early period of labour, yet, as labour progresses, such presentations are almost always converted into those of the upper extremity.

Cases in which the, long diameter of the fœtus does not correspond with that of the uterus.

For all practical purposes we may confine ourselves to a consideration of *shoulder* presentations; the further subdivision of these into *elbow* or *hand* presentations being no more necessary than the division of pelvis presentations into breech, knee, and footling cases, since the mechanism and management are identical, whatever part of the upper extremity presents.

Practically these may be discussed under shoulder presentations.

There is this great distinction between the presentations we are now considering and those already treated of, that, on account of the relations of the fœtus to the pelvis, delivery by the natural powers is impossible, except under special and very unusual circumstances that can never be relied upon. Intervention on the part of the accoucheur is, therefore, absolutely essential, and the safety of both the mother and child depends upon the early detection of the abnormal

Delivery by the natural powers is quite exceptional.

position of the fœtus; for the necessary treatment, which is comparatively easy and safe before labour has been long in progress, becomes most difficult and hazardous if there have been much delay.

Position of the fœtus.

Presentations of the upper extremity or trunk are often spoken of as '*transverse presentations*' or '*cross-births;*' but both of these terms are misleading, as they imply that the fœtus is placed transversely in the uterine cavity, or that it lies directly across the pelvic brim. As a matter of fact, this is never the case, for the child lies obliquely in the uterus, not indeed in its long axis, but in one intermediate between its long and transverse diameters.

Fig. 115.

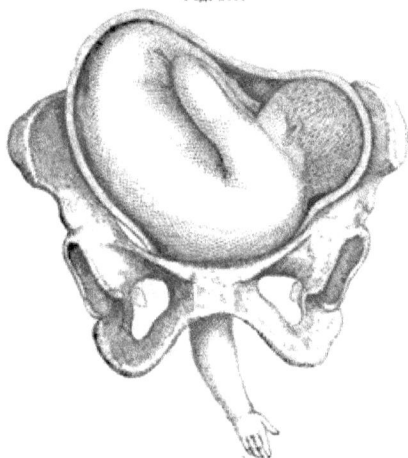

DORSO-ANTERIOR PRESENTATION OF THE ARM.

Divided into dorso-anterior and dorso-posterior positions.

Two great divisions of shoulder presentations are recognised: the one in which the back of the child looks to the abdomen of the mother (fig. 115), and the other in which the back of the child is turned towards the spine of the mother (fig. 116). Each of these is subdivided into two subsidiary classes, according as the head of the child is placed in the right or left iliac fossa. Thus in dorso-anterior positions, if the head lie in the left iliac fossa, the right shoulder of the child presents; if in the right iliac fossa, the left. So in dorso-posterior positions, if the head lie in the left iliac fossa, the left shoulder presents; if in the right, the

right.   Of the two classes the dorso-anterior positions are
more common, in the proportion, it is said, of two to one.

The causes of shoulder presentation are not well known.
Amongst those most commonly mentioned are prematurity
of the fœtus, and excess of liquor amnii; either of these, by
increasing the mobility of the fœtus in utero, would probably
have considerable influence.   The fact that it occurs much
more frequently amongst premature births has long been
recognised.   Undue obliquity of the uterus has probably
some influence, since the early pains might cause the pre-

Fig. 116.

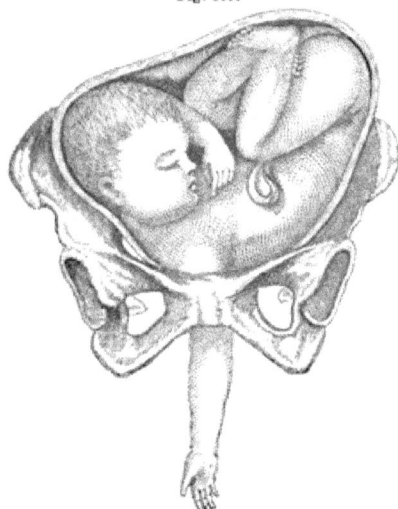

DORSO-POSTERIOR PRESENTATION OF THE ARM.

senting part to hitch against the pelvic brim, and the shoulder
to descend.   An unusually low attachment of the placenta to
the inferior segment of the uterine cavity has been mentioned
as a predisposing cause.   In consequence of this the head
does not lie so readily in the lower uterine segment, and is
apt to slip up into one of the iliac fossæ.   This is supposed to
explain the frequency of arm presentation in cases of partial
or complete placenta prævia.   Danyau and Wigand believe
that shoulder presentations are favoured by irregularity in
the shape of the uterine cavity, especially a relative increase
in its transverse diameter.   This theory has been generally
discredited by writers, and it is certainly not susceptible of

proof; but it seems far from unlikely that some peculiarity of shape may exist, not capable of recognition, but sufficient to influence the position of the foetus. How otherwise are we to explain those remarkable cases, many of which are recorded, in which similar malpositions occurred in many successive labours? Thus Joulin refers to a patient who had an arm presentation in three successive pregnancies, and to another who had shoulder presentation in three out of four labours. Certainly, such constant recurrences of the same abnormality could only be explained on the hypothesis of some very persistent cause, such as that referred to. Pinard [1] states that shoulder presentations are seven times more common in multiparæ than in primiparæ, in consequence, as he believes, of the laxity of the abdominal walls in the former, which allows the uterus to fall forwards, and thus prevents the head entering the pelvic brim in the latter weeks of pregnancy. It is probable that merely accidental causes have most influence in the production of shoulder presentation, such as falls, or undue pressure exerted on the abdomen by badly fitting or tight stays. Partially transverse positions during pregnancy are certainly much more common than is generally believed, and may often be detected by abdominal palpation. The tendency is for such malpositions to be righted either before labour sets in, or in the early period of labour; but it is quite easy to understand how any persistent pressure, applied in the manner indicated, may perpetuate a position which otherwise would have been only temporary.

Prognosis and frequency.

According to Churchill's statistics, shoulder presentations occur about once in 260 cases, that is, only slightly less frequently than those of the face. The prognosis to both the mother and child is much more unfavourable; for he estimates that out of 235 cases 1 in 9 of the mothers, and half the children were lost. The prognosis in each individual case will, of course, vary much with the period of delivery at which the malposition is recognised. If detected early, interference is easy, and the prognosis ought to be good; whereas there are few obstetric difficulties more trying than a case of shoulder presentation, in which the necessary treatment has been delayed until the presenting part has been tightly jammed into the cavity of the pelvis.

[1] *Annal. d'hyg. pub. et de méd.* Jan. 1879.

Bearing this fact in mind, the paramount necessity of an accurate diagnosis will be apparent ; and it is specially important that we should be able not only to detect that a shoulder or arm is presenting, but that we should, if possible, determine which it is, and how the body and head of the child are placed.   The existence of a shoulder presentation is not generally suspected, until the first vaginal examination is made during labour.   The practitioner will then be struck with the absence of the rounded mass of the fœtal head, and, if the os be open and the membranes protruding, by their elongated form, which is common to this and to other mal-presentations.   If the presenting part be too high to reach, as is often the case at an early period of labour, an endeavour should at once be made to ascertain the fœtal position by abdominal examination.   This is the more important as it is much more easy to recognise presentations of the shoulder in this way than those of the breech or foot ; and, at so early a period, it is often not only possible, but comparatively easy, to alter the position of the fœtus by abdominal manipulation alone, and thus avoid the necessity of the more serious form of version.   The method of detecting a shoulder presentation by examination of the abdomen has already been described (p. 121), and need not be repeated.   The chief points to look for are, the altered shape of the uterus, and two solid masses, the head and the breech, one in either iliac fossa. The facility with which these parts may be recognised varies much in different patients.   In thin women, with lax abdominal parietes, they can be easily felt, while in very stout women it may be impossible.   Failing this method, we must rely on vaginal examinations ; although, before the membranes are ruptured, and when the presenting part is high in the pelvis, it is not always easy to gain accurate information in this way.   The difficulty is increased by the paramount importance of retaining the membranes intact as long as possible.   It should be remembered, therefore, that when a presentation of the superior extremity is suspected, the necessary examinations should only be made in the intervals between the pains when the membranes are lax, and never when they are rendered tense by the uterine contractions.

As either the shoulder, the elbow, or the hand may present, it will be best to describe the peculiarities of each

[margin note:] Diagnosis.

[margin note:] Shoulder presenta- tions can often be detected by abdo- minal palpation.

separately, and the means of distinguishing to which side of
the body the presenting part belongs.

Peculiari-
ties of the
shoulder.

1. The *shoulder* is recognised as a round smooth prominence, at one point of which may often be felt the sharp
edge of the acromion. If the finger can be passed sufficiently
high, it may be possible to feel the clavicle, and the spine of
the scapula. A still more complete examination may enable
us to detect the ribs and the intercostal spaces, which would
be quite conclusive as to the nature of the presentation, since
there is nothing resembling them in any other part of the
body. At the side of the shoulder, the hollow of the axilla
may generally be made out.

Mode of
diagnos-
ing the
position of
the child.

In order to ascertain the position of the child we have to
find out in which iliac fossa the head lies. This may be done
in two ways: 1st, the head may be felt through the abdominal parietes by palpation; and 2nd, since the axilla always
points towards the feet, if it point to the left side the head
must lie in the right iliac fossa, if to the right, the head
must be placed in the left iliac fossa. Again, the spine of
the scapula must correspond to the back of the child, the
clavicle to its abdomen; and, by feeling one or other, we
know whether we have to do with a dorso-anterior or dorso-
posterior position. If we cannot satisfactorily determine the
position by these means, it is quite legitimate practice to
bring down the arm carefully, provided the membranes are
ruptured, so as to examine the hand, which will be easily
recognised as right or left. This expedient will decide the
point; but it is one which it is better to avoid, if possible,
for it not only slightly increases the difficulty of turning,
although perhaps not very materially, but the arm might
possibly be injured in the endeavour to bring it down.

Differen-
tial dia-
gnosis
of the
shoulder.

The only part of the body likely to be taken for the
shoulder is the breech: but in that its larger size, the groove
in which the genital organs lie, the second prominence formed
by the other buttock, and the sacral spinous processes are
sufficient to prevent a mistake.

The elbow.

2. The *elbow* is rarely felt at the os, and may be readily
recognised by the sharp prominence of the olecranon, situated
between two lesser prominences, the condyles. As the elbow
always points towards the feet, the position of the fœtus can
be easily ascertained.

3. The *hand* is easy to recognise, and can only be con- The hand.
founded with the foot. It can be distinguished by its
borders being of the same thickness, by the fingers being
wider apart and more readily separated from each other than
the toes, and above all by the mobility of the thumb, which
can be carried across the palm, and placed in apposition with
each of the fingers.

It is not difficult to tell which hand is presenting. If the   Mode of
hand be in the vagina, or beyond the vulva, and within easy   detecting
reach, we recognise which it is by laying hold of it as if we   hand is
were about to shake hands.   If the palm lie in the palm of   present-
the practitioner's hand, with the two thumbs in apposition,   ing.
it is the right hand; if the back of the hand it is the left.
Another simple way is for the practitioner to imagine his
own hand placed in precisely the same position as that of the
fœtus; and this will readily enable him to verify the previous
diagnosis.   A simple rule tells us how the body of the child
is placed, for, provided we are sure the hand is in a state of
supination, the back of the hand points to the back of the
child, the palm to its abdomen, the thumb to the head, and
the little finger to the feet.

It is perhaps hardly proper to talk of a mechanism of   Mechan-
shoulder presentations, since, if left unassisted, they almost   ism.
invariably lead to the gravest consequences.   Still, nature is
not entirely at fault even here, and it is well to study the
means she adopts to terminate these malpositions.

There are two possible terminations of shoulder presen-   The two
tation.   In one, known as '*spontaneous version*,' some   possible
other part of the fœtus is substituted for that originally pre-   tions of
senting; in the other, '*spontaneous evolution*,' the fœtus   shoulder
is expelled by being squeezed through the pelvis, without   presenta-
the originally presenting part being withdrawn.   It cannot   natural
be too strongly impressed on the mind that neither of these   powers.
can be relied on in practice.

Spontaneous version may occasionally occur before, or   Spon-
immediately after, the rupture of the membranes, when the   taneous
fœtus is still readily moveable within the cavity of the uterus.   version.
A few authenticated cases are recorded in which the same
fortunate issue took place after the shoulder had been engaged in the pelvic brim for a considerable time, or even after
prolapse of the arm; but its probability is necessarily much

lessened under such circumstances. Either the head or the breech may be brought down to the os in place of the original presentation.

The precise mechanism of spontaneous version, or the favouring circumstances, are not sufficiently understood to justify any positive statement with regard to it.

Cazeaux believed that it is produced by partial or irregular contraction of the uterus, one side contracting energetically, while the other remains inert, or only contracts to a slight degree. To illustrate how this may effect spontaneous version, let us suppose that the child is lying with the head in the left iliac fossa. Then if the left side of the uterus should contract more forcibly than the right, it would clearly tend to push the head and shoulder to the right side, until the head came to present instead of the shoulder. A very interesting case is related by Geneuil,[1] in which he was present during spontaneous version, in the course of which the breech was substituted for the left shoulder more than four hours after the rupture of the membranes. In this case the uterus was so tightly contracted that version was impossible. He observed the side of the uterus opposite the head contracting energetically, the other remaining flaccid, and eventually the case ended without assistance, the breech presenting. The natural moulding action of the uterus, and the greater tendency of the long axis of the child to lie in that of the uterus, no doubt assist the transformation, and much must depend on the mobility of the fœtus in any individual case.

That such changes often take place in the latter weeks of pregnancy, and before labour has actually commenced, is quite certain, and they are probably much more frequent than is generally supposed. When spontaneous version does occur, it is, of course, a more favourable event; and the termination and prognosis of the labour are then the same as if the head or breech had originally presented.

Sponta-
neous
evolution.

The mechanism of spontaneous evolution, since it was first clearly worked out by Douglas, has been so often and carefully described that we know precisely how it occurs. Although every now and then a case is recorded in which a living child has been born by this means, such an event is of

[1] *Ann. de Gynécologie*, v. v. 1876.

extreme rarity ; and there is no doubt of the accuracy of the general opinion, that spontaneous evolution can only happen when the pelvis is unusually roomy and the child small ; and that it almost necessarily involves the death of the fœtus, on account of the immense pressure to which it is subjected.

Two varieties are described, in one of which the head is first born, in the other the breech ; in both the originally presenting arm remained prolapsed.   The former   is   of

Fig. 117.

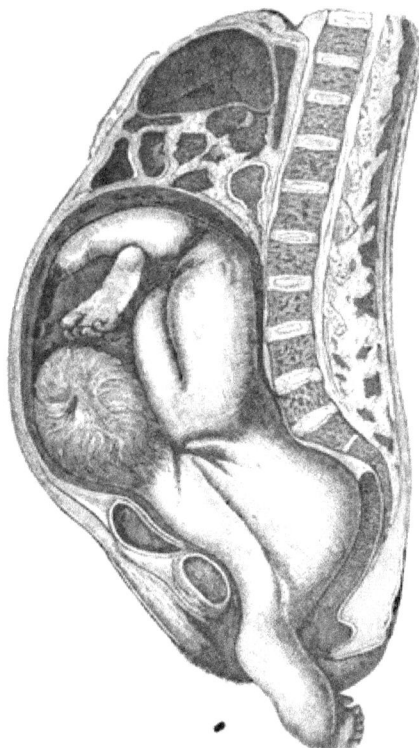

SPONTANEOUS EVOLUTION.   (After Chiara.)

This drawing was made from a patient who died undelivered, the body being frozen, and bisected.

extreme rarity, and is believed only to have happened with very premature children, whose bodies were small and flexible, and when traction had been made on the presenting arm.   Under such circumstances it can hardly be called a natural process, and we may confine our attention to the latter and more common variety.

What takes place is as follows : The presenting arm and shoulder are tightly jammed down, as far as is possible, by the uterine contractions, and the head becomes strongly flexed on the shoulder. As much of the body of the fœtus as the pelvis will contain becomes engaged, and then a movement of rotation occurs, which brings the body of the child nearly into the antero-posterior diameter of the pelvis (fig. 117). The shoulder projects under the arch of the pubis, the head lying above the symphysis, and the breech near the sacro-iliac synchondrosis. It is essential that the head should lie forwards above the pubes, so that the length of the neck may permit the shoulder to project under the pubic arch, without any part of the head entering the pelvic cavity. The shoulder and neck of the child now become fixed points, round which the body of the child rotates, and the whole force of the uterine contractions is expended on the breech. The latter, with the body, therefore, becomes more and more depressed, until, at last, the side of the thorax reaches the vulva, and, followed by the breech and inferior extremities, is slowly pushed out. As soon as the limbs are born the head is easily expelled.

The enormous pressure to which the body is subjected in this process can readily be understood. As regards the practical bearings of this termination of shoulder presentations, all that need be said is, that, if we should happen to meet with a case in which the shoulder and thorax were so strongly depressed that turning was impossible, and in which it seemed that nature was endeavouring to effect evolution, we should be justified in aiding the descent of the breech by traction on the groin, before resorting to the difficult and hazardous operation of embryotomy or decapitation.

Treat-
ment.
It is unnecessary to describe specially the treatment of shoulder presentation, since it consists essentially in performing the operation of turning, which is fully described elsewhere. It is only needful here to insist on the advisability of performing the operation in the way which involves the least interference with the uterus. Hence if the nature of the case be detected before the membranes are ruptured, an endeavour should be made—and ought generally to

succeed—to turn by external manipulation only. If we can succeed in bringing the breech or head over the os in this way, the case will be little more troublesome than an ordinary presentation of these parts. Failing in this, turning by combined external and internal manipulation should be attempted ; and the introduction of the entire hand should be reserved for those more troublesome cases in which the waters have long drained away, and in which both these methods are inapplicable.

Should all these means fail, we must resort to the mutilation of the child by embryulcia or decapitation, probably the most difficult and dangerous of all obstetric operations. In nine cases in the United States the Cæsarean section has been performed under these circumstances, with a successful result to the mother in six.[1]

There are various so-called *complex presentations* in which more than one part of the fœtal body presents. Thus we may have a hand or a foot presenting with the head, or a foot and hand presenting simultaneously. The former do not necessarily give rise to any serious difficulty, for there is generally sufficient room for the head to pass. Indeed, it is unlikely that either the hand or foot should enter the pelvic brim with the head, unless the head was unusually small, or the pelvis more than ordinarily capacious. As regards treatment, it is, no doubt, advisable to make an attempt to replace the hand or foot by pushing it gently above the head in the intervals between the pains, and maintaining it there until the head be fully engaged in the pelvic cavity. The engagement of the head can be hastened by abdominal pressure, which will prove of great value. Failing this, all we can do is to place the presenting member at the part of the pelvis where it will least impede the labour, and be the least subjected to pressure ; and that will generally be opposite the temple of the child. As it must obstruct the passage of the head to a certain extent, the application of the forceps may be necessary. When the feet and hands present at the same time, in addition to the confusing nature of the presentation from so many parts being felt together, there is the risk of the hands coming down, and converting the case into one of arm presentation. It is the obvious duty of the

Complex presentations.

Foot or hand with head.

Hands and feet together.

---

[1] Harris, note to 3rd American edition.

accoucheur to prevent this by insuring the descent of the
feet, and traction should be made on them, either with the
fingers or with a lac, until their descent, and the ascent of
the hands, are assured.

Dorsal displacement
of the arm.   In connection with this subject may be mentioned the
curious dorsal displacement of the arm first described by Sir
James Simpson,[1] in which the forearm of the child becomes
thrown across and behind the neck.   The result is the formation of a ridge or bar, which prevents the descent of the
head into the pelvis by hitching against the brim (fig. 118).

Fig. 118.

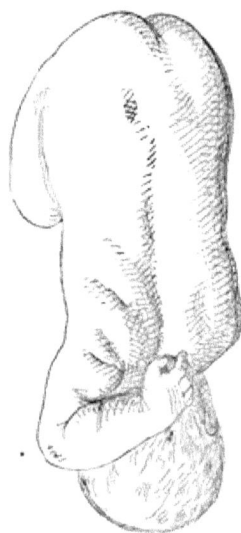

DORSAL DISPLACEMENT OF THE ARM.

The difficulty of diagnosis is very great, for the cause of obstruction is too high up to be felt.   But if we meet with a
case in which the pelvis is roomy and the pains strong, and
yet the head does not descend after an adequate time, a full
exploration of the cause is essential.   For this purpose we
would naturally put the patient under chloroform, and pass
the hand sufficiently high.   We might then feel the arm in
its abnormal position.   That was what took place in a case
under my own care, in which I failed to get the head through
the brim with the forceps, and eventually delivered by turn-

[1] *Selected Obstet. Works*, vol. i.

ing. The same course was adopted by my friend Mr. Jardine Murray in a similar case.[1] Simpson advises that the arm should be brought down so as to convert the case into an ordinary hand and head presentation. This, if the arm be above the brim, must always be difficult, and I believe the simpler and more effective plan is podalic version. A similar displacement may cause some difficulty in breech presentations, and after turning (fig. 119). Delay here is easier of

Fig. 119.

DORSAL DISPLACEMENT OF THE ARM IN FOOTLING PRESENTATIONS.   (After Barnes.)

diagnosis, since the obstacle to the expulsion will at once lead to careful examination. By carrying the body of the child well backwards, so as to enable the finger to pass behind the symphysis pubis and over the shoulder, it will generally be easy to liberate the arm.

It occasionally happens that the umbilical cord falls down past the presenting part (fig. 120), and is apt to be pressed between it and the walls of the pelvis. The consequence is

Prolapse of the umbilical cord.

[1] Med. Times and Gaz. 1861.

that the fœtal circulation is seriously interfered with, and the death of the child from asphyxia is a common result. Hence prolapse of the funis is a very serious complication of labour in so far as the child is concerned.

Fortunately it is not a very frequent occurrence. Churchill calculates that out of over 105,000 deliveries it was met with once in 240 cases, and Scanzoni once in 254. Its frequency varies much under different circumstances, and in different places. We find from Churchill's figures a remarkable difference in the proportional number of cases

Fig. 120.

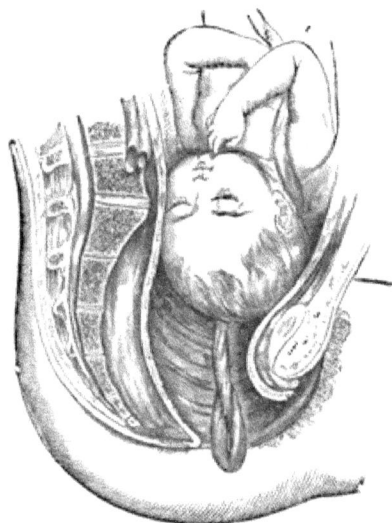

PROLAPSE OF THE UMBILICAL CORD.

observed in France, England, and Germany—viz. 1 in 446½, 1 in 207½, and 1 in 156 respectively. Great as is the proportion referred to Germany in these figures, it has been found to be exceeded in special districts. Thus Engelman records 1 case out of 94 labours in the Lying-in Hospital at Berlin, and Michaelis 1 in 90 in that of Kiel. These remarkable differences are at first sight not easy to account for.

Explana-
tion of its
increased
frequency
in certain
countries. Dr. Simpson suggests, with considerable show of probability, that the difference in frequency in England, France, and Germany may depend on the varying positions in which lying-in women are placed during labour in each country.

In France, where, although the patient is laid on her back, the pelvis is kept elevated, the complication occurs least frequently; in England, where she lies on her side, more often; and in Germany, where she is placed on her back with her shoulders raised, most often. The special frequency of prolapsed funis in certain districts, as in Kiel, is supposed by Engelman[1] to depend on the prevalence of rickets, and consequently of deformed pelvis, which we shall presently see is probably one of the most frequent and important causes of the accident.

With regard to the danger attending prolapsed funis, as far as the mother is concerned, it may be said to be altogether unimportant; but the universal experience of obstetricians points to the great risk to which the child is subjected. Scanzoni calculates that 45 per cent. only of the children were saved; Churchill estimated the number at 47 per cent.; thus, under the most favourable circumstances, this complication leads to the death of more than half the children. Engelman found that out of 202 vertex presentations only 36 per cent. of the children survived. The mortality was not nearly so great in other presentations; 68 per cent. of the cases in which the child presented with the feet were saved, and 50 per cent. in original shoulder presentations. The reason of this remarkable difference is, doubtless, that in vertex presentations the head fits the pelvis much more completely, and subjects the chord to much greater pressure; while in other presentations the pelvis is less completely filled, and the interference with the circulation in the cord is not so great. Besides, in the latter case, the complication is detected early, and the necessary treatment sooner adopted.

The fœtal mortality is considerably greater in first labours; a result to be expected on account of the greater resistance of the soft parts, and the consequent prolongation of the labour.

The causes of prolapse of the funis are any circumstances which prevent the presenting part accurately fitting the pelvic brim. Hence it is much more frequent in face, breech, or shoulder than in vertex presentations, and is relatively more common in footling and shoulder presenta-

Marginal notes: Prognosis. Relative fœtal mortality in different presentations. The fœtal mortality is greater in first labours. Causes. Circumstances interfering with the adaptation of the presenting part to the pelvis.   tions than in any other. Amongst occasional accidental predisposing causes may be mentioned early rupture of the membranes, especially if the amount of liquor amnii be excessive, as the sudden escape of the fluid washes down the cord; undue length of the cord itself; or an unusually low placental attachment. Engelman attaches great importance to slight contraction of the pelvis, and states that in the Berlin Lying-in Hospital, where accurate measurements of the pelvis were taken in all cases, it was almost invariably found to exist. The explanation is evident, since one of the first results of pelvic contraction is to prevent the ready engagement of the presenting part in the pelvic brim.

Pelvic deformity.

Diagnosis.   The diagnosis of cord presentation is generally devoid of difficulty; but if the membranes are still unruptured, it may not always be quite easy to determine the precise nature of the soft structures felt through them, as they recede from the touch. If the pulsations of the cord can be felt through the membranes, all difficulty is removed. After the membranes are ruptured, there is nothing that it can well be mistaken for.

Import-ance of de-termining the pulsa-tions of the cord.   The important point to determine in such a case is whether the cord be pulsating or not; for if pulsations have entirely ceased, the inference is that the child is dead, and the case may then be left to nature without further inter-ference. It is of importance, however, to be careful; for, if the examination be made during a pain, the circulation might be only temporarily arrested. The examination, therefore, should be made during an interval, and a loop of the cord pulled down, if necessary, to make ourselves absolutely certain on this point.

Amount of cord prolapsed.   The amount of the prolapse varies much. Sometimes only a knuckle of the cord, so small as to escape observation, is engaged between the pelvis and presenting part. Under such circumstances the child may be sacrificed without any suspicion of danger having arisen. More often the amount prolapsed is considerable; sometimes so as to lie in the vagina in a long loop, or even to protrude altogether beyond the vulva.

Treat-ment.   In the treatment the great indication is to prevent the cord from being unduly pressed on, and all our endeavours must have this object in view. If the presentation be de-

tected before the full dilatation of the cervix, and when the
membranes are unruptured, we must try to keep the cord
out of the way; to preserve the membranes intact as long
as possible, since the cord is tolerably protected as long as
it is surrounded by the liquor amnii; and to secure the complete dilatation of the os, so that the presenting part may
engage rapidly and completely.

Much may be done at this time by the postural treat- Postural
ment, which we chiefly owe to the ingenuity of Dr. T. Gail- treatment.
lard Thomas, of New York, whose writings familiarised the
profession with it, although it appears that a somewhat
similar plan had been occasionally adopted previously.   Dr.
Thomas's method is based on the principle of causing the
cord to slip back into the uterine cavity by its own weight.
For this purpose the patient is placed on her hands and
knees, with the hips elevated, and the shoulders resting on
a lower level (fig. 121).   The cervix is then no longer the

Fig. 121.

POSTURAL TREATMENT OF PROLAPSE OF THE CORD.

most dependent portion of the uterus, and the anterior wall
of the uterus forms an inclined plane down which the cord
slips.   The success of this manœuvre is sometimes very
great, but by no means always so.   It is most likely to
succeed when the membranes are unruptured.   If, when
adopted, the cord slip away, and the os be sufficiently
dilated, the membranes may be ruptured, and engagement
of the head produced by properly applied uterine pressure.

Sometimes the position is so irksome that it is impossible to resort to it. Postural treatment is not even then altogether impossible, for by placing the patient on the side opposite to that of the prolapse, so as to relieve the cord as much as possible from pressure, and at the same time elevating the hips by a pillow, it may slip back. Even after the membranes are ruptured, postural treatment in one form or another may succeed ; and, as it is simple and harmless, it should certainly be always tried. Attempts at reposition, by one or other of the methods described below, may also occasionally be facilitated by trying them when the patient is placed in the knee-shoulder position.

Artificial reposition. Failing by postural treatment, or in combination with it, it is quite legitimate to make an attempt to place the cord beyond the reach of dangerous pressure by other methods. Unfortunately reposition is too often disappointing, difficult to effect, and very frequently, even when apparently successful, shortly followed by a fresh descent of the cord. Provided the os be fully dilated, and the presenting head engaged in the pelvis (for reposition may be said to be hopeless when any other part presents), perhaps the best way is to attempt it by the hand alone. Probably the simplest and most effectual method is that recommended by McClintock and Hardy, who advise that the patient should lie on the opposite side to the prolapsed cord, which should then be drawn towards the pubes as being the shallowest part of the pelvis.

Reposition by the fingers. Two or three fingers may then be used to push the cord past the head, and as high as they can reach. They must be kept in the pelvis until a pain comes on, and then very gently withdrawn, in the hope that the cord may not again prolapse. During the pain external pressure may very properly be applied to favour descent of the head. This manœuvre may be repeated during several successive pains, and may eventually succeed. The attempt to hook the cord over the fœtal limbs, or to place it in the hollow of the neck, recommended in many works, involves so deep an introduction of the hand that it is obviously impracticable.

Instruments used for reposition. Various complex instruments have been invented to aid reposition (fig. 122), but even if we possessed them they are not likely to be at hand when the emergency arises. A simple instrument may be improvised out of an ordinary

male elastic catheter, by passing the two ends of a piece of
string through it, so as to leave a loop emerging from the
eye of the catheter. This is passed through the loop of
prolapsed cord, and then fixed in the eye of the catheter by
means of the stilette. The cord is then pushed up into the
uterine cavity by the catheter, and liberated by withdrawing
the stilette. Another simple instrument may be made by
cutting a hole in a piece of whalebone. A piece of tape is
then passed through the loop of the cord
and the ends threaded through the eye
cut in the whalebone. By tightening
the tape the whalebone is held in close
apposition to the cord, and the whole is
passed as high as possible into the uterine
cavity. The tape can easily be liberated
by pulling one end. If preferred, the
cord can be tied to the whalebone, which
is left in utero until the child is born.
Nothing need be said as to the various
other methods adopted for keeping up
the cord, such as the insertion of pieces
of sponge, or tying the cord in a bag
of soft leather, since they are generally
admitted to be quite useless.

Fig. 122.

BRAUN'S APPARATUS FOR
REPLACING THE CORD.

It only too often happens that all
endeavours at reposition fail. The subsequent treatment must then be guided
by the circumstances of the case. If the
pelvis be roomy, and the pains strong,
especially in a multipara, we may often
deem it advisable to leave the case to
nature, in the hope that the head may
be pushed through before pressure on the
cord has had time to prove fatal to the child. Under such
circumstances the patient should be urged to bear down, and
the descent of the head promoted by uterine pressure, so as
to get the second stage completed as soon as possible. If
the head be within easy reach, the application of the forceps
is quite justifiable, since delay must necessarily involve the
death of the child. During this time the cord should be
placed, if possible, opposite one or other sacro-iliac synchon-

Treatment
when re-
position
fails.

drosis, according to the position of the head, as the part of the pelvis where there is most room, and where the pressure would consequently be least prejudicial. If we have to do with a case in which the head has not descended into the pelvis, and postural treatment and reposition have both failed, provided the os be fully dilated, and other circumstances be favourable, turning would undoubtedly offer the best chance to the child. This treatment is strongly advocated by Engelman, who found that 70 per cent. of the children delivered in this way were saved. There can be no question that, so far as the interests of the child are concerned, it is, under the circumstances indicated, by far the best expedient. Turning, however, is by no means always devoid of a certain risk to the mother, and the performance of the operation, in any particular case, must be left to the judgment of the practitioner. A fully dilated os, with membranes unruptured, so that version could be performed by the combined method without the introduction of the hand into the uterus, would be unquestionably the most favourable state. If it be not deemed proper to resort to it, all that can be done is to endeavour to save the cord from pressure as much as possible, by one or other of the methods already mentioned.

END OF THE FIRST VOLUME.

LONDON : PRINTED BY
SPOTTISWOODE AND CO., NEW-STREET SQUARE
AND PARLIAMENT STREET

.